DARWIN'S
BIOLOGICAL WORK

Solanum rostratum Dun. Two flowers of the racemose inflorescence showing the asymmetry of the reproductive parts. The style curves downwards and always lies between the lower enlarged stamen and the main axis of the inflorescence so that adjacent flowers of the inflorescence are mirror images of each other. See note on p. ix.

DARWIN'S BIOLOGICAL WORK

SOME ASPECTS RECONSIDERED

BY

P. R. BELL J. CHALLINOR
J. B. S. HALDANE
P. MARLER H. L. K. WHITEHOUSE
J. S. WILKIE

EDITED BY P. R. BELL

CAMBRIDGE
AT THE UNIVERSITY PRESS
1959

CAMBRIDGE UNIVERSITY PRESS
Cambridge, New York, Melbourne, Madrid, Cape Town,
Singapore, São Paulo, Delhi, Tokyo, Mexico City

Cambridge University Press
The Edinburgh Building, Cambridge CB2 8RU, UK

Published in the United States of America by Cambridge University Press, New York

www.cambridge.org
Information on this title: www.cambridge.org/9781107697157

First published 1959
First paperback edition 2011

A catalogue record for this publication is available from the British Library

ISBN 978-1-107-69715-7 Paperback

CONTENTS

LIST OF PLATES

PREFACE

On 24 May 1859, reviewing the papers which had been read before the Linnean Society of London in the preceding session, the President, Thomas Bell, is recorded in the history of the Society as having regretted that the session had passed without being marked 'by any of those striking discoveries which at once revolutionize, so to speak, the department of science on which they bear', or which 'shall produce a marked and permanent impress on the character of any branch of knowledge'. Nevertheless, one contribution was certainly 'to produce a marked and permanent impress', for, on 1 July 1858, the communications of Charles Darwin and Alfred Wallace *On the Tendency of Species to form Varieties* had been read to the Society. Few judgments can have proved so erroneous as that of Thomas Bell.

The story of the publication in the following year of *The Origin of Species* and of the controversy which ensued has often been told. It is less well known that feeling was intense even within the Linnean Society and several Fellows withdrew from it because Darwin was not summarily expelled. Yet, within ten years, the Council of the Linnean Society was more rigorous in its interpretation of 'Darwinism' than Darwin himself. This humble man, who, in diffidently sending specimen chapters of *The Origin of Species* to his publisher, requested that after perusal they be sent to a 'lady being excellent judge of style [who] is going to look out for errors for me', was indeed beginning to effect a revolution.

Yet we should not think of Darwin solely in relation to evolution and natural selection. He was in fact one of the foremost geologists of his day and his early papers made noteworthy contributions to that science. His work came to encompass the whole field of biology, but was never superficial. Several of his publications, for example *The Various Contrivances by which Orchids are fertilized by Insects*, have not yet been superseded. It is true, of course, that evolution came to be a unifying principle in all Darwin's scientific work, but such unifying principles are a great stimulus to biological progress.

Preface

The contributions of Darwin to experimental biology have received less attention than his descriptive and theoretical work, but they were by no means unimportant. For example, he recognized the importance of plant physiology and pressed for the establishment of laboratories for its study. Starved though he was of equipment, his penetrating observations have been the origin of some of the most active branches of contemporary biology.

For a century and more biologists have drawn freely and profitably upon Darwin's work and his flashes of originality continue to stimulate his successors. It is for these reasons, not because of the endless and often ignorant controversies which have become popularly associated with his name, that Darwin occupies a unique place in the history of biology. It is to put this point of view that *Darwin's Biological Work* is being published at this time when the name of Darwin is again before the public. All of the contributors are actively engaged in biological research (and research into the development of ideas in biology is considered to be an essential part of the science). Our hopes are the same as Darwin's when he submitted the manuscript of *The Origin of Species*: 'the book *ought* to be popular with a large bag of scientific and semi-scientific readers'.

ACKNOWLEDGEMENTS

Acknowledgements are gratefully made to the following: to Mr G. Atkinson, for the frontispiece; to the Controller, H.M. Stationery Office, for figs. 2 and 3 in the article by J. Challinor (fig. 12 from British Regional Geology, *The Welsh Borderland*, and fig. 25 from *The Hampshire Basin and Adjoining Areas*, respectively), and to Professor T. N. George and Messrs C. A. Watts and Company Limited for fig. 4 in the same article; to the late Dr W. J. Arkell, F.R.S., and the Council of the Palaeontographical Society for Pl. I; to Dr H. B. D. Kettlewell and Mr Michael Lyster, Department of Zoology, University Museum, Oxford, for Pl. II, and to Dr H. B. D. Kettlewell and Mr John Haywood for Pl. III, both plates illustrating the work of Dr Kettlewell discussed by J. B. S. Haldane; to Dr J. W. S. Pringle, F.R.S., for Pl. IV, taken from his article in the *Journal of Experimental Biology*; and to Mrs M. P. Whitehouse, A.R.P.S., for Pl. V.

The Editor would personally like to thank Mr John G. Murray for allowing him to see unpublished letters of Charles Darwin, the Cambridge University Press for the care they have devoted to this volume, and in particular Elizabeth Bell for undertaking much of the preparation of the manuscript and other invaluable assistance.

NOTE

Superior figures in the text refer to the Notes and References at the end of the book (p. 311).

1

P. R. BELL

THE MOVEMENT OF PLANTS IN RESPONSE TO LIGHT

The Power of Movement in Plants[1] is not one of Charles Darwin's better known books, nor is it one of his most readable. The gradual development of a grand theme and the sustained excitement of *The Origin of Species* are lacking, and all but the most persevering reader would to-day be discouraged by the lengthy descriptions of experiments and their results. Nevertheless, at its publication in 1880, *The Power of Movement* excited considerable interest and *The Times*, despite its previous antipathies, took the occasion to eulogize its author and his work.[2]

One of the investigations reported upon in *The Power of Movement* concerned the movements of plants in response to differences in illumination, a phenomenon then referred to as *heliotropism** since its occurrence in nature was frequently related to movements of the sun. It was in this field that Darwin made significant discoveries which affected the whole development of the subject and which contributed ultimately to the discovery of plant growth substances and the opening of a new and vigorous chapter of plant physiology. It is the purpose of this essay to trace the understanding of the part played by light in causing the movement of plants, to show Darwin's work in relation to that of his contemporaries, and to discuss the developments that have followed from his investigations. We shall, as Darwin, confine ourselves here to the reactions of the higher plants, although the effects of light on the lower, such as the fungi and the algae, are no less interesting and pose problems no less perplexing. By concentrating upon a familiar

* Those movements of plants in which the direction of the movement has a definite relation to the direction of the agent causing it are referred to as *tropisms*.

and apparently circumscribed biological problem we shall, too, be the better able to appreciate the significance of the changes of outlook that have occurred in half a century of research.

The Recognition of Light as a Cause of Plant Movement

The movements of shoots, leaves and flowers in relation to light must have been amongst the earliest observations of civilized man. Theophrastus, writing in the third century B.C., was familiar with the turning of leaves towards the sun,[3] and Varro in the first century B.C. records similar phenomena and also the existence of flowers which follow the course of the sun throughout the day.[4] Pliny, a century later, describes how the leaflets of *Trifolium* (clover) close together at the approach of a storm.[5] In later years Acosta, writing of medicinal plants, mentions the conspicuous folding together of the leaflets and leaves of *Tamarindus indica* L. (tamarind) at evening and their opening at dawn.[6] Other plants, all members of the Leguminosae, among them species of *Acacia*, *Sesbania sesban* (L.) Merr. and *Cassia absus* L., whose leaves show these 'sleep' movements are described by Alpinus in the sixteenth century.[7]

The first general discussion of movements of this kind and speculation as to their cause is not to be found until the end of the seventeenth century. By this time it had already been shown that the opening of the flowers of *Anemone* could be brought on by heat in the absence of light.[8] In Oxford, a Dr Sharrock, 'very knowing in vegetables and all pertaining therunto', had grown plants before an open window and, by reversing them after an interval of time, caused their stems to assume S-shaped curves, or bend to any position which his friends cared to indicate.[9] These experiments were known to John Ray, the illustrious British botanist, and he was led to regard the bending of stems towards the light as being caused, not by a difference in illumination of the two sides of the stem, but by a difference in temperature. The more strongly illuminated side is that closer to the fresher air; it follows, he argued, that its temperature will be lower and its growth consequently slower than that of the shaded.[10]

Light as a Cause of Plant Movement

The reluctance shown by Ray to ascribe heliotropic curvature to a direct effect of light upon the plant persisted throughout the eighteenth century, possibly because light had not been one of the Aristotelian elements which played so important a role in medieval philosophy. Even Stephen Hales, for example, who occupies an honoured place in the history of botany for his experiments in the ascent of sap, attributed this curvature to a greater loss of water by evaporation from the illuminated side of a stem than from the shaded,[11] causing shrinkage of the former. Linnaeus, aware of the movements of flowers[12] and leaves[13] in relation to light, observed that the 'sleep' movements of leaves at dusk were not caused by a fall in temperature, since they also took place in the more or less constant temperature of a conservatory,[14] but he was at a loss for an alternative explanation. The Genevan naturalist Bonnet,[15] in the middle of the eighteenth century, appears to have been the first to carry out extensive experiments on the nature of plant movements, although, of course, by modern standards these experiments are extremely confused. His conclusions were that movements were principally determined by warmth and moisture, particularly the latter, and in arriving at this position, which his experiments hardly justify, he was clearly elaborating the earlier views of the French Academician Dodart.[16] From Bonnet's own experiments, Duhamel,[17] a more penetrating observer, appears to have concluded that heliotropic curvatures were, in fact, dependent upon light alone, but his explanation of the manner in which these curvatures were caused had no greater validity than Bonnet's. He envisaged the existence of 'vapours' within the plant, the quantity and flow of which could be influenced by external factors, and it was to these that the movements of the plant could be ascribed.

These early attempts to explain the behaviour of plants in light, though imperfect, were, nevertheless, the beginnings of an attempt to link observable phenomena with tangible causes. Refined methods of experimentation, together with advances in the physical sciences, led to much more rapid advances in the nineteenth century. Although it would be illegitimate to attempt to divide the history of science into centuries, it is true that here the difference in outlook between the eighteenth and nineteenth is profound. Whereas before the turn of the century methods were crude and thinking confused, from the beginning of the nineteenth century to the present there can be traced the steady development of fruitful ideas and profitable experimentation. In the study of the effects of light on vegetation, the reluctance to consider light itself as an active factor, so conspicuous in Ray and Bonnet, disappeared with the eighteenth century.

De Candolle, one of the first in the nineteenth century to experiment on the effects of illumination, and, incidentally, the first to use the term 'heliotropism', regarded the bending of unequally illuminated stems as part of the general phenomenon of 'etiolation'.[18] A plant grown in darkness lacks chlorophyll and its stems are usually extremely elongated. (It is now known that these two effects, although commonly associated, are not necessarily related.) According to de Candolle, to put his view in modern terms, in an unequally illuminated stem the metabolism on the illuminated side is more intense; as a consequence, the illuminated cells mature faster and extend less than the shaded and a curvature results. The shaded side can, in fact, be regarded as partially etiolated. This theory, although it explained curvature towards light (now referred to as *positive phototropism*), would not account for curvature away from light (*negative phototropism*), such as is seen in those roots which react at all to light, and in certain stems, such as underground rhizomes.

Shortly after its publication, Dutrochet challenged the theory, on other grounds, as the result of an ingenious experiment.[19] Taking a stem of *Medicago sativa* L. (lucerne) which had bent towards the light, he split it into two longitudinally, perpendicular to the plane of bending. The side adjacent to the light curved still more, while that away from it straightened, whereas, argued Dutrochet, if elongation of the shaded side was, as maintained by de Candolle, the cause of curvature, the reverse result should have been obtained. Dutrochet consequently concluded that light was affecting only the illuminated side and that it was the shaded that was passively involved. His theory to account for phototropic bending supposed a contraction of the contents of the illuminated cells, leading to a contraction of the cells themselves, arising from metabolic changes attributable to light, and a difference in the sizes of the cells in different regions of the plant. In the stem, for example, he envisaged a diminution in the sizes of the cells from the outside towards the centre. The contraction of the larger cells was greater than that of the smaller and, in unilateral illumination, curvature towards the illuminated side was inevitable. In roots the gradation in cell size was supposed to be reversed, and consequently the curvature. But there was no anatomical support for this view and, ingenious though Dutrochet's theories were, they had to be discarded. In fact, as a result of careful measurements, it had become clear by the middle of the nineteenth century that unilateral illumination caused a difference in the growth of the two sides of the stem. Curvature was, in consequence, confined to the extending regions, a fact which had been demonstrated unwittingly by Sharrock two centuries earlier. De Candolle's interpretation of the manner in which light acted upon the stem was, in essence, generally accepted at this time, even by Sachs, although his view changed radically later. It was, nevertheless, also clear, largely as a result of the work of Sachs and his students, that the question was more complex than imagined by de Candolle. Etiolation had, in fact, several different aspects, only one of which was the effect of light on the extension of individual cells.

By 1874 the writings of Sachs[20] were turning attention to the possibility of a direct effect of light on the extensibility of the cell wall, or of changes in the protoplasm of the cell which might affect the properties of the wall. In the growing region of a stem, he suggested, light promoted the growth in thickness of the cell walls and consequently the extension of the cells by the absorption of water was impeded. Sachs's student, de Vries, later maintained that in phototropic curvature there was, apart from any effect upon the cell wall, an actual increase in turgor on the shaded side contributing to its greater extension.[21] There was, too, a new element entering Sachs's writings at this time, for he had noticed that the more nearly perpendicular was the direction of light entering the cells of a normally upright stem to the longitudinal axis of those cells (and consequently of the stem), the greater was its phototropic effect. This, he pointed out, was quite similar to the effect of gravity, in that both agents cause the longitudinal axes of the cells they act upon to come into alignment with their own line of action. In addition, Francis Darwin, Charles's third son, and his assistant in all his botanical experiments towards the end of his life, working in

Sachs's laboratory in 1881 had shown that the roots of *Sinapis alba* L. (white mustard) grew slower in the light than in the dark.[22] But these roots were indisputably negatively phototropic and, according to Candollian ideas, should have shown the reverse behaviour to account for the receding curvature.* The behaviour of stems and roots in relation to light thus appeared the precise reverse of that in relation to gravity and Sachs, developing an idea of Frank,[25] came to regard phototropism and geotropism as similar phenomena, differing only in the nature of the initial stimulus.[26]

It is to the credit of Wiesner, working in Vienna at this time, that he was the first to investigate the relation between the amount of light falling on a shoot and the subsequent curvature.[27] He was also the first to show that in the absence of oxygen no curvature occurred at all,[28] indicating that the bending process required a supply of energy from within the plant in addition to any received from the light. The mechanical cause of the bending he considered to be changes in turgescence of the cells in the growing region, probably brought about by light, coupled with an increase in the extensibility of the cell walls. Wiesner's quantitative approach was later developed extensively by the school of F. F. A. C. Went at Utrecht.

Pfeffer must also be mentioned here for, although he himself did little directly concerned with phototropism, it is to him that the elaboration of the Darwins' experiments is largely due. Having earlier worked with Sachs at Würzburg, he studied, amongst other physiological problems, those concerned with the periodic movements of leaves, such as those of *Desmodium gyrans* DC. (telegraph plant) and other members of the Leguminosae. At this time he attributed the phototropic responses of higher plants to changes in the turgescence of the cells, due to osmotic effects. He did not share with Sachs his later view of the all-importance of the direction of the light, nor did he deny, as did Sachs, the possibility of one and the same structure displaying both positive and negative phototropism according to the intensity of the light, a property first

* A suggestion that the tissues of the root, being translucent, acted as a lens, so that the light was concentrated on the shaded side, was later shown to be correct.[23] Even so, this concentration of light is sharply limited and the total illumination of the shaded side is less than that of the illuminated.[24]

demonstrated in *Lepidium sativum* L. (cress) by Müller.[29] Later, in 1887, he settled at Leipzig and founded the school which was to have such a profound influence on the study of plant movements. Influenced, as we shall see, by Darwin, he, more than any of his contemporaries, now developed the view that tropisms were manifestations of Irritability, a fundamental property of living matter of all kinds. The property of Irritability, which encompasses those processes by which living matter reacts to disturbances in such a way as to maintain its equilibrium, was well known as a philosophical notion,[30] but for Pfeffer became one of the bases of physiology. In the enlarged edition of his *Pflanzenphysiologie* of 1897,[31] Pfeffer formulated the principles of Irritability with a clarity and precision which, even though the value of the concept is now questioned, are conspicuous in the writings of the period.

That phototropism, along with geotropism, was a response of this kind was envisaged by Dutrochet as long ago as 1824,[32] but subsequently ignored by him and overlooked until noticed by Pfeffer. By 1880 it had become general to regard tropic responses to light as a manifestation of Irritability; even Sachs who, like de Candolle, at first saw the response to light as due to physical changes forced on to the plant by the external agent, came to accept the notion without demur. Wiesner, however, remained a conspicuous exception.

The Originality of Darwin's Work

We are now in a position to consider Darwin's contribution to the study of phototropism. *The Power of Movement in Plants* was the result of some five years' experimenting, in which Charles was assisted by Francis Darwin, just down from Cambridge.* It was, in Darwin's own words, 'a tough piece of work' which followed directly from *The Movements and Habits of Climbing Plants*. It is clear from the preface and footnotes that Darwin was familiar with the work of Sachs at Würzburg, with the current theories of growth of cells and with the experiments of Wiesner in Vienna. These, however, are mentioned only incidentally; Darwin's whole approach to the problem of movement was different from that of the Conti-

* The writing of the book appears to have been entirely the work of Charles Darwin.[33]

nental workers and his inspiration lay elsewhere. Whereas Sachs, Wiesner and Pfeffer were seeking to explain the behaviour of plants in terms of physical and chemical causes, Darwin turned to plant movements as yet a further demonstration of the interrelatedness of living plants and their common origin. Although he accepted fully that growth curvatures were an expression of Irritability, for him it was sufficient to demonstrate that these phenomena had advantageous consequences. He envisaged the response as having become linked to the stimulus in the course of evolution, so producing an organism better fitted in the struggle for existence.

It will be recalled how in his study of climbing plants Darwin had demonstrated that the apices of these plants, when not in contact with a support, described remarkable sweeping movements. 'In accordance with the principles of evolution', wrote Darwin in 1881, 'it was impossible for climbing plants having been developed in so many widely different groups unless all kinds of plants possess some slight power of movement of an analogous kind'.[34] His experiments do, in fact, reveal that growing stems, roots and leaves show in general irregular rotatory movements (*circumnutation*). That the striking movements of the stems of climbing plants were a development of this common property was and still is an acceptable hypothesis. Darwin, however, made a further claim, namely that the movements associated with light and gravity were also modified forms of circumnutation, a generalization that was not adequately supported by his evidence, as Francis Darwin was forced to admit in later years.* It was, nevertheless, probably Darwin's preoccupation with the property of circumnutation that led him to pay close attention to the apices of stems and roots in his experiments on phototropism, although Ciesielski's demonstration[36] some years earlier that the roots of *Pisum* would turn downwards in the normal way only if their tips were intact may also have been in his mind.

Darwin used as material for his research either young seedlings, where the absence of a mature stem simplifies observation, or germinating seeds of cereals and grasses. In the latter plants, belonging to the family Gramineae of the Monocotyledons, the stem, unlike that of most Dicotyledons, shows little elongation until flowering. That which passes for a stem in the vegetative state

* According to Blackman.[35]

consists principally of a tube formed of sheathing leaf bases. This peculiarity is reflected even in the embryo, and the growing point of the young stem is contained within a tubular structure called the *coleoptile*. As the seed of a cereal or grass germinates, the coleoptile behaves like a stem and its summit, solid and pointed, is pushed up through the soil to the surface (fig. 1). At this stage the first normal leaf, formed from the enclosed growing point, soon ruptures the coleoptile and leaves it a withered remnant at its base. In its actively growing stem-like phase, the coleoptile shows the phototropic responses of a stem and it forms, as Darwin found, a very convenient experimental object, since it is easily obtained and consists of little more than a tapering hollow cylinder of tissue, uncomplicated at its tip by rudimentary leaves or branches. Moreover, the way in which it grows can be easily seen; there is initially an increase in the number of cells forming the coleoptile, but after it has reached approximately 1 cm. in length, all subsequent growth is exclusively by cell elongation, only the mature cells at the base remaining unchanged.[37] The coleoptile is not circular in transverse section, but oval, with one broad side adjacent to the seed. In critical phototropic experiments it is necessary to know the orientation of the coleoptile in relation to the light, because the broad side will absorb significantly more radiation than the narrow.

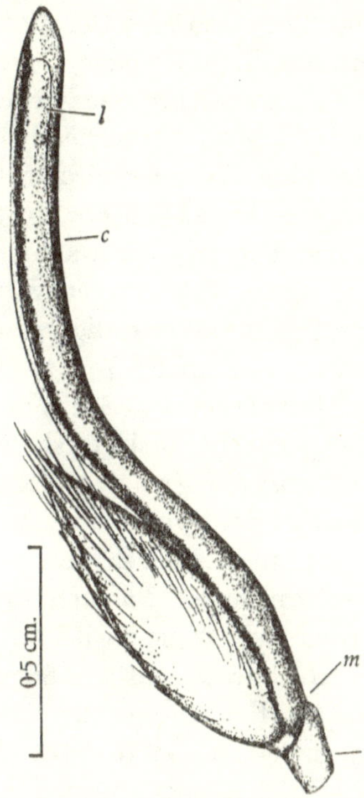

Fig. 1. The germinating seed of *Avena sativa*, the glumes (husk) removed. *c*, coleoptile; *l*, first leaf enclosed within the coleoptile; *m*, mesocotyl; *r*, coleorhiza containing first root. The coleoptile has become erect through the action of light and gravity. Note the flat side adjacent to the seed.

The Darwins' first experiments on phototropism were concerned with the effect of unilateral illumination on the circumnutation of seedlings and coleoptiles. They demonstrated that the movement towards the source of the light was superimposed upon the circumnutation, suppressing the latter completely when the light was strong. They also tried the effect of blackening one half of a coleoptile longitudinally and placing it so that one of the vertical boundaries of the blackened area was in the middle of the illuminated side, the other boundary, diametrically opposite, forming the mid-line of the shaded side. In these circumstances the plane of the curvature of the coleoptile was deflected by approximately 45° from the window. This result clearly throws doubt on Sachs's directional theory for, according to Sachs, the curvature of such an organ should be such as to bring its axis into alignment with the radiation, which here must have been principally from the window. Darwin was either unaware of this latter theory of Sachs or chose not to comment upon it, using this experiment solely to show that light acts over the whole stem and not just along a narrow longitudinal strip adjacent to the source.

The first observations which suggested that the response to light was not entirely local, but in part transmitted, were made upon the coleoptiles of *Phalaris canariensis* L. (canary grass).[38] The Darwins were struck with the way in which these coleoptiles (as other coleoptiles and seedlings), when exposed to light, bent first at the tip, the bending then travelling down the coleoptile for some 2 cm. in such a way that the alignment of the tip in relation to the light remained unchanged. They demonstrated that the mechanical transmission of the bending was not an essential part of the phenomenon, for, where the upper parts of the coleoptiles were constrained within narrow glass tubes, the bending continued to appear in the free part below.

Two further experiments led them to the view that the upper part of the coleoptile determined the bending of the lower. In the first, approximately 1·5 to 4·0 mm. were cut from the tips of coleoptiles; with the smaller decapitations there was diminished sensitivity to light, but removal of 2·5 to 4·0 mm. destroyed the sensitivity altogether. In the second, to combat the reasonable objection that wounding from the amputation might have interfered with the sensitivity, the apex was shielded from light by blackened tubes or caps of thin tinfoil. These experiments led to the conclusion that in *Phalaris* exclusion of light from the upper 2·5 mm. had no strong influence in diminishing the curvature of the lower part, but exclusion from about 4 mm. or more completely prevented it. When the procedure was reversed and only the tips of the coleoptiles were illuminated, the remainder being shielded from the light by immersion in damp dark sand, the bending took place as if the whole coleoptile were illuminated. Similar results were obtained with the coleoptiles of *Avena sativa* L. (oat) and some seedlings.

Here then was demonstrated for the first time that curvature could occur in regions of the coleoptile not acted upon directly by light. The experiments with caps suggested that in normal coleoptiles the curvature of the basal region was determined by influences from the apex, and that in the absence of these influences, even if the base was illuminated, no curvature occurred. Darwin did not speculate upon the nature of these influences except to speak of 'some matter in the upper part [of the coleoptile] which is acted upon by light, and which transmits its effects to the lower part'.[39] He later compared the phenomenon in general terms, as had Sachs compared transmitted turgor changes, with the nervous system of animals. It is doubtful whether the quantitative aspects of these responses interested Darwin very much. He states that there did not appear to be a 'close parallelism between the amount of light which acts on a plant and its degree of curvature', but did observe that plants which had been brightly illuminated the previous day showed much more sluggish phototropic responses than those which had been kept in the dark.[40]

Darwin was very anxious about the reception of his work on the Continent. Sachs's response, such as it was, was one of contempt. He regarded the Darwins as rather poor experimenters,[41] and it must be admitted that Charles Darwin's way of writing about them and, one suspects, the experiments themselves had not the precision associated with Sachs. Such, too, was the autocratic nature of Sachs and so convinced was he at this time of the correctness of his directional theory (which, he believed, 'as soon as it was but generally understood, will be accepted on all sides'),[42] that he probably had little time for work not directly supporting it.*

* According to Goebel, Sachs's conviction of the correctness of his theory diminished towards the end of his life.[43]

Wiesner, on the other hand, although critical, charmed Darwin by his response. He compiled and presented to Darwin a reasoned argument against the interpretations he had placed on his results,[44] but written in so generous a spirit that, wrote Darwin, 'No man was ever vivisected in so sweet a manner before'.[45] He rejected Darwin's claim of the universality of circumnutation and refused to be convinced by the evidence for a transmission of stimulus in the phototropic response. The bending of the lower part of coleoptiles and seedlings following the curvature of the upper part was attributed by Wiesner to growth effects brought about by the unequal distribution of weight following the preliminary upper curvature. Pfeffer alone of the leading physiologists recognized the full significance of the Darwins' work. Proofs of the German edition of *The Power of Movement* reached Pfeffer in time for him to incorporate the main results in the first edition of *Pflanzenphysiologie*. It is clear that for him evidence of the spatial separation of perception and action in tropic responses was of profound importance and opened an entirely new field of profitable experimentation. The discoveries which followed proved to have an importance far beyond the elucidation of the phenomenon of phototropism.

The Development of the Study of Phototropism since Darwin

For the first part of the twentieth century it is possible to trace two main streams of research in phototropism. The first, associated particularly with the school of F. F. A. C. Went at Utrecht and drawing inspiration from Wiesner, remained antagonistic to Pfeffer's approach. They took advantage of the refined equipment becoming available to study in particular the quantitative aspects of phototropic curvatures. Their view, put forward particularly by Blaauw, was that the effect of light was a direct one upon the irradiated cells and that it was neither necessary nor desirable to envisage a mediated response or to invoke notions of Irritability. The other, associated with Pfeffer, concentrated upon the transmission of the phototropic stimulus and attempted to determine its course and nature. In both, fundamental discoveries were made which remain valid today and which form the basis of modern research.

THE CONFIRMATION OF THE SPECIAL SENSITIVITY OF APICAL REGIONS TO LIGHT

After the publication of Darwin's work, despite Pfeffer's appreciative reception of it and the provocation of Wiesner's reasoned attack, little immediate interest was taken in his claims for the remarkable phototropic properties of the apices of stems and coleoptiles. Nevertheless, evidence for the transmission of phototropic stimuli was coming from other sources, for Vöchting showed that phototropic movements of the petioles of members of the Malvaceae (mallows and allied plants) took place when the leaf blade alone was illuminated.[46] A repetition of the Darwins' experiments under strict laboratory conditions was clearly called for, and the establishment of Pfeffer at Leipzig provided the opportunity. One of his first students there, Rothert, repeated and extended the work of the Darwins in a series of well designed and critical experiments.[47] The claims of the Darwins were vindicated with only minor modifications. Rothert's conclusions were: (1) The capacity for perceiving the light stimulus was widely distributed in plant tissues, but not universal. (2) In stems and coleoptiles the capacity for perceiving the light stimulus was very pronounced at the tip, but was nevertheless present, although much diminished, below. (3) The capacity for perceiving the light stimulus was independent of growth; numerous growing organs lacked the capacity, yet others, although mature, still possessed it. (4) The capacity for bending in response to the light stimulus was distinct from the perceiving of it and, in contradistinction to the latter, was entirely dependent upon growth. In consequence the degree of bending was the resultant of two separate properties in the plant, namely, the sensitiveness to the stimulus and the intensity of growth. Rothert confirmed that removal of the upper 5 mm. of the *Avena* coleoptile eliminated the phototropic response, but discovered that within 24 hours the sensitivity had returned and was distributed within the stump in the same way as in the intact coleoptile.[48] The responses of these stumps in which the sensitivity had been regenerated were almost as strong as in intact control plants.

The question of the distribution of the sensitivity to light within the intact coleoptile received the attention of Rothert, Wilschke and others, but careful and quantitative investigations were first made by Sierp and Seybold[49] and Lange.[50] Their results showed clearly that the sensitivity fell away very sharply below the apex. Lange used a more effective method of limiting the unilateral illumination to a definite zone of the coleoptile than Sierp and Seybold, and greater accuracy in calculating the amount of light falling upon it. He also placed his plants on a klinostat* after illumination, so that any responses should not be masked by an opposed geotropic reaction. The results of the two investigations are shown in Table I: although there is seen to be broad agreement, those of Lange show that the relative sensitivity does not fall off quite so sharply as thought by Sierp and Seybold, but both found that the sensitivity had fallen to unity or less at 2·5 mm. below the apex. The differences may arise from physiological differences between the material of the two investigations. Lange suggested that the basal region was, in fact, completely insensitive to light, the curvature following its illumination being caused by some of the light scattered in the cells of the base reaching the sensitive apex.

TABLE I. *The distribution of the sensitivity to unilateral light in the apical 2 mm. of the coleoptile of* Avena sativa (*from Lange*)

Zone (in mm. from tip)	Relative sensitivity to unilateral light (Reciprocals of energy required to produce curvature): Sierp and Seybold	Lange
0–0·5	33,948	38,870
0·5–1·0	564	2,870
1·0–1·5	34·7	253
1·5–2·0	8·42–24·20	24·7

THE QUANTITATIVE ASPECTS OF PHOTOTROPISM

The Effect of Light of Different Wavelengths. The manufacture of light filters permitting the passage of almost pure colours at last made it possible to determine precisely the relationship between the colour (or wavelength) of light and its effectiveness in promoting curvature. It was by now realized that, for the results to be valid, the intensity of the light falling on the experimental plants must be the same whatever the colour. Neglect of this precaution and the

* The klinostat is a device for rotating plants in a vertical plane so that the effects of gravity upon them are neutralized. It is customary to have the longitudinal axis of the plant horizontal.

absence of good light filters almost certainly account for the contradictory results of the early workers. Blaauw, using *Avena* coleoptiles grown in the dark, showed that red light produced little or no phototropic response, but as the wavelength diminished a sharp increase in response occurred in the blue range of the spectrum, falling off again on entering the violet end.[51] These results were substantially the same as those obtained earlier by Payer with *Lepidium*.[52] More extensive measurements have now shown that the curve of sensitivity is not as regular as supposed by Blaauw,

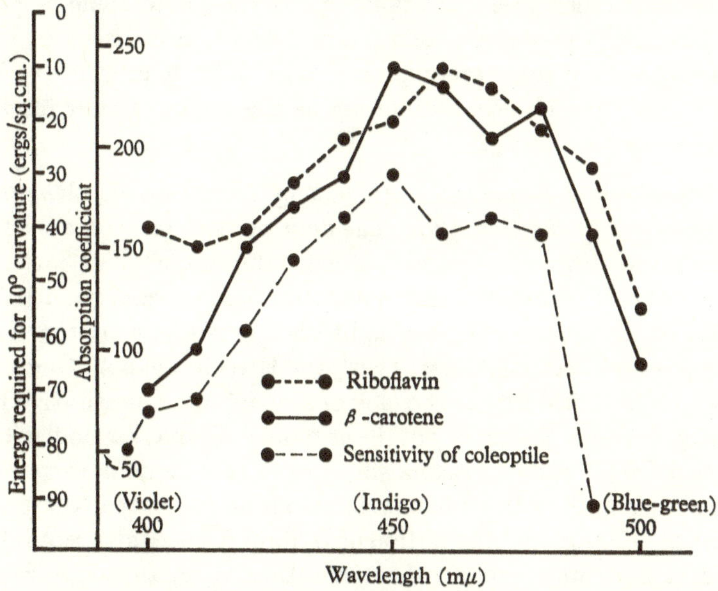

Fig. 2. The sensitivity of the coleoptile of *Avena sativa* (the narrow side facing the light) to light of different wavelengths. The absorption of light of different wavelengths by the pigments β-carotene and riboflavin is shown for comparison. (All data from Galston and Baker.[53]) Note: 1mμ = 1/10⁶ mm. = 10 Å. (Similar information for the root of *Sinapis alba* has been given by Kohlbecker.[54])

especially between the wavelengths 450 and 500 mμ.[53] These irregularities have special significance since they are also found in the absorption spectrum of β-carotene (fig. 2). Outside the visible spectrum both the tip and the base of the coleoptile show a positive response to ultraviolet light.

When the very low sensitivity to red light of plants grown in the dark was clearly established, this was frequently used when preparing plants for phototropic investigations, and blue light of known wavelength and intensity used for the actual experimental illumination. It should be noted, however, that there is now incontrovertible evidence that red light may cause phototropic curvature in certain green plants.[55] In *Zea mays* L. (maize), for example, the green coleoptile of a seedling grown in the light shows a phototropic response to red light, but the etiolated coleoptile is insensitive. Green and etiolated seedlings of *Lepidium sativum* show a similar difference, but the coleoptile of *Avena* shows little or no response to red light whether green or etiolated. If the phototropic response depends upon the presence of a photoreceptor, as is discussed later, the sensitivity of some green plants to red light in addition to blue may follow from chlorophyll acting as the photoreceptor for light of the longer wavelength.

The Relationship between the Amount of Light and the Magnitude and Direction of the Curvature. The development of accurate equipment for measuring the intensity of light, the lack of which Sachs had bemoaned half a century earlier, was a stimulus to careful quantitative studies of the several effects of light upon growth, an approach that became associated particularly with the Utrecht School. One of the principles of Irritability is the lack of any definite quantitative relationship between the energy of the stimulus and that expended in the response.[56] If such a relationship can in fact be demonstrated in phototropism, then this phenomenon can be compared to a photochemical reaction and the notion of Irritability becomes redundant. The first experiments to test these quantitative relationships, carried out by Blaauw[57] at Utrecht and Fröschel[58] in Vienna concerned the length of time for which it was necessary to subject the *Avena* coleoptile to unilateral illumination before the curvature began to appear, the period known as the *presentation time*. Their results appeared to show that no curvature of the coleoptile occurred before a definite quantity of light, approximately 20 metre-candle-seconds,* had been transmitted to the coleoptile. Such a threshold had been expected by analogy with the stimulus-response mechanisms of animals, but, as pointed out by Arisz,[59] the figure obtained

* The metre-candle (or lux) is a measure of light intensity; the metre-candle-second (m.c.s.) (or lux-second) a measure of light energy. 1 metre-candle = 0·0929 foot-candles.

by Blaauw has no simple physiological interpretation, since the first recognition of the curvature by the human eye is an intermediate event in the bending process and subject to considerable variation with different observers.

Changes in the rate of growth following uniform illumination had been detected by Wiesner,[60] but more detailed investigations were made by Vogt,[61] using the *Avena* coleoptile, and Blaauw,[62] using the *hypocotyl* (the stem below the cotyledons) of *Helianthus globosus* L. In these experiments the plants were returned to darkness or dim red light after the illumination and the growth rate measured until it became stable. The usual result was to obtain first a depression and then an acceleration of the rate, the whole effect, lasting an hour or more, being referred to as the *light-growth reaction* (*Photowachstumsreaktion*). Blaauw, following Wiesner, maintained that phototropic curvatures were an inevitable result of light-growth reactions of different extent on the two sides of an unequally illuminated plant, or, in other words, phototropism was a secondary phenomenon, the primary being the light-growth reaction. He rejected the notion and terminology of Irritability. The plant cells were not 'excited' by differences in amount of illumination or its direction (this Sachsian view was still receiving attention).[63] No 'perception' was involved, but solely a direct effect of light, becoming detectable in a few minutes, upon the growth of the living cell.

These results were based upon the illumination of the whole length of the coleoptile. Sierp was the first to suggest that the light-growth reaction following a short burst of illumination consisted of two successive components, one resulting from the illumination, the other from its cessation.[64] It later became clear that the apex and the subapical portion of the coleoptile behaved differently in this respect. As a result of illuminating different zones, van Dillewijn distinguished two reactions proceeding at different rates, namely, a 'long' reaction, following irradiation of only the tip of the coleoptile and reaching its maximum in one to two hours, and a 'short' reaction following irradiation of any growing part and occurring within about thirty minutes.[65] Only very small quantities of light were sufficient to bring about the 'long' reaction at the tip, but the 'short' reaction required greater quantities and higher intensities. The different 'long' reactions of the light and dark sides were considered responsible for phototropic bending, and the 'short' reactions for preliminary oscillatory movements.

Many of the fundamental physical relationships of the phototropism of the *Avena* coleoptile were determined by Arisz whose paper forms one of the outstanding contributions of the Utrecht laboratory.[66] He showed conclusively that the relation between

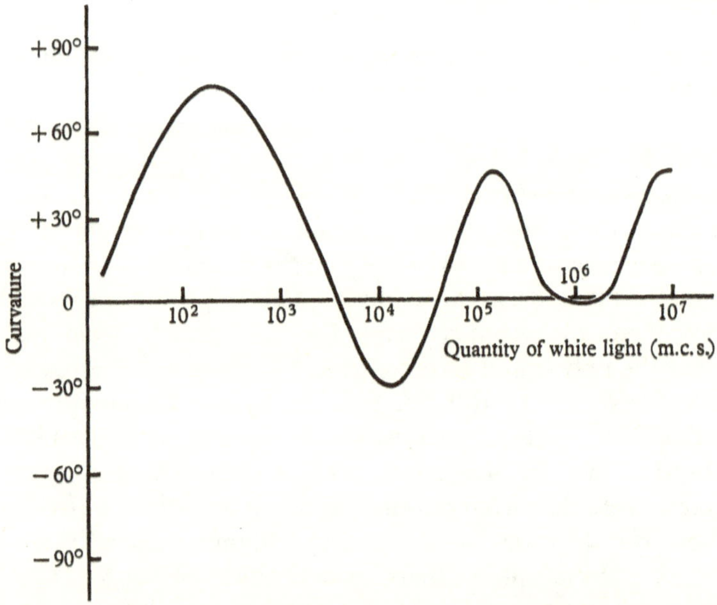

Fig. 3. The curvatures of the coleoptile of *Avena sativa* in response to different quantities of light. Semi-diagrammatic; the data from du Buy and Nuernbergk.[67]

the amount of bending and the quantity of light energy received by the plant was by no means simple. With small quantities of light the amount of bending increases with the quantity of light used, but not in simple proportion. At a fairly low level of light energy (between 100 and 250 m.c.s.) no further increases in bending occur, and above this level the maximum response of the coleoptile begins to decline. With still greater amounts of light, the coleoptile bends away from the light source. As a result of these experiments and others by du Buy and Nuernbergk (in which blue light was used) there is now evidence for the existence in the oat coleoptile of three ranges of positive response, alternating with two of negative (fig. 8). They follow illumination of the whole coleoptile or of the tip alone, illumination of the base alone giving only a positive curvature. Examination of Arisz's results in detail shows that the nature of the response does not depend entirely upon the amount of light energy received by the coleoptile, but also upon the intensity at which it is delivered. For example, 6000 m.c.s. of light delivered in 60 seconds produces a negative curvature, but if delivered in 20 minutes a positive curvature precedes a later negative. If the intensity is yet lower and the illumination is prolonged, no negative curvature occurs at all. The interpretation of these results is therefore difficult and there is no simple photochemical parallel such as was suggested by the earlier work of Blaauw.

TABLE II. *The effect of preliminary uniform illumination upon the amount of light required to produce the first positive curvature in the coleoptile of* Avena sativa (*from Arisz*).

Quantity of light delivered in preliminary illumination (m.c. × sec.)	Quantity of light delivered unilaterally (m.c. × sec.) 12 × 5	12 × 10	100 × 5	100 × 10
	Curvature developing within 2 hours of second unilateral illumination (+, definite; ++, strong)			
0	++	++	++	++
25 × 10	Not done	+	+	+
100 × 10	0	+	+	+
25 × 100	0	0	?	?
100 × 100	0	0	0	0

It had of course been known for some time that a preliminary general illumination diminished the sensitivity of coleoptiles and seedlings to subsequent unilateral illumination, but Arisz was the first to investigate this effect quantitatively.[68] A selection of his results is shown in Table II. Within the range of the first positive curvature, preliminary illumination clearly reduces the effect of the subsequent unilateral.

One of the last investigations at the Utrecht laboratory directly concerned with the explanation of phototropism along the lines of Blaauw's theory was that of van Dillewijn.[69] His approach, too, was quantitative and attempted, as had Sierp's earlier,[70] to explain the various phototropic curvatures in terms of the light-growth reaction. His experiments followed from his estimate that approximately one-thirtieth of the light falling upon the body of the *Avena* coleoptile from one side penetrated to the other, the remainder being dispersed or absorbed. First, the light-growth reactions were determined for seven different quantities of light delivered at intensities of from 15 to 2400 m.c., and supplied uniformly around the coleoptile. Secondly, the reactions were determined for one-thirtieth of each of these seven quantities, giving seven pairs of quantities of light, the members of each pair being in the proportion

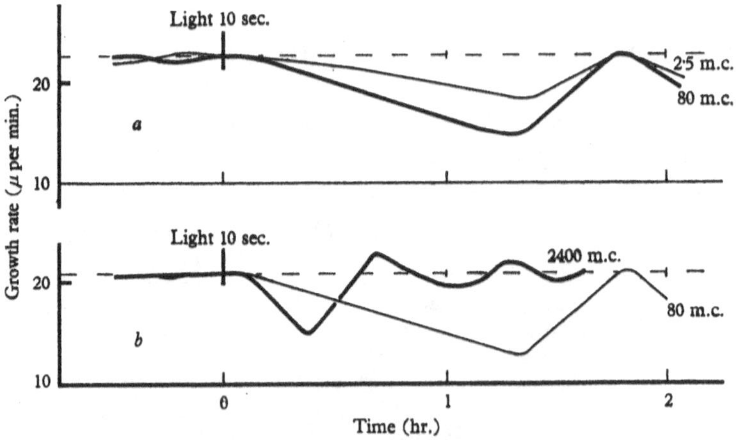

Fig. 4. Two pairs of light-growth reactions in which the ratio of the quantities of light is or approaches 30:1. It is seen that in *a*, a positive curvature would be expected if the coleoptile were illuminated with 800 m.c.s. from one side; and in *b*, a negative if the 24,000 m.c.s. were supplied in this way. For further explanation, see text. Note: 1μ = 1/1000 mm. (Data from van Dillewijn.)

30:1 and delivered in the same time. The two growth reactions for the quantities in each pair were then superimposed upon a graph. It was now possible to predict, assuming that the quantities of light reacting on the two sides of a coleoptile illuminated from one side were also in the proportion 30:1, the direction and magnitude of the curvature which would have resulted from supplying the larger quantity of light of each pair from one side only. The predicted results agreed closely with those obtained by Arisz, giving Blaauw's theory considerable support (fig. 4).

THE PATH AND DIRECTION OF THE 'PHOTOTROPIC STIMULUS'

The passage of a stimulus from the tip to the lower region of seedlings and coleoptiles had been conclusively demonstrated, but how and where was this stimulus conducted? In the *Avena* coleoptile Rothert had shown that conduction was not confined to the vascular system.[71] Only two vascular bundles ascend the coleoptile, one at each margin, and they do not branch or connect with one another. When they were severed and the region above illuminated, bending took place as usual below the cuts. This question was taken further by Fitting, earlier a student at Leipzig. He showed that transverse cuts made in growing coleoptiles 0·5 to 1 cm. below the tip, although causing small variable curvatures due to wounding, did not permanently impair the rate of growth or phototropic responses.[72] No matter what the orientation of the cuts in relation to the direction of the light, the basal part still continued to respond when the tip alone was illuminated. Even two overlapping cuts from opposite sides of the coleoptile did not prevent the response; only when small plates of mica or tin-foil were inserted into the cuts was there a conspicuous reduction in the amount of bending. This last effect was attributed to an interference with the water supply of the tip which, partially wilted, reacted only slowly to the light. It appeared, therefore, that no matter by what course the stimulus reached the basal part of the coleoptile, the correct curvature resulted. To account for this, Fitting imagined some form of polarization of the basal cells, but, although subsequent discoveries have confirmed a polarity in the cells of the coleoptile, Fitting's original conception was too involved to commend itself generally. Regarding the mode of transmission of the stimulus, he rightly disposed of facile comparisons with the animal nerve, but admitted that it appeared to be different from any other form of transmission in plants then known.

Before Fitting's work was published Boysen Jensen had begun at Copenhagen similar experiments concerned with the effect of transverse incisions on the transmission of the stimulus.[73] In his account he pointed out that Fitting had considered no possibility other than the passage of the stimulus in the living matter of the coleoptile. If one admitted the possibility of the transmission continuing in non-living matter, then other interpretations of Fitting's experiments became possible. Boysen Jensen showed that if the experiments were repeated in a saturated atmosphere, as used by Fitting, he obtained Fitting's results, but in a drier atmosphere (having a saturation deficit equivalent to about 7 mm. mercury) a cut on the side of the coleoptile away from the light prevented the transmission of the stimulus, while a cut on the side adjacent was without effect. If a mica plate were inserted into the incision on the side away from the light, then even in a saturated atmosphere there was no transmission of the stimulus below the incision, but a thin transverse slice of the stem of *Calamus* (a palm with a thin climbing stem, the tissue of which is perforated by small holes, the vessels) similarly placed did not impede it. To explain these apparent contradictions Boysen Jensen called attention to the fact that when the normal turgid coleoptile was cut there was an emission of sap and a droplet formed between the lips of the cut. This drop persisted in a saturated atmosphere. If transmission continued across this bridge then both Fitting's and Boysen Jensen's results were consistent with wholly longitudinal transmission of the stimulus confined to the side away from the illumination. To prove conclusively that conduction could occur across a discontinuity of living matter, Boysen Jensen removed the upper centimetre of the coleoptile and replaced it upon a ring of gelatin solution, sealing the wound with cocoa-butter. These coleoptiles showed almost unimpaired phototropic responses.

Boysen Jensen's experiments did not escape criticism. Wolk[74] in Went's laboratory pointed out that Boysen Jensen had ignored the tropic effects of wounding and carried out modified experiments the results of which, he claimed, supported Fitting. The point remained in dispute, but later results at Copenhagen appeared to confirm that the major part of the influence of the apex upon the base was by way of the shaded side.[75]

The early experiments also clearly established that the stimulus causing the curvature was able to pass only from the apex to the base of the coleoptile, but whether illumination of the base can cause a change in the sensitivity of the apex or the extent of its response to subsequent unilateral illumination, as claimed by Wolk and others, was disputed and this important point has never been finally settled.[76]

THE INTERPRETATION OF PHOTOTROPISM IN TERMS OF GROWTH-REGULATING SUBSTANCES

The idea that the control exerted by the tip of a coleoptile or seedling upon the curvature of the lower part was brought about by the movement of some substance is, as we have seen (p. 11), to be found in embryo in *The Power of Movement.* Rothert did not deal with this aspect of the problem, but made the interesting speculation, largely substantiated by subsequent discoveries, that the means by which the stimulus was transmitted was probably identical whatever the nature of the tropism.[77] The possibility of the unequal distribution of a growth-regulating substance being involved in phototropism was, according to Boysen Jensen, first proposed by him,[78] but the concept is by no means clearly expressed in his first writings. In 1911, discussing the interpretation of his experiments described above, Boysen Jensen attributed the curvature solely to an acceleration in the rate of growth on the shaded side. In considering how this effect might be brought about at a distance from the tip, he dismissed as unlikely physical means of transmission, such as changes of pressure, and concluded that 'la transmission de l'irritation est de nature chimique'.[79] The substance envisaged was capable of travelling across a non-living barrier.

Meanwhile Fitting had discovered that the morphological changes associated with fruiting in certain orchids could be brought on by applying dead pollen or even extracts of it to the stigma.[80] In his view, since confirmed, the stimulus causing these morphological changes could only be chemical. His argument was inspired, as those of Pfeffer in the preceding century, by advances in animal physiology, where it had been discovered that substances circulating in the blood (*hormones*) were capable of controlling the functioning of the body.[81] The term hormone was introduced into botany by Fitting in the discussion of his results and it has come into common, although by no means universal, usage. It should, however, be noticed in passing that so great is the difference between the organization of higher animals and that of plants, and so much less clearly separated and localized are the various metabolic activities of the latter, that there is probably no good analogy between the animal hormones and the so-called hormones of plants. The work of Fitting is important because it drew the attention of botanists more forcibly than that of Boysen Jensen to the existence of influences, the nature of which could only be chemical, initiating and controlling growth in higher plants. Such substances were, in fact, in direct line with the *Bildungsstoffe* of Sachs.[82]

The work of Paál, carried out at Leipzig, continued directly from that of Boysen Jensen and is noteworthy for continuing the first clear formulation of the notion of a growth-regulating substance.[83] Going further than Boysen Jensen, he showed that the phototropic responses continued even if a disc of *Calamus* stem, 0·05 to 0·10 mm. thick, impregnated with 10 per cent gelatin solution were inserted between the tip and the base. In other experiments the tip of the coleoptile was replaced as before, but a slip of mica or tin-foil was inserted across half the area; these coleoptiles showed a curvature even when kept in the dark. Since the controlled coleoptiles, which had been operated on in a precisely similar manner, except that the tip had not been replaced, remained straight, the curvature was in all probability produced by some influence from the apex and not from the wound. In *Coix lachryma* L. (Job's tears), a grass in which a short coleoptile surmounts the solid mesocotyl, it was possible to detach the coleoptile and replace it with gelatin cement so that it was displaced laterally a distance equal to its radius, leaving a segment of coleoptile, less than half the total area of the cross-section, in contact with the stump. A strong curvature resulted, the side bearing the coleoptile being convex.

To explain these results Paál supposed the continuous and uniform production in the apex of a growth-promoting substance. Further, the amount of this substance was supposed to be affected by light, either by its depressing the formation of the substance at the apex or by inactivating it directly through some photochemical reaction. Unequal illumination of the two sides of a stem would then lead to a greater reduction in growth on the light than on the shaded side and consequently to the phototropic curvature. The *Coix* experiment showed how a similar curvature followed the mechanical displacement of the growth-promoting substance.

Confidence in the existence of growth-regulating substances continued to grow as a consequence of experiments not directly related to phototropism, but for some time it was not clear whether the effect of these substances was to promote or to depress growth.[84] Assuming that their function was to depress, Brauner proposed a theory of phototropism which was credible, but soon shown to be false.[85] It was based upon his discovery that, were decapitated coleoptiles illuminated from one side, transferred to complete darkness and the tips (which had remained in complete darkness) replaced, the reconstituted coleoptiles developed a strong positive curvature. Brauner also demonstrated that the initial effect of illuminating plant cells was to increase their permeability; in *Avena* he estimated an increase of 25 per cent within the first 30 minutes. The steps envisaged by Brauner in the phototropic reaction were, in order of occurrence, an increase of permeability on the illuminated side leading to an enhanced downward movement on this side of a growth-inhibiting substance (the production of which was possibly promoted by light), and lastly, as a consequence, a bending towards the illuminated side.

This theory collapsed when it was shown that the substance leaving the apex did in fact promote growth. The proof was due to Söding[86] who compared the rate of growth, both in darkness and in more natural conditions in a glasshouse, of intact coleoptiles, decapitated coleoptiles with the tip replaced, and the stumps alone. Growth was greatest in the intact coleoptiles, weaker in the decapitated coleoptiles with the tip replaced and weakest in the stumps. If a gelatin disc were inserted between the tip and the stump, growth was still promoted in the base. Consequently, asserted Söding, the growth is promoted by a hormone moving from the apex towards the base. Nielsen,[87] working in Jensen's laboratory, arrived at the same conclusion. He was able to repeat Paál's experiment of replacing the tip of the coleoptile to one side with *Avena* instead of *Coix*, and obtained the same result, that is, curvature away from the side bearing the tip.

There were, of course, those whose scientific caution caused them to be sceptical of the whole concept of natural growth-regulating substances, especially in view of the heterogeneous assemblage of materials known to produce responses. Priestley,[88] for example, noting the ready exudation from the cut surfaces of the coleoptile (which often interfered with experiments requiring the replacement of the tip) and finding it difficult to reconcile the idea of the downward movement of growth substances with a vigorous upward movement of sap, sought to explain the results of Continental workers as the effects of blocking or maintaining (by means of gelatin) the loss of water from the coleoptile. The 'regeneration' of the properties of the apex at the summit of the stump was, he claimed, nothing more than the consequence of the sealing of the cut surface by drying and oxidation. Phototropic curvature of the *Avena* coleoptile was explained by an increase in the resistance to stretching on the illuminated side and a general enhancement of the entry of water and movement of sap following illumination.

Nevertheless, the accumulating evidence clearly indicated the soundness of attempting to interpret phototropic curvatures in terms of growth substances.[89] The vindication of this view came with the isolation of a natural growth substance from the coleoptile of *Avena*.

THE ISOLATION OF A GROWTH-PROMOTING SUBSTANCE AND THE SYNTHESIS OF THE QUALITATIVE AND QUANTITATIVE APPROACHES

By 1927 it had become clear that light-growth reactions in themselves were inadequate to explain phototropic curvature and the conclusion became general that any correspondence between light-growth reactions and phototropic bending was not obligate, but accidental, and indicated no causal relationship between these two phenomena.[90]

Although F. F. A. C. Went continued to protest vigorously against the approach of Pfeffer and his followers, especially the application of 'the principles of human physiology to stimulation in plants'[91] (presumably referring on this occasion to Fitting's introduction of the hormone concept into botany), it was nevertheless true to say that the continued study of the quantitative physical aspects of phototropism was becoming sterile. It reached its climax in Dillewijn's work, but already by this time interest was shifting to the discovery of some causal mechanism within the plant. The bridge between the quantitative and qualitative approaches was fittingly provided by Went's son, F. W. Went, in his work upon the coleoptile of *Avena*.[92] In his first paper he described how, inspired by Paál's experiments, he placed a number of coleoptile tips close together, their cut surfaces downwards, on a thin layer of gelatin. After about an hour, he cut the gelatin into small cubes and placed a cube asymmetrically on each of the original coleoptiles (which had not been left sufficiently long for the apical properties to have been 'regenerated'). In a control series pure gelatin was used which had not been in contact with the tips. Whereas pure gelatin (and also gelatin on which rings cut from the bases of coleoptile tips had stood) gave no regular curvatures, the gelatine which had borne the tips caused pronounced curvatures away from the side of the stump bearing it. The growth-promoting substance or substances of Paál and Söding must, therefore, have diffused into the gelatin and thence asymmetrically into the stumps. Went recognized that here lay the basis of a method for quantitative estimation, for, as the experiments were done under standard conditions, the extent of the curvature would be a measure of the amount of growth-promoting substance present in the gelatin.[93]

It was also possible to imitate phototropic curvatures by placing on to a decapitated coleoptile two gelatin cubes, one on each side, containing the growth substance from apices which had received different amounts of illumination. The equivalents of both positive and negative phototropic curvatures were produced in this way.[94]

Went's extended and masterly account of these experiments appeared in 1928, an account which will undoubtedly form one of the classics of botany.[95] His precise and uniform techniques and the statistical soundness of his experiments preclude lack of confidence in his results and the existence of growth-regulating substances was removed from doubt and uncertainty. In his later experiments, in which he used agar blocks in place of gelatin, he was able to establish an exact relationship between the amount of growth substance diffusing into the base of a coleoptile and its growth (measured by the angular deflection produced by placing a block unilaterally on the base under uniform conditions). At low concentrations there was simple proportionality, but at higher, the amount of growth approached its maximum value at a diminishing rate as the concentration of growth substance increased. A relationship of this kind, familiar to plant physiologists, suggests that the growth substance can be looked upon as a 'limiting factor' of growth in Blackman's sense.[96] Furthermore, measurement of the rate of diffusion of the growth substance from one agar block to another indicated that its molecular weight lay between 350 and 400. To account for the evident rapid longitudinal movement of the growth substance, about 200 times that which would be expected by simple diffusion, Went envisaged its being carried by protoplasmic streaming within the cells and by diffusion between cells.

One of Went's most important experiments from the point of view of phototropism was his demonstration of the effect of light on the distribution of the growth-promoting substance. Coleoptiles were illuminated from one side only with 1000 m.c.s. of light, the tips removed and placed on two agar blocks separated by a razor blade, just penetrating the base of the coleoptile and dividing it exactly into the illuminated and unilluminated halves. When the amount of growth substance in the agar blocks was estimated in the standard manner, he found that, although the total amount of growth substance coming from the illuminated tips was about 16 per cent less than that from the controls kept in the dark, nevertheless, the amount coming from the shaded side of the former was more than twice that coming from the lighted, and was therefore more than half that coming from the controls. This suggested that the main effect of illumination was to cause lateral transport of the growth substance from the light to the dark side, and that destruction of the growth substance by light took place in the conditions of this experiment only to a small extent.

These experiments have been described at length because they form a landmark in the study of growth and tropisms. They were followed by intense activity, both in the study of normal growth and of growth curvatures. At this point the older notion of Irritability, the respectability of which had been declining since the work of Paál, ceased to have any utility. Even if the process by which light exerted its initial influence remained obscure, once the influence had been imparted to the plant all subsequent activity could be interpreted in terms of a tangible substance. Moreover, it was no longer necessary to imagine a specific 'phototropic hormone', for all that was involved was the fundamental process of the regulation of growth, the mechanism of which was likely to be common to all higher plants. Went also proposed for geotropism an explanation similar to that for phototropism. Cholodny, whose experience was mainly of geotropism, arrived independently at the same conclusion.[97] Their view, often referred to as the *Cholodny–Went theory*, was that all growth curvatures resulted from an unequal distribution of a growth-promoting substance. In phototropism the unequal distribution was brought about by the transverse movement of the substance away from the illuminated side; in geotropism, in the direction of the field of force.*

THE CHEMICAL NATURE OF THE GROWTH-PROMOTING SUBSTANCES

Many substances, such as malt extract and saliva, were known to be capable of promoting the growth of the *Avena* coleoptile,[99] but the first to isolate pure chemicals with this property were Kögl and his collaborators at Utrecht.[100] Two were isolated, the first from human urine and the second from malt and maize germ oil; both were found to be complex organic compounds. Kögl proposed for them the general term *auxin*, referring to the first as auxin *a* and the second as auxin *b*. Subsequently yet another compound of similar activity was isolated from urine and named *heteroauxin*. This proved to be indole-3-acetic acid (II), already known from certain fungal fermentations, and closely related to the amino acid tryptophane (I), which occurs generally throughout the plant kingdom.

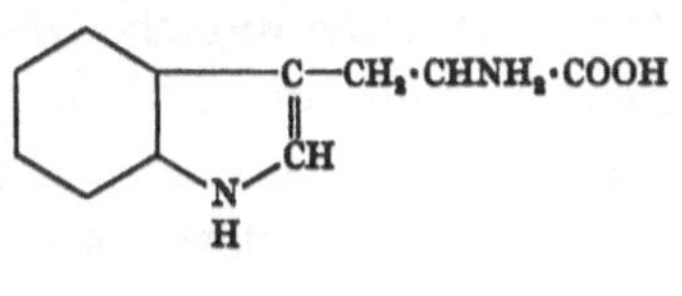

(I)

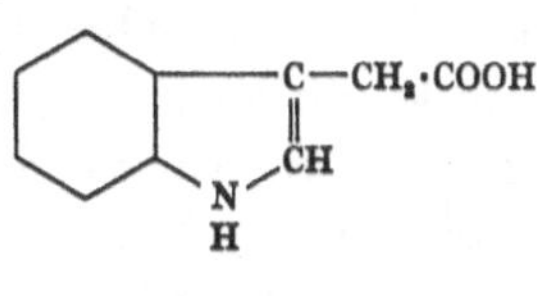

(II)

The term *auxin*, although originally used for the chemically pure compounds isolated at Utrecht, has come to be a convenient general term for all substances, both artificial and natural, having a growth-promoting property. It will be used here with this broad meaning.

* It should be noted that the simple Cholodny–Went explanation of geotropism has recently been seriously challenged.[98]

The identity of the naturally occurring auxin or auxins involved in phototropism became the subject of immediate inquiry. Went's estimate of the molecular weight of the *Avena* auxin agreed well with that of auxin *a*, and the auxin from the coleoptiles of *Zea mays* (maize) and the 'regenerated' tip of *Avena* coleoptiles seemed to be similar. In the early investigations the auxin of coleoptiles appeared to be destroyed in alkaline solution, a property of auxin *a*, but not of indole-3-acetic acid. These observations indicated that the *Avena* auxin might be identical with Kögl's auxin *a*, a view that was strengthened when it was shown that growth curvatures produced in the decapitated coleoptile by the unilateral application of auxin *a* and natural auxin were both diminished to a similar extent if the coleoptiles were uniformly illuminated.[101] It appeared that both the natural auxin and the auxin *a* were suffering photo-inactivation, and there was already available some evidence that auxin *a* was in certain conditions unstable in the light. In the isolation of auxin *a*, Kögl had evidence that it was in equilibrium in solution with its lactone. This lactone also stimulated growth, but was converted by ultraviolet light into an inactive lumi-auxin *a* lactone. Certain pigments, known as carotinoids, occur in coleoptiles and absorb light very strongly in the blue-violet region of the spectrum. It was thought that these pigments, acting as sensitizers, might bring about photo-inactivation of auxin *a* within the cells of the coleoptile. To add strength to this conjecture it was later shown by Kögl and Schuringa that β-carotene did in fact sensitize the photo-inactivation of pure auxin *a* lactone in visible light.[102] It was already known that not only did the *Avena* coleoptile, although apparently colourless, contain carotene pigments,[103] but also that β-carotene was more abundant in the region most sensitive to light than elsewhere, except that it was more or less absent at the extreme tip.[104] For a time, therefore, the evidence that natural auxin was identical with auxin *a* was very strong and the identity appears to have been accepted without question by some of the workers at Utrecht. However, in addition to the contradictory evidence from recent studies, it has not been possible to confirm that β-carotene sensitizes the photo-inactivation of natural auxin.[105]

Notwithstanding the weight of the evidence in favour of the

occurrence of auxin *a* in the *Avena* coleoptile, more recent research has strengthened the claims of indole-3-acetic acid, usually referred to as IAA (in German, IES), as the principal naturally occurring auxin. In higher plants this substance was first isolated from maize meal by alkaline hydrolysis. It was subsequently obtained from immature maize grains (in the 'milk' stage, about 15 days after fertilization of the ovule) by simple extraction with a mixture of ethyl alcohol and water.[106] Auxin obtained in this way is regarded as 'free', that is, it is capable of movement within the plant or of diffusing into agar, whereas the auxin which can only be obtained by alkaline hydrolysis or by prolonged extraction with solvents is referred to as 'bound' auxin. Recently Wildman and Bonner[107] have estimated that the auxin which can be extracted from the *Avena* coleoptile by cold ether represents about one-fifth of the total diffusible auxin. They further show that this auxin is almost entirely confined to the upper 5 mm. of the coleoptile, that this distribution is directly related with that of an enzyme system capable of transforming the amino acid tryptophane into IAA, and that the auxin itself displays the reactions of IAA. Moreover, if auxin collected by diffusion into agar in the normal manner is subsequently purified by extraction with ether, it is found to have a diffusion constant indicating a molecular weight of 206. This, although a little above the expected value for IAA (175), is considerably closer to that than to the molecular weight of auxin *a* (328). These authors concluded that the bulk of the auxin in the *Avena* coleoptile was IAA, allowing the possibility of the presence of another, possibly auxin *a*, in small quantity.

Other workers in Germany, however, found no auxin in the coleoptiles of *Avena* and *Zea* which remained stable in alkali, although such an auxin, presumably IAA, occurred in the seed. Von Guttenberg maintained, in fact, that the action of IAA was not to influence growth directly, but to bring about the liberation of auxin *a*.[108] This view was supported by experiments of two kinds. In the first, certain dicotyledonous plants treated with IAA appeared to contain in consequence an increased amount of an auxin having the properties of auxin *a*.[109] In the second, it was shown that in *Avena* coleoptiles whose content of natural auxin had been very much reduced by two successive decapitations growth curvatures were produced much more rapidly by natural auxin than by IAA, even though when tested on coleoptiles decapitated only once, the amounts of natural auxin and IAA were such that they produced equal responses.[110] These latter results do not, of course, prove that the natural auxin used in these experiments consisted of auxin *a*; it may have been IAA, but enhanced in its action by other substances accompanying it. Recent work, in fact, is wholly in support of Wildman and Bonner. Reinert, for example, estimated that 90 to 95 per cent of the auxin extracted from *Avena* coleoptiles by cold ether was identical with IAA.[111] The deficiency of some 5 per cent was not identified but did not appear to be auxin *a*.*

The view that auxins are fundamental metabolites, unlikely to show differences between species of the higher plants, has received no serious challenge. In some early experiments it was found that if the apices were interchanged between the coleoptiles of grasses of different species, the phototropic response was in general decreased according to the systematic distance between the species.[113] There were, however, exceptions; for example, when the tip of *Avena sativa* was placed on the stump of *Triticum polonicum* L., the curvature was actually greater than that produced by the same quantity of light in normal *Triticum* coleoptiles. These results can be interpreted just as well in terms of differences in amount of auxin coming from the different apices as of differences in kind. Söding, studying the activity of extracts from a wide range of plants, concluded that one growth substance was common to them all, but present in varying amounts.[114] Again, as techniques for the identification of IAA in plant extracts were improved, the occurrence of IAA was found to be widespread.[115] By 1952 it was possible to state with confidence that IAA was 'a principal native auxin of higher plants'.[116] Evidence has also accumulated that substances closely related to IAA, namely indole-3-acetaldehyde (IAc) and indole-3-acetonitrile (IAN), also occur naturally in some plants and are either auxins themselves or are capable of being rapidly transformed into IAA in living cells. Other substances which accelerate and depress growth, some of unknown chemical nature, are also known to occur in some plants in addition to IAA. Their isolation has been facilitated

* Note, too, that von Guttenberg, using modern techniques of identification, is now prepared to accept as IAA auxins formerly thought not to be this.[112]

by paper chromatography, a technique which depends upon the fact that substances of different molecular weights in solution are carried across absorbent paper at different rates.[117] It should be remembered of course that it remains a matter of judgment to decide to what extent substances identified in plant extracts actually enter into the metabolism of the living cells.

In addition to the auxins occurring naturally a wide range of synthetic substances is now known to have similar properties. They and their applications in horticulture and agriculture have been lucidly described by Audus.[118]

THE TRANSPORT OF AUXIN IN THE PLANT

The means by which natural auxins are transported within the plant remains an unsolved problem. That it could be accounted for by movement of protoplasm within the cells appeared unlikely when it was shown that the speed of movement of auxin, unlike the streaming of protoplasm, was independent of temperature.[119] According to Bottelier, however, the moving of protoplasm is markedly depressed by light and the effectiveness of light of different wavelengths in producing this depression closely parallels the effectiveness in producing phototropic curvature.[120] It is tempting to suggest that this close parallel indicates a causal relationship, but there is no direct evidence that such exists. In fact, the transport of IAA in coleoptiles in which the movement of the protoplasm has been suppressed by saponin remains unimpaired. On the other hand, sodium glycocholate in low concentrations suppresses the transport of IAA, but not the movement of protoplasm. Incidentally, coleoptiles treated in this way continue to show normal phototropic and geotropic responses.[121]

It appears that the movement of auxin normally takes place in the direction away from the apex in the coleoptile and from actively growing tissues of aerial stems generally. If a segment of one coleoptile is inserted into another near the tip, the phototropic response is propagated into the base only if the inserted portion is in an upright position; if inverted, no curvature appears below.[122] Auxin cannot be made to move through segments of coleoptile in a reverse direction, even if the difference in concentrations above and below the segment is very great. In these conditions auxin may even be transported from the region of low concentration to that of high, a process the reverse of diffusion. This strict polarity in the movement of auxin does not appear to occur in roots, nor does it apply to the movement of the precursor of auxin which ascends into the coleoptile during germination. It should be noted, too, that some deny that the transport of IAA is strictly polar.[123]

The Part Played by Auxin in Phototropism

THE INTERPRETATION OF THE OLDER EXPERIMENTS

The Cholodny–Went theory immediately allowed a satisfactory explanation of the older decapitation experiments. Stark and Dreschel, for example, had subjected *Avena* coleoptiles to unilateral illumination for 1 hour, at the end of which the coleoptiles were bending towards the light. The light was then extinguished and the tips of the coleoptiles removed and transferred to freshly decapitated coleoptiles which had been raised and kept in total darkness. Very soon after the transfer the bending was propagated from the tips into the foreign base.[124] This can clearly be interpreted as a consequence of the unequal distribution of auxin in the illuminated tip.

The interpretation of the incision experiments of Boysen Jensen and others is not quite so simple. If an incision on the side away from the illumination completely prevents the descent of auxin on that side, then, if the coleoptile is to remain straight, auxin must also be absent from the illuminated side, but this is not in conformity with Went's results. In later experiments at Copenhagen the coleoptiles with the incision on the shaded side showed a small, but definite, positive curvature. One must either assume that a 'leakage' of auxin took place around the incision, or that the small amount of auxin descending on the illuminated side and passing the incision was redistributed below it. In any event the interpretation of all experiments involving unilateral incisions is hazardous because of the lack of any certain knowledge of the extent to which the wounding disturbs the physiological symmetry of the coleoptile.

The demonstration of the strong polarity of the movement of auxin explained the early failure to obtain curvature of the apex of the coleoptile by illumination of the base. It appears doubtful whether the claims (see p. 22) to have modified the responses of the apex by a previous illumination of the base can be correct, but the point merits further attention, especially since von Guttenberg has questioned the strictly polar movement of IAA in the coleoptile.

Van Overbeek[125] showed that the system revealed by Went existed also in dicotyledonous seedlings. In *Raphanus sativus* L. (radish) auxin flows from the cotyledons down into the hypocotyl. Unilateral light causes a redistribution of this flow in a way precisely similar to that occurring in the *Avena* coleoptile.

As is usual in scientific research, the new theory, although bringing order where there was confusion, posed as many new questions as it solved. It soon became clear that the amount, distribution and activity of auxin in the plant might be influenced in a number of different ways by light. In consequence, a phototropic curvature, although mediated by auxin, might follow from a simple intermediary process or a complex. It will assist the analysis if we consider the possible actions of light and the evidence for their occurrence separately before attempting an assessment of their relative importance in the phenomenon of phototropism.

THE INACTIVATION OF NATURAL AUXIN BY LIGHT

Although the growth substance extracted by Went in his original experiments was stable in the light (as is IAA), nevertheless, both the uniform and unilateral illumination of the apex of the coleoptile appeared to show a definite reduction in the amount of auxin leaving it. These experiments were extended by others who used a wider range of lighting. Their results, particularly at high light intensities, showed a marked decrease in the growth-promoting properties of the apices of both *Avena* and *Zea* coleoptiles following illumination.[126] In addition, and contrary to Went's findings, the effectiveness of agar blocks into which natural auxin had diffused was also reduced by light. On the other hand, van Overbeek, in his experiments with *Raphanus*, found no evidence for the photo-inactivation of auxin in the cotyledons; on the contrary, the output of auxin from green cotyledons transferred to the dark rapidly decreased.[127] By 1936, however, it seemed that auxin *a*, as distinct from IAA, suffered photo-inactivation in *Avena*, since the growth of decapitated coleoptiles to which it was supplied was clearly depressed in the light. Unfortunately, the value of the work of the Utrecht school at this time was lowered by the assumption that experiments involving auxin *a* were representative of the natural system. The significance of the results of Koningsberger and Verkaaik, who also provided convincing evidence for the photo-inactivation of auxin *a* in de-seeded and twice decapitated coleoptiles (which can be assumed to be more or less free of natural auxin) is doubtful for the same reason.[128]

In 1940, Stewart and Went showed that the auxin extractable by ether from the *Avena* coleoptile was less after illumination.[129] Since this auxin would now be considered to be a measure of the 'free' auxin,[130] these results can be considered to confirm the earlier ones in which the auxin was obtained by diffusion. Oppenoorth also carried out extensive experiments of a similar kind, illuminating the upper part of the coleoptile unilaterally with blue light (wavelength 436 mμ).[131] According to his results, with small quantities of light (less than 150 ergs* per sq.cm.) no destruction of auxin occurs, but with 330 ergs per sq.cm., at which the first positive curvature is developed to a maximum extent, the amount of auxin extractable by ether diminishes by about one-third and this reduction occurs equally on the lighted and shaded sides of the coleoptile. With 3000 ergs per sq.cm., sufficient to produce the first negative curvature, no reduction in the amount of auxin can be detected, but increasing quantities of light again diminish the amount of auxin in the tip.

When the naturally occurring auxin was thought to be the auxin *a* of Kögl, there was evidence that photo-inactivation might be brought about by carotene, but no such mechanism was known for IAA. This remained an argument against IAA being the auxin involved in phototropism, until it was shown in 1949 that another pigment, riboflavin, could cause the oxidation of IAA and the closely related amino acid, tryptophane, in visible light.[132] It was later shown that the disappearance of IAA from ground-up tissue of *Pisum* (pea) containing the enzyme IAA-oxidase was accelerated by light. This effect again appeared to be associated with the presence of riboflavin.[133] Riboflavin is certainly present in the *Avena* coleoptile, although, unlike β-carotene, it is not present in

* The erg is a measure of energy. 1 erg per sq.cm. per sec. of light of wavelength 436 mμ can be taken as equivalent to an intensity of 25 m.c. of white light.

greater quantity at the apical region. The auxin diffusing from *Avena* coleoptiles, if incorporated into an agar block in the presence of riboflavin, is inactivated in the light, strong evidence in favour of its being IAA. It seems safe to conclude that a system involving riboflavin and capable of inactivating IAA in visible light is present in the *Avena* coleoptile and other plants. Other evidence is now accumulating that the presence of carotene may influence the action of riboflavin, possibly by absorbing the wavelengths of light by which the riboflavin is activated.[134]

THE EFFECT OF LIGHT UPON THE SYNTHESIS OF AUXIN

The synthesis of auxin must depend upon the availability of the precursor and the presence of an enzyme system. If tryptophane be the precursor, then the experiments mentioned in the preceding section show that light may influence the synthesis by causing the destruction of the precursor. On the other hand, the experiments of van Overbeek, in which the amount of auxin diffusing into agar from the cotyledons of *Raphanus* declined in the dark, suggest that here light may promote the synthesis. The same holds for the cotyledons of *Lepidium sativum* (cress) and possibly for the apical portions of certain other plants,[135] but there is no general agreement that the production of auxin is directly dependent upon light.[136]

Oppenoorth has attempted to follow the effects of unilateral illumination on the synthesis of auxin. He submitted a series of *Avena* coleoptiles to 3000 ergs per sq.cm. of blue light unilaterally and measured the changes in the amount of auxin extractable by ether on the illuminated and shaded sides of the tips for a period of some 4 hours after the illumination.[137] An increase in the amount of auxin, predominantly on the illuminated side of the coleoptile, appeared to begin immediately after the illumination and to continue for the duration of the experiment. That the most marked increase should have been on the illuminated side agrees well with the fact that this quantity of light produced a negative curvature. Evidence for an enhanced synthesis of auxin was also obtained with 330 ergs per sq.cm. of blue light and 500 m.c.s. of white light, both of which in these experiments produced the maximum first positive curvature. Here, however, the increased synthesis occurred only after an initial fall in the amount of auxin present.

At the present time it is not safe to draw any general conclusion about the effect of light on the synthesis of auxin in the aerial parts of plants. It is clearly very difficult to distinguish between the inactivation of auxin and the suppression of its synthesis in intact living tissue during or immediately following illumination. Whether suppression of synthesis occurs must remain for the moment an open question.

THE REDISTRIBUTION OF AUXIN IN UNEQUAL LIGHTING

In addition to Went's original experiments, van Overbeek provided evidence that lateral transport of auxin took place in the apical region of the seedling of *Raphanus* in similar conditions.[138] Later experiments upon *Avena*, using greater quantities of light, have shown that in the range of the first negative curvature the amounts of auxin diffusing from the apex are higher on the lighted than on the shaded side, and in the range of the second positive curvature the relationship of the amounts is again reversed. The results of Miss Wilden[139] are shown in Table III. Her figures were obtained by placing the whole tip after illumination with either its light or its dark side directly upon the decapitated coleoptile used for the quantitative estimation, the object being that natural movement of auxin could continue while the estimate of the redistribution was being made. This method is open to some objections and the results should not be regarded too critically; nevertheless, the proportions obtained for the first negative curvature are of the same order as those obtained earlier by Asana[140] using Went's agar method.

TABLE III. *The amounts of auxin diffusing from the illuminated and shaded sides of the upper 2 mm. of coleoptiles of* Avena sativa *after unilateral illumination (from Wilden)*

Amount of light (m.c. × sec.)	Nature of curvature	Ratio of amounts of auxin* (illuminated : shaded)
100 × 15	+	17:83
380 × 30	−	62:38
3,000 × 50	+	36:64

* Determined from the test curvatures developing 130 to 140 minutes after the illumination.

Boysen Jensen also carried out experiments with *Avena* designed to demonstrate active lateral movement of auxin. He placed agar blocks containing the auxin from the fungus *Aspergillus niger* (probably IAA) on to the tops of decapitated coleoptiles illuminated from one side.[141] The phototropic curvatures of these coleoptiles were much stronger than in a control series bearing blocks of pure agar alone. The experiment is not conclusive, for the unequal growth of the two sides of the first series of coleoptiles might have followed from the inactivation of the auxin or its reduced transport on the illuminated side. Since the transported auxin was not collected at the base of the coleoptile, it is not possible to eliminate these alternative explanations.

There is, however, evidence of another kind in support of the lateral transport of auxin. In 1928 Boysen Jensen longitudinally bisected the apex of the *Avena* coleoptile and inserted a thin glass coverslip. When the illumination of the apex was perpendicular to the coverslip only very little curvature occurred, but if the slip were removed and the two halves of the apex kept in contact the coleoptile behaved almost normally.[142] The effect of the glass plate, which interfered hardly at all with the gradient of light through the apex, could be ascribed to its preventing the lateral movement of auxin. Studies of the exact course of the phototropic bending also give direct support to the idea of lateral transport. In the apical zone there is only a very slight increase in the growth rate of the shaded side, but growth of the lighted side is sharply retarded so that the tip bends abruptly to the light.[143] In the lower zones there is both acceleration on the dark side and retardation on the light side. These effects would follow a redistribution of auxin, assuming that in the apical zone an increase of auxin on the shaded side would produce little or no acceleration in growth, since it would already be present in adequate amount, whereas in the lower zones, where growth would be directly dependent upon the amount of auxin, both a decrease in growth on the lighted side and an increase on the shaded would be possible.

Despite these experiments and observations many remained sceptical of the existence of active lateral transport of auxin,[144] finding it necessary to envisage nothing more than inactivation of auxin or a change in its rate of production. Very recently, convincing evidence has come in support of lateral movement from experiments on *Zea*.[145] The upper 4 mm. of the coleoptiles were used in these experiments. First, it was shown that if the tip was longitudinally bisected and the halves kept in contact, the amount of auxin diffusing into agar in 3 hours in the dark was not significantly less than that coming from entire tips (fig. 5, *a* and *b*). Secondly, if coleoptiles were illuminated from one side at an intensity of 21·5 m.c. at their surface for the same period, there was no significant change in the amount of auxin entering the agar (fig. 5, *c* and *d*). Thirdly, if the agar block and the coleoptile at its base were bisected by a thin glass coverslip perpendicular to the direction of light, then a much larger amount of auxin was obtained on the shaded side. If the

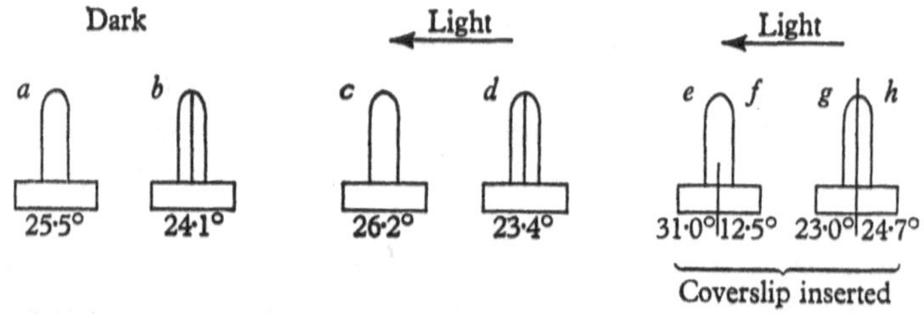

Fig. 5. Auxin diffusion experiments on the tips of coleoptiles of *Zea mays*. *a*, three intact tips kept in darkness; *b*, three tips completely split longitudinally, also in darkness; *c*, three intact tips illuminated; *d*, three tips completely split longitudinally and illuminated; *e*, six partially split tips, dark side; *f*, the same, light side; *g*, six tips, completely split, dark side; *h*, the same, light side. The numbers indicate amount of auxin obtained by diffusion into agar blocks, expressed as degrees of curvature of the coleoptiles used for estimation in the standard way. (From Briggs, Tocher and Wilson.)

coverslip extended to the tip of the coleoptile, however, this concentration of auxin on the shaded side was prevented (fig. 5, *e*, *f* and *g*, *h*). These experiments are complementary to that of Boysen Jensen in 1928 and, since the results rule out a suppression of the synthesis of auxin or its inactivation by the lighting used, lateral transport of auxin in certain conditions seems to be firmly established.

It should be noticed, however, that if radioactive IAA is supplied to the summits of *Avena* coleoptiles from which the upper 0·2 mm. have been removed, there is no evidence that light of low intensity from one side causes its lateral transport.[146] It may, of course, be that the removal of the uppermost tissue, although minute, disturbs the normal system.

Auxin in Phototropism

THE INFLUENCE OF LIGHT ON THE LONGITUDINAL TRANSPORT OF AUXIN

Some of Went's early results suggested a direct effect of light upon the transport of auxin.[147] Further evidence in support of this was produced by du Buy,[148] but his results were contradicted by van Overbeek's experiments in which pure auxin *a* was used.[149] There is, as yet, no general agreement upon the effect of light upon transport. Some have maintained that in the *Avena* coleoptile light is inhibitory, while others hold that in the hypocotyl of *Helianthus* light is essential for transport. The polarized movement of auxin must demand a source of energy and light may affect the availability of this energy. Treatments which interfere with respiratory processes, for example, usually diminish the transport of auxin as well.[150] Photo-inactivation of auxin may also involve the production of substances which interfere directly or indirectly with transport.

THE INFLUENCE OF LIGHT ON THE SENSITIVITY OF CELLS TO AUXIN

There is no doubt that the sensitivity of cells to auxin varies. This variation is one of the difficulties in using the *Avena* coleoptile for quantitative estimation of auxin. Coleoptiles of genetically pure lines of *Avena*, raised under identical conditions, may vary in their response to the same amount of IAA and it has not yet been possible to explain these differences in behaviour satisfactorily.*

Evidence in favour of light bringing about a change in the sensitivity of cells to natural auxin is gradually accumulating. It has been shown that in *Pisum* the response of the cells of the stem to auxin is diminished by illumination, although the content of auxin remains unchanged. Similar results have been obtained with the hypocotyl of *Cucumis sativus* L. (cucumber).[151] A reduction in sensitivity of cells to auxin would account well for the results of Brauner's experiment, upon which he based his theory of phototropism (p. 25), if we assume that a difference in the sensitivity

* It is now known that it is possible to set up by simple means cyclical variations in the growth rate of the *Avena* coleoptile. The cycle has a period of about 24 hours and continues without further stimulation in a constant environment. The extent of the response to IAA, and possibly to light and gravity, of such coleoptiles may depend upon the point in the cycle at which these stimuli are given.

of the cells of the light and dark sides of the base of the coleoptile persisted after the illumination. Similarly, it has been found that the uniform application of IAA to the summit of the hypocotyl of *Cucumis* after it has been illuminated from one side considerably increases the subsequent curvature in the dark.[152]

THE MECHANISM OF PHOTOTROPISM

There seems every reason to believe that inactivation and lateral transport of auxin, and a change in the sensitivity of cells to it are all involved in phototropism, but the degree to which each of these processes participates in producing the curvature varies according to the intensity of the illumination. Curvatures appear at very low intensities at which, as we have seen, there is no evidence to suggest inactivation of auxin. Moreover, coleoptiles bending towards a weak light continue to increase in volume at the same rate as those remaining straight in the dark.[153] Increase in volume is a measure of cell expansion, and if this in turn is directly related to the supply of auxin, this is additional evidence that inactivation plays no part in these curvatures. A marked change in the sensitivity of the cells at these low intensities seems equally unlikely, so the cause must lie almost entirely in the redistribution of auxin from the light to the dark side, for which strong evidence has now been provided.

At higher light intensities, inactivation of auxin is almost certainly involved. Brauner, for example, has designed a model which shows how photo-inactivation can bring about curvature.[154] The cotyledons were removed from a seedling of *Helianthus* and a glass capillary containing a solution of IAA and riboflavin was attached to each stump (fig. 6). When one of the capillaries alone was illuminated the hypocotyl began to bend strongly towards the side bearing it, indicating that a phototropic curvature could follow from a similar mechanism. Nevertheless, the extent to which inactivation enters into the mechanism of curvature in general is not known, although that caused by ultraviolet light in *Avena* may result from the direct inactivation of a compound of IAA in the affected cells. In *Avena* in visible light, Oppenoorth's results suggest that the inactivation occurring at intensities producing the first positive curvature is of the same extent on both the light and dark sides of the coleoptile. If this be a true picture of the conditions in the coleoptile, then the curvature must follow from the subsequent redistribution of auxin or the different rates of synthesis on the light and dark sides which his results indicate. It should be mentioned that the amount of blue light used by Oppenoorth to produce the first positive curvature appears extremely high. The figures given by du Buy and Nuernbergk[155] show that Oppenoorth's maximum 'first positive curvature' was obtained at intensities which would be expected to produce the first negative.

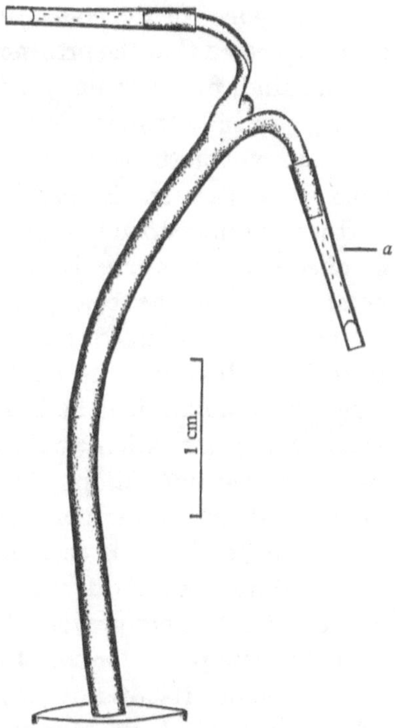

Fig. 6. Brauner's experiment. The capillary *a* has been illuminated, but not the other. For explanation, see text.

Little evidence is available of the extent to which a change in the sensitivity of the cells to auxin enters into phototropism, although it seems likely that it does so at higher light intensities. It has been shown that illumination of the cells of the growing zone of the hypocotyl of *Helianthus* with intense light (3024 m.c.) diminishes their capacity for both reversible stretching (*elasticity*) and irreversible stretching (*plasticity*), due to changes in the wall.[156] These changes may not, of course, be produced directly by the light, but follow from its effect upon the supply of auxin in the cells, since auxin is known to affect the plasticity of cell walls.[157] The fact that preliminary uniform illumination diminishes the phototropic sensitivity of seedlings and coleoptiles may be caused by either a general lowering of the level of mobile auxin through photo-inactivation, or by a change in the sensitivity of the cells, or possibly by both. The same would hold for the reduction in the phototropic response brought about by uniform lighting after the unilateral.

The mechanism by which the lateral transport of auxin is brought about is obscure. In *Avena* it has for long been held that the strength of the phototropic curvature depends upon the steepness of the gradient of light across the coleoptile.[158] It now seems likely that the steepness of this gradient may influence the curvature by controlling the extent of the lateral movement of auxin, for, if the interior of the coleoptile be filled with an opaque substance such as indian ink, the gradient of light is steepened and the phototropic response of the coleoptile is markedly enhanced.[159] It has been suggested that the presence of carotene, acting as a light screen of this kind, may account for the greater phototropic sensitivity of the tip generally than the lower region of the coleoptile,[160] but this does not explain the very pronounced sensitivity of the extreme tip, from which carotene is absent.

It is not understood how the actual lateral transport of the auxin is effected, but there is evidence that a transverse electrical polarization of the coleoptile may be involved.[161] When the coleoptile is illuminated from one side, the side of the apex adjacent to the light becomes electrically negative in relation to the shaded side. Since auxin, if it be IAA, probably exists in the cells as an anion, this polarization would promote the movement of the auxin to the shaded side. Biophysical investigations of this kind are still in their infancy and important discoveries may emerge from them in the future. There is certainly reason to think that the method of lateral transport is different from the longitudinal, since inhibition of the latter does not prevent the onset of phototropic curvature.[162]

If the time taken for the *Avena* coleoptile to respond to light be plotted against its intensity, there is a sharp break in the continuity of the curve at about 54 m.c. (fig. 7).[163] This seems to be due to there being two overlapping responses, one associated with the upper 1·5 mm. of the coleoptile, the other with that part of the coleoptile below the uppermost 3 mm. It is therefore possible that the redistribution of auxin, which we have considered to be responsible for the curvature at low intensities of light, is a process more or less confined to the tip; whereas the processes responsible for curvature at intensities from about 50 m.c. up to about 1000 m.c. are to be found principally below the uppermost 3 mm. Above this intensity the situation may change yet again. The two responses occurring at intensities below 1000 m.c., besides showing spatial separation, also show different sensitivities to light of different wavelengths. Although there are considerable divergences, the curve of sensitivity of the tip is closer to the absorption curve of riboflavin than to that of β-carotene, but the significance of this

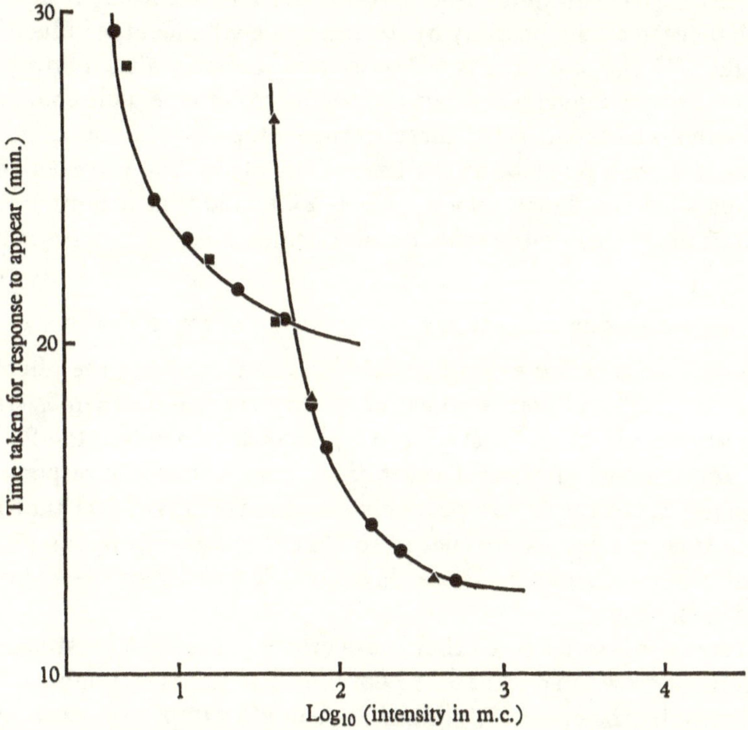

Fig. 7. The relation between the intensity of light (white) and the time taken for the first positive curvature to appear in coleoptiles of *Avena sativa*. The exposure to the light in all experiments was one second. ●, unshielded coleoptiles; ■, all but uppermost 1·5 mm. shielded; ▲, uppermost 3 mm. only shielded. (Data from Haig.)

must remain doubtful while the role of this pigment in phototropism is undecided. Nevertheless, the differences in spectral sensitivity support the notion that two processes or groups of processes are involved in phototropism, spatially separated and coming into play at different intensities of light. There is a clear parallel with the two light-growth reactions discussed earlier (p. 17). The light-growth reactions in themselves proved incapable of explaining phototropic curvature, but there is no reason to doubt that they and the latter are aetiologically similar.

In the study of phototropism most experiments have been concerned with stems and coleoptiles, although similar movements of leaves and flower stalks are, as we have seen, amongst the earliest recorded. It seems that these movements, too, are brought about by changes in auxin, possibly by its displacement under the influence of light.[164] The movements of leaves, and probably also of flowers, are not growth movements, but are produced by reversible changes in volume of certain cells, often grouped together in the case of leaves to form a *pulvinus* at the base of the blade. The leaves of the members of the Leguminosae which attracted the attention of Alpinus and others all possess pulvini of this kind.*

THE EFFECT OF LIGHT ON NATURAL VEGETATION

Although we now have considerable information about the phototropic behaviour of plants under laboratory conditions, there is still little known about the part played by light in controlling the form and development of natural vegetation. The phototropic responses of leaves, for example, are probably responsible for the fact that in many trees the leaves are placed so that they overlap to the least possible extent, and that those leaves which do overlap are distant from each other.

In stems it is well known that those grown in the light are stockier than those grown in darkness or poor light and this is no doubt to be explained by the effect of light either on auxin within the stem or on the sensitivity of the cells to it. Seedlings nearing the surface of the soil will respond to weak light penetrating the crevices, and the curvatures developed in these conditions will probably be equivalent to the first positive curvatures of the laboratory. In full daylight, however, the position is less simple, because the intensity of light may approach 100,000 m.c., that is to say within the range producing the second positive curvature of the *Avena* coleoptile. It

* Experiments have shown that the daily movements of the leaves of some Leguminosae persist for a time in continuous darkness, suggesting that they may arise from some cyclical process in the plant. Nevertheless, it seems clear that in normal conditions light, and to a certain extent temperature, determine the period of the movement.

may be that the form of vegetation at different latitudes is in part determined by the quantity and intensity of the light to which it is subjected. For example, certain plants which are normally prostrate in exposed conditions at equatorial latitudes become upright when they are partially shaded.[165] This may be because in full illumination their stems show a negative phototropic response and so are maintained in a prostrate condition, whereas in reduced illumination they are either indifferent or respond positively. Differences in habit of this kind between species may be explained by their containing different amounts of auxin, or by their cells differing in their sensitivity to it. Reversals of phototropic response, such as we see in the pedicel of *Cymbalaria muralis* Baumg. (ivy-leaved toadflax), which is initially positively, and finally negatively phototropic, may be explained by similar differences coming about in an organ as it matures. Research into these topics has hardly yet begun.

Conclusion

In almost a century of research there have been great advances in our understanding of phototropism, but much remains obscure. Blaauw's assertion, 'Surely there is no problem in phototropism itself', has proved to be unduly optimistic. It is true that phototropism is now known to be intimately connected with those general problems of growth to which Blaauw was directing attention. Nevertheless, phototropism has its own problems, such as the precise means by which light energy is able to enter into the functioning of the apical cells and cause the lateral displacement of auxin which now seems, in certain conditions, to be proved. There seems, too, to be some amplification mechanism within the plant, since the energy of the incident light may be less than that involved in the response. Explanations of these perplexing phenomena will require a study of the biophysics of cells as well as their biochemistry.

Not only has research shown that there is no specific 'phototropic hormone', but it is now clear that auxin itself is less specifically concerned with the extension of cells than was formerly assumed. Auxin is now thought to be involved in processes as diverse as those determining the sex of flowers and controlling the outgrowth of lateral buds in a shoot. The discovery of naturally occurring antiauxins, substances capable of reducing or neutralizing the effect of IAA on elongating cells, is a further complication which suggests that the simplicity of the Went–Cholodny hypothesis may be deceptive. Moreover, there is a growing belief that the expansion of cells is a much more active process than formerly thought, involving an increase in their protein content and metabolic activity. This implies that the morphological effects of auxin are more likely to follow from its entry into the fundamental metabolism of cells than from its action on some isolated feature, such as the plasticity of the wall.

We are, in fact, approaching the position where not only tropisms, but also the general control of form and development in plants, are to be ascribed to auxin or a balance of auxin and inhibitors. If these substances are to have functions so multifarious and so fundamental in organization, they must be in the nature of essential metabolites playing a particularly crucial and delicate part in the biochemical systems of plant cells. We are then back at the basic question of how it is that light and gravity can affect certain plant cells in such a way as to disturb their metabolism so drastically at this vital point. It seems unlikely that these two agents act in precisely the same way since, for example, the distribution of the sensitivity to light in the *Avena* coleoptile, although similar to that of gravity, is not identical with it, and the nodes of the stems of grasses respond to gravity, but not to light (unless they have first been stimulated by gravity). Also climbing plants, such as *Humulus lupulus* L. (hop) are said not to show phototropic responses, but it is unlikely that auxin does not enter into the metabolism of these plants.

How little we know about what causes plants to grow into the shapes that they do is often lost sight of by the physiologist, but it constantly distresses the morphologist. It is in the hope that some will be stimulated to look into these basic problems of growth and development, of which tropisms form a part, that this essay has been written. They must, of course, be prepared to find, as Darwin, that originality is likely to be met with scorn from many of the scientifically orthodox of their day. They must also remember that the successful elucidation of the processes underlying apparently simple phenomena will in all probability depend upon an awareness of the advances in our knowledge of the chemistry and physics of living matter. It has been shown that the study of phototropism, like that of many other aspects of botanical research, has become intimately concerned with the basic problems of cell physiology. For this reason the tacit acceptance, not uncommon, of botany as consisting of two more or less independent disciplines, morphology and physiology, is neither sound nor likely to be profitable. Such a dichotomy would have astonished Darwin, for all his compelling interest in evolution. May it astonish his successors no less.

2

J. CHALLINOR

PALAEONTOLOGY AND EVOLUTION

The General Showing of Palaeontology

The lack of a thoroughly satisfactory harmony between the fossil record and the theory of evolution has presented a most intractable problem ever since Darwin so clearly exposed it in the tenth and eleventh chapters of *The Origin of Species*. He well realized that evolution, as a general proposition, apart from any particular explanation of it, must stand or fall on the record of life in the past as revealed by fossils; and his candour forced him to admit, indeed to insist, that this record, as then known, was a 'most obvious and serious objection', one which could justly be urged against the whole foundation of his beliefs. But he met the objection with determination and reason, explaining the disharmony as being due to 'the extreme imperfection of the geological record'.[1]

During the last hundred years an enormous amount of new material has been brought to light and studied by the palaeontologist, but this has by no means resulted in the record being smoothly reconciled with the theory. Much of the new detailed and precise information only shows that some of the obstacles, by remaining unmoved, are all the more formidable; but, on the other hand, many of the seeming contradictions are found to be less flagrant. In any case, Darwin's contention that it is the record that is at fault, not the theory, is an even more obviously sound one now than it was then.

The purpose of this essay is to inquire into the nature and extent of the evidence provided by palaeontology on the question of evolution. We shall attempt this by briefly considering theoretical matters and by taking general views of the facts. As to theory, we shall try to examine first principles, however many truisms this may involve; and as to facts, the British record is so extraordinarily rich and representative that we need hardly go beyond it for our illustrations.

Towards the end of the eighteenth century it began to be realized that there was such a thing as 'geological time', of which the rocks were a witness, stretching back into ages vastly more remote than the beginnings of humanly recorded history and, presumably, of the human race itself. To catch some glimpse of the life (if any) of those times seemed impossible. Yet that is indeed what we are able to do, and, though we now realize that this is no miracle when we consider the nature of those particular rocks which record the passage of geological time, we may well be astonished at the amount we can see.

Do students of evolution always make the best use of their good fortune in having this priceless treasure, the fossil record? One might well think that the great theme of the palaeontological evidence of evolution would have been so thoroughly and so often explored and expounded that there could now be no room for any further dissertation, unless it were to bring with it some new significant facts. Yet, surprisingly enough, there is little to be found.[2] It may well be that Darwin's chapters were not only the first, but are still the best, review of the whole matter. In reading those weighty pages we are not likely to forget that Darwin, apart from his fame as a biologist, was one of the foremost geologists of his day.[3]

The occasion of the centenary of the publication of *The Origin of Species* will be a timely reminder that we need to re-examine the whole question of what fossils have to show us. This is not intended to be an extended treatise, but a brief contribution to a general review of this fundamental problem.

It seems rather curious that, at its publication, *The Origin of Species* did not at once cause a rush among palaeontologists to find more evidence as to the manner in which new fossil species appeared in the rocks. One might have thought that attempts in evolutional palaeontology would have been very much to the fore in the eighteen-sixties, but it was not till nearly the end of the century that deliberate studies began to be made in Britain.

What, and where, are the data for any and every discussion in palaeontology? The fossils themselves, in the rocks, are the primary data; the different kinds of fossils and their occurrence in the different strata. Fossil specimens in a museum are samples (not altogether random) of these kinds; their provenance, geological position and other information having been noted (once for all) at the time of collecting. A piece of palaeontological research gives the results, in writing and picture, of the study of a relatively small number of specimens in as much detail as practicable. These accounts are the secondary data upon which a wider inquiry can be based, producing firstly generalizations and secondly theories and speculations (which may or may not be evolutionary) as to the significance of these generalizations. Secondary data, while having the demerit of being but partial excerpts from the facts of nature transcribed on to paper through the fallible agency of the human eye, hand and brain, have at the same time the merit of being a condensed and classified epitome of these facts. Nevertheless, for an inquiry into evolution, concerned with the different kinds of fossils and their relationships in space and time, the secondary data themselves are vast in extent. A complete bibliography would include nearly everything ever written on palaeontology. A selection of the significant works is included in the notes to this essay; they, in turn, will usually be found to contain lists which open up paths ramifying far and wide through the vast territory of palaeontological literature.

From estimates of the ages of rocks by means of the observed transformation in them of certain radioactive minerals, it now seems that the age of the earth, as a planet with a rocky crust, may be about 4500 million years. But geological history is chiefly concerned, and the history of life is almost entirely concerned, with the last 500 million years or so, that is, from the beginning of Cambrian times (see Appendix to this essay). Rocks of this latter part of geological time occupy the greater part of the earth's surface, even excluding the areas covered by superficial Pleistocene deposits; but it is the fact that they contain fossils, while the rocks of the preceding eras do not, that enables a geological history and, at the same time, a history of life to be read.

The complete stratigraphical succession over a wide area, and ultimately over the whole world, can be put in order by piecing together local successions, strata being approximately correlated from one region to another, chiefly by means of the particular fossils they contain. The rocks, arranged in a general time-sequence, thus reveal the time-sequence of fossils of all the different kinds they contain. So we obtain an historical record of life in the past, both on the grand scale, and, in places, with extraordinarily fine detail.

It is at once clear, on taking a general view of the kinds of fossils occurring in the successive stratigraphical formations, that there have been great changes. That is the first generalization. Then we have to inquire whether these changes seem to have been continuous and more or less gradual, or whether they seem to have been disconnected and sudden. If the former, then we can only think of new forms of life having arisen out of the old; that is, we have evolution. If the latter, separate and independent creation of new forms of life is suggested and, if considered universal, all distinct forms must in turn be considered immutable.

On an unprejudiced view, we may say that the evidence partly supports evolution and partly separate and independent creation. Now it is hardly likely that both processes are at work together. Is there any evidence that is conclusive for the one and quite inconsistent with the other? Is there any positive proof, from any part of the evidence, that evolution has, or has not, occurred? There is no visible proof, nor any kind of certain proof, either way, anywhere. But it is easier to explain the sudden appearance of new forms among fossils by invoking the imperfection of the geological record than to explain the cases of apparent continuous orderly change by invoking chance coincidence, accidental circumstances or the working of some process added to, or involved in, that of separate creation, simulating the effects of evolution.

Apart from that, evolution as the means of change in the organic world seems, nowadays, the only reasonable theory, while the separate creation of every species, or any separate creation at all within the flow of life, seems to the last degree unlikely. 'It is possible to put into words the proposition that all the animals and plants of each geological epoch were annihilated and that a new set of very similar forms was created for the next epoch; but it may be doubted if any one who ever tried to form a distinct mental image of this process of spontaneous generation on the grandest scale, ever really succeeded in realizing it'[4]. The evidence regarding evolution is still very much a matter 'of anxious inquiry', but hardly now one 'for warm and tedious controversy'.[5]

We may say, therefore, that palaeontology confirms the theory of evolution, but that it does so, as it were, somewhat grudgingly. One might have hoped for something more thoroughly convincing.

Palaeontology itself can hardly be expected to throw any light on the causes or mechanism of evolution. The palaeontologist can infer something about the environment from the matrix of the fossil, the lithology of the stratum and the general geological conditions, but he cannot see the relation between organism and environment to anything like the extent that the student of present-day life, the neontologist, can. Nevertheless, everything that palaeontology has to show strengthens the assumption that the organism was as perfectly adapted to its surroundings and its way of life at all times in the past as it is at the present day, and that the pressure of competition must have been as effective then as now. Everything, in fact, strengthens the conviction that evolution is a matter of 'descent with modification through variation and natural selection', but the palaeontologist must seek from the neontologist information as to the manner in which inheritable variations arise and so become permanent modifications.

The Imperfection of the Geological Record of the History of Life

THE PARTIALITY OF FOSSILIZATION

The first and most obvious reason for the imperfection of the geological record of the history of life is that only the 'hard parts' of organisms can be fossilized at all.[6] Very exceptionally, membranous or very delicate skeletal structures may be preserved, and these are especially important as giving us glimpses, at various times, of that part of the world of life (and it must be by far the greater part) that is normally denied to the palaeontologist. Tracks, trails and burrows, in so far as these are genuinely made by animals (which the evidence does not always certainly tell us) offer indications of life, but a footprint is a certain indication of an individual existence.

Most animal fossils, then, are the remains and impressions of more or less robust shelly and skeletal parts. The original material, mineral or organic, may be almost unaltered; partly decayed (the substance becoming more friable); reinforced with, or partly or completely replaced by, new mineral matter; or be entirely lost, only a space remaining. An initially hollow shell may be perpetuated as three fossils; namely, the shell itself in some form of representation, an internal cast (in rock-material or, less commonly, in a crystallized mineral) and an external impression on the surrounding rock matrix. A solid structure has the counterpart of its various aspects impressed as a 'negative' on the matrix.

In the normal fossilization of plants, the original material is usually more or less carbonized. It may be replaced by mineral matter. Hollow spaces give internal casts, and there will be external impressions. Plant fossils are usually detached parts, such as seeds, leaves and stems, and it is often difficult to say which parts went together in the living state.

'Petrification' is a surprising and fortunate kind of plant fossilization resulting in perfect preservation of microscopic internal structures, so that detailed anatomical comparisons can be made. This kind of preservation is due to chemical precipitation within the individual cells. The cell walls may remain, more or less carbonized, or they may themselves be replaced by mineral substance.

The same species of organism, animal or plant, may be represented by very different-looking individual specimens. Thus hollow or strongly curved shells are often compressed and flattened and this is especially so with most of the ordinary carbonized plant remains. Fossils may be distorted by the forces of earth pressure. They may present various views according to the position in which they lie exposed in the matrix, or the direction in which they happen to have been compressed.[7]

Specimens are usually to some extent broken and fragmentary; an individual fossil seldom reveals the complete structural morphology of the hard parts of the original individual organism, even after careful cleaning and, perhaps, dissection.

We may well marvel at the processes of fossilization, which have resulted in such a wealth of biological material and which have incidentally made of geology an exact historical science; but we must remember at the same time that, because these processes are inevitably highly selective, most of the scenes in the drama of life are unrecorded in the rocks, as are most of the living elements in those scenes of which we do get some view.

THE DEGREE OF COMPLETENESS OF THE STRATAL RECORD

The geological record is written in rocks of both igneous and sedimentary origin; but we are concerned here only with the sedimentary, or stratal, record as this is the only one that incorporates the record of life. Nor do the igneous rocks, in any case, exhibit within themselves any definite and detailed time-sequence.[8]

Over a small area of a few square miles the stratified rocks appearing at the surface may show a detailed succession which is a certain small part of the whole general geological column. It is obvious that the wider the region we survey the more extensive (vertically) is likely to be the stratal record. When we take the whole of Britain the stratal record is found to be nearly complete; remarkably complete, for nowhere else over the earth's surface is so much of the geological column exposed to view in so small a space. Within this island every system is found to pass, somewhere, with hardly a break, into the next and every system is itself apparently complete. That is to say, geological time, from the Cambrian onwards, appears to be continuously represented by strata. There are two relatively small exceptions, for the latest part of Cretaceous time, together with the earliest part of the Eocene, and most of the Miocene, are without deposits here.[9]

The stratal record is even more nearly complete if we include the adjacent continental areas, and it might be thought that over the whole surface of the Earth it would be firmly established. But to what extent does a perfectly conformable sequence of strata really represent the continuous flow of time? The layers in a stratum, the strata in a formation, represent the net gain in sedimentation over removal and nearly contemporaneous erosion. Any thin layer in the stratum may have been deposited on the layer now immediately below it almost at once, or after a pause, or after quite a long period of time, during which, very likely, other layers have been laid down on this lower layer and removed. A local succession, perfectly continuous structurally, may, nevertheless, contain stratal non-sequences, or time-gaps. These time-gaps, individually, will usually be indetectable, because they are not likely to have been longer than a fraction of the time taken in depositing the smallest of the stratal units that are, within the limitations of our methods of correlation, generally recognizable over a wide area. In sum, however, they might amount to a considerable part of the total time represented by the whole local succession.[10] Here again, the wider the area the more will the gaps be filled in. Continuous deposition of shallow water deposits, with no non-sequences, could only occur in a region of continuous subsidence. Most of the strata in the geological record appear to have been deposited in shallow water. It may be noted in passing that the actual thickness of beds is hardly any indication of the amount of time taken in their accumulation; a thousand feet of strata in one region may be represented by less than a hundred feet in another not many miles away, the latter as complete as the former, but condensed.

These considerations lead us to certain basic questions; they cannot as yet be answered, but their existence should be remembered. Can we say with any certainty that somewhere in the earth's crust is preserved a continuous stratal record? The confidence of our answer depends on the degree to which we push our conception of continuity. We may regard the record as continuous when viewed against the background of the whole immense stretch of geological time; but is it continuous in the sense that every second of time in the past is represented by at least one speck of mud or grain of sand now present somewhere in the earth's crust? For our purpose a possibly significant unit of time would be the normal life-span of the shortest-lived among the more commonly fossilized creatures. Is every generation of these potentially recordable somewhere in the rocks?

To be of use to the palaeontologist in the ordinary way, a stratum, if of marine origin, must have been uplifted (probably being tilted or folded in the process) and must now be accessible as an exposed outcrop. Rock cores extracted in boring operations may, however, contain abundant micro-fossils, in addition to some of the larger kinds. Thus, for practical purposes, we want continuity of accessible exposure within a continuity of actual strata.

Strata are formed of rocks of different kinds: mudstones, sandstones, limestones and their varieties; that is, every stratum has its particular *lithological facies*. It is pertinent to our general inquiry to consider to what extent there may be continuity (in a vertical sense) of the various facies. Assuming that there is a continuous stratal record, is there a continuous record of the predominant facies or of any one facies? It is conceivable that this question could be answered, but it would involve the analysis and collation of all the information ever published of the sedimentary geology of the world—a Herculean task that has not yet been attempted.

THE FOSSIL RECORD WITHIN THE STRATAL RECORD

Evolution necessarily implies an absolutely unbroken continuity of life from the time it first began. Assuming also that there is a continuous record of strata of some kind, to what extent may this continuity of life be recorded within the continuous stratal record? For there to be a continuous record, two conditions must have been fulfilled. First, the partiality of fossilization necessitates that there should be, within the continuity of life itself, a continuity of fossilizable life. There seems no doubt that that condition has been realized from Cambrian times.

Secondly, there must also have been a continuity of geological and geographical conditions, somewhere, which would allow fossilization to occur. Again, there seems no doubt that that requirement has been met; and from times far earlier than the Cambrian.

But what is most valuable from the point of view of evolution is continuity of one biological group (or several together). This implies vertical continuity of strata containing fossils of this group or assemblage of groups; that is, a continuous *faunal facies*.

Both the lithological and the faunal facies of a stratum, or a stratal formation, are expressions of environment. The kinds of deposit being formed on the sea-floor (or anywhere else) and the kinds of life living there at the same time depend on the environment. Now lithological and faunal facies largely correspond, but not exactly. Thus faunal facies are more refined than lithological in reflecting salinity and temperature (which hardly affect the kind of deposit at all), but are less refined when, for example, planktonic organisms fall to the sea-floor and become incorporated in whatever sediment happens to be forming there.

Thus we have to ask the same sort of question again that we asked, as a preliminary, in connection with the strata themselves; and the question is now a more immediate one.

Assuming that there is a continuous fossil record of life of one kind or another, is there a continuous record for each, some, or at least one kind of environment? To answer this is even more difficult, for it is plain that within the almost certain continuity of a stratal record of some kind and the quite certain continuity of fossilizable life, the evidence for the continuity of the record of an evolving group is likely to be precarious and difficult to follow. At best, we must everywhere endeavour to detect it, isolate it in our observations, and follow it in contemporaneous deposits as well as vertically. We must also remember, as biologists, that changes within an environment may have stimulated, either locally or generally, evolutionary trends within a group or groups. Again, evolution within a group may have occurred not gradually, but in jumps, and a search for evidence of gradualness in the rocks becomes necessarily futile. Variations in rate of evolution may also have occurred and these will increase the difficulty of interpreting the fossil record.

Strata vary greatly in the degree to which they are fossiliferous; thus desert sands are almost entirely unfossiliferous, while some shallow-water limestones may be almost entirely made up of shells. Even within a lithological facies that in general may be said to be fossiliferous, there are all degrees of plenty in the fossils, bed by bed, from none at all to great abundance. Thus the sporadic distribution of fossils, even in a favourable facies (together with probable non-sequences) adds still further to the imperfection of the record and our difficulty in following it.

The immensity of the number of fossils lying in the rocks must yet be but a minute fraction of the total number of potentially fossilizable individuals that have lived. Of that number, what a minute fraction can be within range of the geologist's hammer; and of these, yet again, what a minute fraction have been actually collected! Even if it should be said that our collections between them probably contain representative specimens of most of the broadly defined species of fossilizable animals and plants, thus forming a fair sample, no individuals are exactly alike and the proportion of known forms, from among the infinite variety that have existed, must remain infinitesimal. Fossils of the rarest species or varieties are unlikely to have been found at all; and it may well be that the forms most significant in evolution lie largely among the rarest.

LOCAL EVOLUTION AND MIGRATION

In addition to geological age and physical environment, the particular species of animals and plants that may be fossilized in a stratum depend on the geographical area in which that stratum was deposited; that is, on the palaeobiological province. The existence of biological provinces, to-day and in past times, follows naturally from the consideration that evolution must have place as well as time and the evident fact that the dispersal of organisms is limited.

Local evolution and migration are further biological factors which must be considered in examining the imperfection of the geological record of life. The speed of migration obviously varies enormously in the various groups according to their powers of locomotion or ability to drift, and the extent to which they can migrate also depends on the varying degree of tolerance of conditions among the different kinds of organisms. Thus local evolution and migration will attenuate and entangle still further the evolutionary threads we are trying to follow.

It will be seen from the foregoing that, although a general correspondence between the fossil record and the theory of evolution is immediately apparent, in detail there are likely to be incompletenesses and difficulties for a variety of reasons. Nevertheless, when the foregoing considerations are brought to bear upon them, what at first sight appear as stumbling blocks may, in fact, provide valuable evidence of the course of evolution, the imperfections lying not only in the record, but also in our preconceptions.

The Kinds of Evidence in Evolutional Palaeontology

The complete corpus of fact as to the history of life through the ages can be given by the distribution (in terms of both range and abundance) in time and space of every distinguishable form or, at least, every form that includes individuals so much alike as to be called by the same name; in practice, every species. This still leaves room for the detailed study of the distribution of morphological variations within species. We can place the time-ranges as lines on a diagram in which the time covered by the fossil record, or a part of that time, is shown upwards (the usual way) or sideways. The lines can be expanded into strips (made symmetrical if we please), the width at any part showing abundance (fig. 1*a*). Naturally the species of one genus would be placed together, the genera of one family together and similarly for the higher categories; so that the lines or strips for species could be combined to form similar graphical indications for genera, these for families and so on, up to classes and phyla.[11] Generally, the strips for the groups taper at both ends and widen to a maximum, often somewhere about the middle, thus indicating that the group appears with one or two species, gradually becomes more and more abundant (and varied), as more and more species appear, rises to a maximum and then, unless it is still expanding to-day, gradually declines, perhaps becoming extinct.

The general history of life thus at once shows something of an orderly evolution in that there is a more or less gradual rise and fall of groups. This rise and fall can be described and the genera and species particularized.[12] But what we want to find, for our present purpose, are 'links' between these species, genera and groups; that is, we want at the time of inception of a new species (which may also be initiating a new genus and, perhaps, a still higher group) a form or forms morphologically intermediate between the new species and a pre-existing one.

The extinction of groups is one of the fundamental facts of evolution; without it, the stream of life would be continually swollen. The causes of extinction are a problem in themselves; there is no reason to suppose that they will help us to understand any better the main problem of evolution, which is the origin of new forms of life.

To demonstrate evolutionary series we have first to demonstrate the existence of morphological series. A morphological series may be a very closely graded series such as we could get by sorting out and putting in a row the small variations from within a species or set of merging species. The range of form covered by such a series would be comparatively narrow. Or the series may be loosely graded, or more open, the members forming 'stepping stones' rather than a continuous 'causeway', and the range of form wide. We now introduce the time element; without the facts of succession in time, no series of forms can provide anything but a guess about their evolution. If the morphological series is found to correspond with time to a degree such that mere chance seems to be ruled out, then we may be said to have an evolutionary series. In an evolutionary series the morphological intergrades become transients. (Links, as above defined, are transients and must form part of an evolutionary series.)

Instead of first setting out a morphological series and then comparing it with time we might, of course, first set out the material in time and then see if there were any corresponding morphological sequences. But it is probably more satisfactory generally to do the operation the first way as this makes, in practice, the two stages more completely separate and we are not in danger of being prejudiced by having the time element presented to us from the start. Also this order of procedure seems the logical one.

An evolutionary series is thus a series of species and varieties showing orderly morphological change, usually in several features together, as time goes on.

Evolutionary series, whether of the closer or more open kind, are difficult to detect among fossils; the observed succession of forms within what may be supposed to be one natural genus or family (being bound together by the possession of some stable, persistent characters) usually showing very little, if any, orderly gradation. Species belonging to one genus or family, following one another in time but not falling into evolutionary series, must be supposed to be samples from the successive 'cross-sections' of the evolving genus or family, but on different strands of evolution. There is, however, vast room, in most cases, for more intensive collecting from wider areas, which must surely make less imperfect the palaeontologist's view of the inevitably imperfect geological record.[13] The strands will then appear less as merely disconnected odds and ends. On the other hand, the apparent rarity of closely graded evolutionary series may express a real rarity and be due, as suggested earlier, to evolution usually proceeding by jumps rather than by perfectly smooth gradation.

It is the orderly appearance of the several grades, at successive times, that establishes an evolutionary series. Their disappearance (extinction) need not be in any particular order and some of the earlier members may very likely outlast the later. Consequently, while a morphological series is necessarily a line, an evolutionary series may be more in the nature of a plexus.[14]

Works which refer to the fossil evidence of evolution usually cite a few of the well-known cases of evolutionary series as if they were merely representatives of a host that might have been quoted, instead of stressing the fact that records of such cases are rare. Nevertheless, these exceptional cases are the significant ones, as they demonstrate that gradual change, slow enough to leave a trail of evidence in the rocks, has occurred. Something near proof of evolution being obtained in some cases, we are justified in assuming that most probably it is universal and we must try to explain the general paucity of evidence as best we may. But someone seriously combating the whole idea of evolution might well ask, in some exasperation, what evidence against evolution the evolutionary palaeontologist could not explain away to his own satisfaction.

When we come to review particular evolutionary series we shall see that these themselves form something of a gradation; some are clearly documented in the rocks, others less so and in yet others it is doubtful whether there is anything more than a suggestion. These last grade into the commonest condition where there is no visible sign of orderly evolution in the successive fossils (so far found) of a genus or family.

The detection and setting out of a morphological series of whole individuals implies that if there is more than one variable character these all vary harmoniously together; the other characters being constant. Thus, every individual fossil has a particular position in the linear series, a position that can be expressed as some kind of measurement. If every individual fossil is also stratigraphically dated, we can plot each on a diagram where the horizontal axis represents morphological state and the vertical axis stratigraphical position, or time (fig. 1*b*). We can group the morphological state within arbitrarily limited, as nearly as possible equal, narrow ranges of form; that is, we can allocate each and every specimen to some particular morphological grade. We can similarly group the stratigraphical horizons (probably in any case restricted, by collecting having been confined to certain fossiliferous and accessible bands); these horizons need not be equally spaced, nor are they likely to be in practice. A third variable now appears; namely, the frequency of individuals of a certain morphological grade occurring

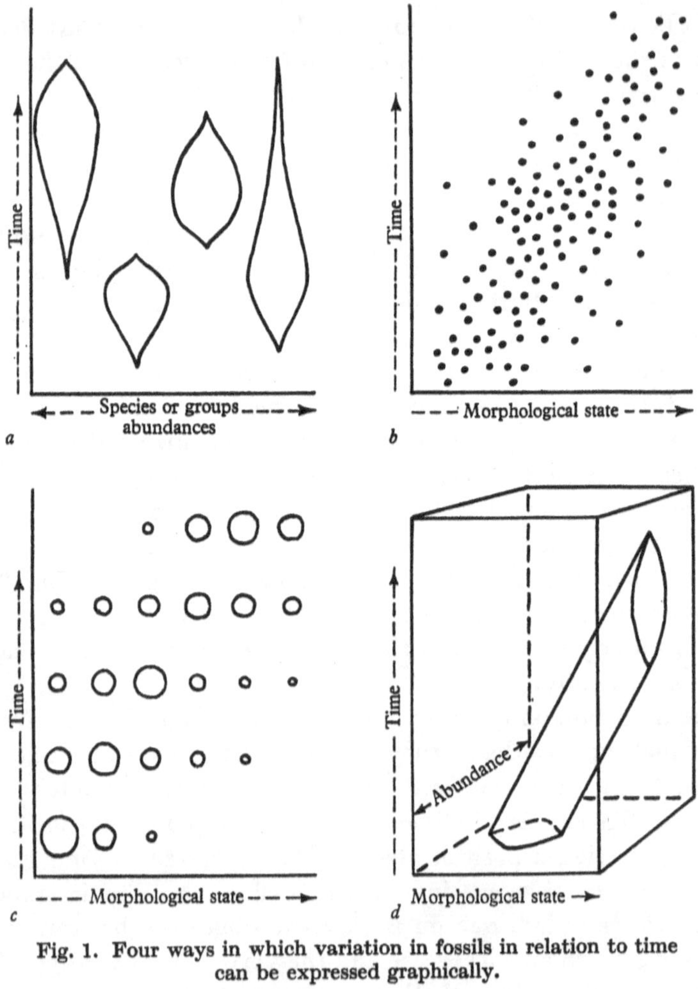

Fig. 1. Four ways in which variation in fossils in relation to time can be expressed graphically.

within the upper and lower limits of a certain stratigraphical group. We can express these three variables on our two-dimensional diagram by indicating abundance by the size of a circle (or square) at the relevant points (fig. 1*c*). The horizontal rows of these circles are almost the same thing as graphical curves; but we really need a three-dimensional frame, if we are to show properly these three variables simultaneously (fig. 1*d*).

Another kind of evolutionary evidence is that of trends, throughout a group, in separate structural characters.[15] For example, we may take one variable character and set out a morphological series to represent its variation, ignoring the other characters. If we now consider the time element it may be found that the whole morphological gradation in that one character corresponds with the onward passage of time. An evolutionary trend in this character is thus demonstrated. Such trends are not very rare.

There may be evolutionary trends in separate structural features within a whole group of any taxonomic size from phylum to genus. There may be, as we have seen above, evolutionary series of varieties, species or genera in which each member is, in all its characters, a link between its neighbours. These are two distinct kinds of evolutionary phenomena. Whereas evolutionary trends within a group may be manifest without any evolutionary series within that group being so, an evolutionary series must necessarily display an evolutionary trend in one or more features (the other features remaining constant). Usually we speak of an evolutionary trend (in one feature) when we are taking an over-all view of a large group, such as an order or class. Species that we might select to exemplify the several stages of a particular trend in such a group would not usually themselves form an evolutionary series of whole individuals.

A very important question, one which has been much debated, is whether the individual organism may itself, in its structural make-up, throw any light on its racial history. The study of ontogeny, the development of the individual organism, has suggested that there may be some correspondence between this process and that of phylogeny, the history of the race. Does the individual, in its early growth, tend to pass through stages which either repeat the later stages of its past evolution or foreshadow the evolutionary history of its descendants? The ontogeny of skeletal structures is in many cases permanently visible to the palaeontologist in the morphology of the complete fossil. Any valid 'law' connecting ontogeny and phylogeny would obviously be of the very greatest importance in detecting evidence of evolution among fossils. The following considerations are very elementary, but it seems worth while mentioning them as they do not, all of them, always seem to be borne in mind.[16]

There are the following categories of ontogenetic tendencies:

1. *Direct development.* This is the tendency for a structure to grow up in the most direct way, that is, the tendency to grow in size gradually, and more or less gradually to assume any adult intricacies of structure. This tendency must always be expressed to some extent.

2. *Changing adaptation.* If the mode of life at the early stages of growth differs from that at a later, one would expect this change to be reflected in ontogeny.

3. *Racial retention or foreshadowing.* The early stages of growth may retain ancestral characters or foreshadow future ones. In contrast to the foregoing, this is not a necessary, nor perhaps a likely, situation. Nevertheless, we have the possibility of recapitulation, a familiar idea, or the converse, where the earliest ontogenetic stage looks to the future instead of to the past. It should be noted that ontogenies may legitimately be placed in the third category only if the simpler and prior first and second categories are convincingly demonstrated to be unsatisfactory.

Finally we have to consider the general question of classification. The evolutional palaeontologist, like any other palaeontologist, has two kinds of observable fact to go on and no more: the morphology and the stratigraphical occurrence of individual specimens. From these facts he seeks to discern relationship, biological affinity, and reconstruct the branches and twigs of the 'evolutionary tree'. He first of all wants to have his fossils sorted out into the different kinds. The essential thing here is that if evolutionary evidence is going to be sought from the classification, evolutionary ideas must not be put into the classification to begin with. The sorting should be done entirely by morphological comparison, by assessing degrees of likeness; all structural features in all parts of the specimen, at all stages of growth, being taken into account. There would doubtless be a number of classifications, differing in detail, which might be said to express almost equally well the degrees of likeness (and nothing else) between groups of organisms, but it is improbable that any one of them would express exactly all the true genetic relationships. The ability to select the most satisfactory and comprehensive classification is a measure of the skill of the investigator.

On this clean generalization and classification of pure morphological fact, together with the data of stratigraphical occurrence, all hypothetical elements being rigorously excluded, the palaeontologist is now perfectly free to reason and to speculate as to biological affinity and evolutionary lines. It would be fatal to freedom in this respect if it were felt that classification and nomenclature had to follow these ideas. The true phylogenetic grouping, the 'natural' grouping, is what we are seeking; but this grouping, which must always remain more or less hypothetical, is best kept apart from a stable body of ascertained and classified fact. The basic morphological classification should be altered and expanded only when errors of fact are found or when new facts come to light which show that the current classification does not adequately represent the state of knowledge about the comparative morphology.

It must also be borne in mind that the distinction between a purely morphological and a natural classification is not always immediately apparent. For example, if a species A gives rise (through transients) to a very different species X, and a species B, very like A, gives rise at the same time to a species Y, very like X, but unlike B, is it more 'unnatural' for A and B to be placed together in one genus, and X and Y in another, than for A to be placed with X, and B with Y? Is a 'horizontal' classification always and inevitably less 'natural' than a 'vertical' classification? Would a genus be 'unnatural' (from the point of view of relationship) if, in fact, its species were derived from separate species of another genus? These questions, which are to a certain extent subjective, must be decided by the palaeontologist in the light of his general training and experience.

Stratigraphical Palaeontology

Stratigraphical palaeontology describes the occurrence of the kinds and assemblages of fossils in the several strata. It deals with that complete corpus of fact we have mentioned above; but 'horizontally', bed by bed, instead of 'vertically', species by species, group by group.

We can take a glance at the kind of facts supplied by stratigraphical palaeontology by holding up a sample for inspection. The Cambrian system is specially interesting because it is the first chapter in the recorded history of life. The outcrops of the Cambrian rocks in Britain are not extensive and the fossiliferous outcrops are even more restricted. This makes the material all the easier to deal with; yet it is not so sparse that the showing is unrepresentative of what the geological record provides. Moreover, very little was known of the life of Cambrian times in Darwin's day; it is the one major period of which the fossil record has been almost entirely discovered during the last hundred years.

We can find compilations in, for instance, Neaverson's treatise[17] or can combine the summaries given in the Geological Survey regional handbooks (and both these works give pictures of the fossils), but the following paragraphs, giving something of the geological, geographical, topographical and historical circumstances of the occurrence and discovery of the different kinds of fossils in these rocks, are the result of an independent perusal of the original sources of information. Reference to them, together with this brief account, will provide all the details so far recorded, and form and substance can be given to the names of group, genus and species from specimens or descriptions in the systematic works. These are examples of that primary practical experience which supplies all the data of evolutional palaeontology. Some of the fossils mentioned in the following outline are illustrated in fig. 2.

THE FOSSIL RECORD OF THE CAMBRIAN SYSTEM

Lower Cambrian. The chief groups are Brachiopoda, Trilobita and small conical shells of doubtful biological place. Of these last, a very few appear to be cap-shaped Gasteropoda; the rest, more tubular in form, the Hyolithidae, are of unknown affinities. Other groups definitely represented among the fossils so far found are Cystidea (phylum Echinodermata) and Conchostraca (phylum Arthropoda). Bryozoa are doubtfully present. What appear to be worm-borings and worm-casts occur. We thus have the surprising fact, abundantly demonstrated from the British fossils alone, that the oldest preserved fauna is made up of a very wide variety of

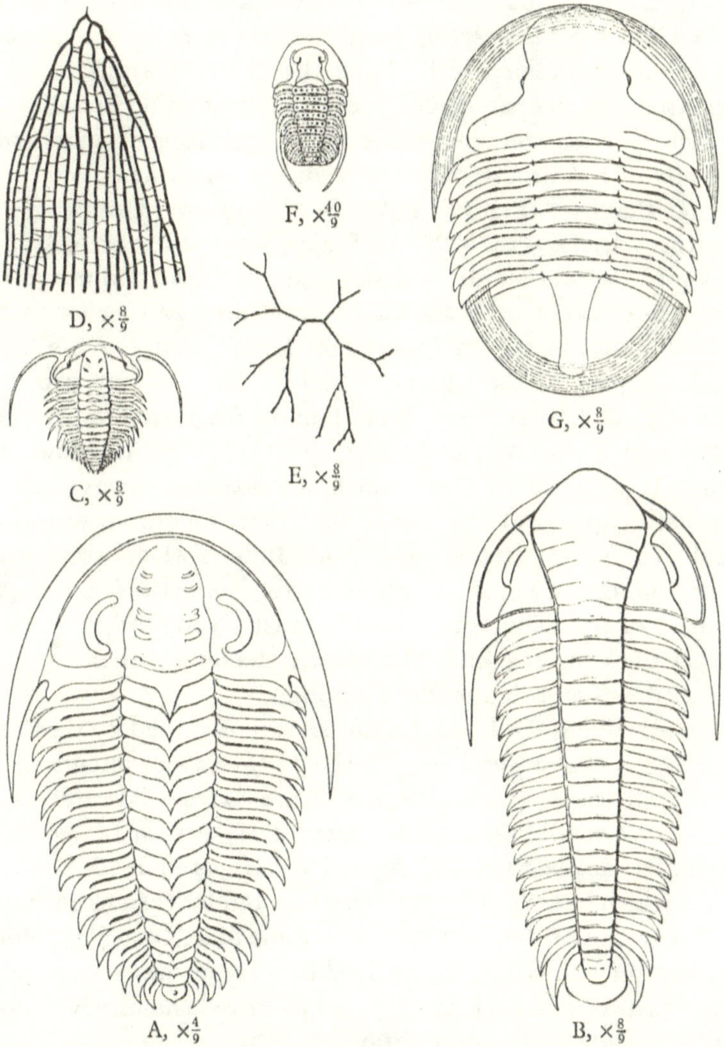

Fig. 2. Cambrian fossils. A, *Callavia callavei* (Lapworth), Lower Cambrian; B, *Para doxides hicksi* Salter, Middle Cambrian; C, *Ctenopyge flagellifera* (Angelin), Dolgelly Beds, Upper Cambrian; D, *Dictyonema flabelliforme* (Eichwald), Tremadoc Beds, Upper Cambrian; E, *Clonograptus tenellus* (Linarsson), Tremadoc Beds, Upper Cambrian; F, *Shumardia pusilla* (Sars), Tremadoc Beds, Upper Cambrian; G, *Asaphellus homfrayi* Salter, Tremadoc Beds, Upper Cambrian. (From R. W. Pocock and T. H. Whitehead, *The Welsh Borderland*, London, 1948.)

quite distinct groups of marine invertebrate animals.* The trilobites are the most interesting group because here there is considerable structural detail which allows the different kinds to be discriminated and compared. The fossils so far found do not allow any satisfactory stratigraphical zones to be established, either within the limits of the British outcrops or over a wider field.

By far the most extensive outcrops of the Lower Cambrian rocks in southern Britain are in North Wales, where they are splendidly exposed in the Merioneth mountains and in the slate quarries (and other exposures) of northern Caernarvonshire. But no fossils have yet been found in any of these rocks except near the top, in the uppermost part of the Llanberis Slates. Here, about two dozen specimens altogether have been found; fragments of trilobites (particularly *Pseudatops viola*) and hyolithids.[19] This shows, incidentally, that a Lower Cambrian fauna does not always, perhaps not usually, appear at the base of the Lower Cambrian series; that is, not everywhere at the same time. Indeed when strata with a Lower Cambrian fauna are conformably underlain by a great thickness of unfossiliferous strata it must be somewhat uncertain whether these lower strata, particularly the lowest of them, should be classed as Cambrian or Pre-Cambrian.

In South Wales the Lower Cambrian is represented by the Caerfai Beds. The rocks are well seen in the St David's promontory, particularly in the sea-cliffs, but fossils are extremely rare. A few specimens of small brachiopods, placed in the genera *Lingulella* and *Discina*, were recorded and figured by Hicks in 1871, and small objects which have been at various times and by various authorities thought to be Ostracoda or the hypostome of an Olenellid trilobite. Very few fossil finds have been added to these of the early searchers. All the fossils have been found in a bed of red shales which occurs about the middle of the formation.[20]

The Shropshire areas of Comley and Rushton, some 10 miles apart, are much the most important for Lower Cambrian fossils in Britain. Natural exposures are few and researches on the stratigraphy and palaeontology have been largely made by cutting trenches and making exposures artificially. The fossils are crowded

* It is noteworthy that the oldest known land flora (from the Silurian of Australia) also consists of widely different forms.[18]

into a few bands in the top 5 or 6 feet of a total thickness of some 500 or 600 feet. They are nearly all small, chiefly trilobites, brachiopods and hyolithids and, further, are usually fragmentary (particularly the trilobites). The famous *Callavia* from the small, now largely overgrown, Comley quarry is an exception as regards size; it has itself never been found except in fragments, but one nearly complete specimen of the allied *Nevadia* was found on a loose stone picked up in the quarry. Our knowledge of this fauna, certainly one of the most interesting in Britain, is chiefly due to Cobbold's work.[21]

Fossils, apparently of much the same species of small brachiopods and hyolithids, occur in both the Malvern Quartzite and the overlying Hollybush Sandstone which crop out at the southern end of the Malvern Hills and are exposed in two or three spots.[22] At Nuneaton, in Warwickshire, the Hartshill Quartzite is well exposed in large quarries, but the only fossils found, chiefly hyolithids, are in 2 or 3 feet of limestone not far from the top of the formation.[23] *Callavia* has been found at the base of the overlying Purley Shales. The Lickey Hills Quartzite, supposedly of Lower Cambrian age, in north Worcestershire, has not yielded any fossils.

The commonest fossils in the Lower Cambrian of the north-west Highlands of Scotland are of a somewhat doubtful nature, being of worm-like kinds. The abundant and characteristic 'pipes' of the Pipe Rock are supposed to be the casts of worm-burrows. The 'fucoids' (the name implying Algae) of the Fucoid Beds are now taken to be flattened worm-casts. Meanwhile, the 'serpulite' ('worm') of the Serpulite Grit (named *Salterella*) is regarded as *incertae sedis*. A stream on the slopes of a mountain just south of Loch Maree exposes tribolite-bearing bands of the Fucoid Beds with, particularly, *Olenellus*.[24]

Middle Cambrian. The chief fossils are the trilobites. The genus *Paradoxides* is characteristic of this portion of the stratigraphical succession in Europe and eastern Canada. With this genus, the most abundant species are those of the Agnostidae; the genus *Agnostus* shows a great number of species, but the species *Eodiscus punctatus* is the most abundant of all Middle Cambrian fossils.

Second in importance to the trilobites are the brachiopods; and among other fossils are the hyolithids and the sponge (Porifera) *Protospongia*. A definite succession of widespread faunas within the series, based first on the species of *Paradoxides*, and secondly on those of *Agnostus*, is more satisfactorily established than in the Lower Cambrian.

In North Wales, in Merioneth, the lower part of the series is presumably represented by the upper beds of the unfossiliferous Harlech Grits. In the succeeding shales a rich fauna of trilobites has been described from the St Tudwal's peninsula, but elsewhere in North Wales the series has not been intensively studied palaeontologically in recent years.[25]

The Middle Cambrian in South Wales comprises the Solva beds (below) and the Menevian beds (above). The well-known fossils of the South Wales Middle Cambrian, which include the largest and the smallest trilobites (*Paradoxides davidis* and the agnostids), were found by Salter and Hicks nearly a hundred years ago, but since their day few more records have been made and good specimens of fossils are difficult to collect.[26]

In Shropshire, natural exposures are few and the structure is complicated, thus masking the time-succession of the forms within the series. Small fossils, trilobites, brachiopods and hyolithids, occur as fragments through what appears to be a considerable thickness of strata.[27]

The most abundant and best-preserved Middle Cambrian fossils so far found in Britain occur in the Nuneaton outcrop. Good exposures are very few, but fossils were suspected in the Abbey Shales, forming the top part of the Middle Cambrian, and were found by cutting carefully sited trenches. Thus was revealed, in the words of their discoverer, Illing, 'a succession of horizons which teem with beautiful fossils'. These are chiefly trilobites (more than fifty species), and of these the agnostids are by far the most abundant. The other fossils are, as usual, brachiopods and hyolithids.[28]

Upper Cambrian (*Lower Division*). Trilobites are the characteristic fossils, species of *Olenus* in the lower part and of somewhat similar, presumably allied, genera in the upper. *Agnostus* is still a common genus. The brachiopod fauna is poor in species, but *Lingulella davisi* is one of the best known of British fossils. Other groups are only locally represented.

North Wales is very much the most important British area for the series; the 'Lingula Flags', with its three divisions, the Maentwrog, the Festiniog and the Dolgelly. Apart from the trilobites, the brachiopods *Lingulella davisi* and *Orthis lenticularis* are characteristic of the Festiniog and Dolgelly divisions respectively, tending to occur abundantly in bands, the *Lingulella*-band, constituting the topmost layers of the Festiniog beds, being particularly distinct and widespread. (The *Lingulella* fossils are usually distorted by cleavage strains so as to look like lamellibranchs.) The Festiniog beds are also characterized by *Hymenocaris vermicauda* (Arthropoda, Phyllocarida) and the peculiar and problematical rope-like *Cruziana semiplicata* (in relief on the undersides of the flaggy layers).[29]

In South Wales the rocks are poorly exposed and fossils are rare, except *Lingulella davisi*.[30] In Shropshire a stream-section has been found and fossils, chiefly small trilobites, occur in calcareous nodules.[31] At the southern end of the Malvern Hills are the White-leaved Oak Shales in which the commonest fossil is the little trilobite, *Sphaerophthalmus alatus*.[32] At Nuneaton, the series is represented by the Oldbury Shales, but little work has been done on the fossils of these rocks since 1898.[33]

Upper Cambrian (*Upper Division*). (This series, outside Britain, is placed as the lowest series of the Ordovician system.) The lower part is characterized by graptolites, which are absent from the upper part. The upper part is characterized by trilobites, but these fossils also occur in the lower part. The other groups, which include brachiopods, hyolithids and cystids, are sporadically distributed. The graptolitic part of the succession falls into two zones, that of *Dictyonema flabelliforme* below and that of *Clonograptus tenellus* above. Of these, the former zone is especially well defined, constant and easily recognized, because of the vertical restriction, widespread occurrence and abundance of this distinctive fossil. There are a number of trilobite species, of which *Asaphellus homfrayi* (the most abundant), *Angelina sedgwicki* and *Shumardia pusilla* are perhaps the best known. Two species of *Lingulella* and *Acrotreta sabrinae*, are widespread brachiopod members of the fauna. Hyolithids and cystids also occur.

In North Wales, the series (widely known, far beyond Wales, as the Tremadocian) is represented by the Tremadoc Slates which crop out round the outer rim of the Harlech Dome (Tremadoc, Arenig and Dolgelly areas) and in the St Tudwal's peninsula. Many well-preserved fossils, particularly trilobites (*Angelina sedgwicki, Asaphellus homfrayi, Niobella homfrayi* and *Cheirurus frederici* are among the chief), were collected by the official geological surveyors and the Portmadoc amateurs David Homfray and Frederic Ash from the neighbourhood of that town about 100 years ago. Fossils are now not so easy to find; except the dendroid graptolite *Dictyonema flabelliforme*, which occurs very abundantly in a particular band.[34]

It is in the Shineton Shales of Shropshire that the greatest number of species, and the most perfectly preserved specimens are to be found. Here also, the faunal succession is most distinct. These rocks with their fossils have been described in two well-known papers, just 50 years apart. They form the dingle country and undulating land to the south of the Wrekin and the little cliff-section of Shineton Brook is a famous collecting ground. Most of the Tremadocian species, except *Angelina sedgwicki*, are to be found in the Shineton Shales.[35] The Habberley Shales of the Shelve country of west Shropshire also contain some characteristic species.[36]

Tremadocian species have been found in the Bronsil Shales of the Malvern district,[37] in shales at Pedwardine in Herefordshire,[38] in the Breadstone Shales at Tortworth in Gloucestershire[39] and in the Merevale Shales of Nuneaton.[40] Parts of the Highland Border series in Scotland contain a few fossils, some of which appear to be of Upper Cambrian, and others of Middle Cambrian, age. The unfossiliferous part of the Durness Limestone, in the north-west Highlands, between its lowest part, of Lower Cambrian age, and the upper part which contains a peculiar fauna of supposedly Ordovician age, presumably ranges through the Middle and Upper Cambrian.

From the picture that emerges it will be seen that we have many different groups, genera and species distributed in little apparent serial order. The story to be read from the facts is a very disconnected one. If we turn to Neaverson's book for an account of the Cambrian faunas in other parts of the world we find successions that run more or less parallel to the British succession; there are biological provinces, zoogeographical regions. But in none of them separately, nor when we combine them, do we find much more evidence of an evenly connected flow within the changing scene; there is the same baffling lack of evolutionary evidence.

A COMPARISON WITH THE CRETACEOUS SYSTEM

Lest it should be said that it is hardly fair to take the several series of the Cambrian system as representative of what the record as a whole can show, let us go to that stratigraphical series which, of all others, is most favoured in its conditions for giving a steadily moving picture of the history of life: the Chalk. This is entirely of marine origin and was laid down on a gradually sinking sea-floor.

> The conditions under which the Chalk was formed were apparently uniform over wide areas; moreover, deposition appears to have been almost continuous, no appreciable gaps comparable with the frequent non-sequences in Jurassic strata having been detected, though the possibility of their occurrence cannot be denied. Consequently the succession of beds, though subject to local variations in lithology and other characters, is fairly regular over wide areas, and the fossils reveal a true and almost unbroken succession of faunas. Indeed, no other formation, except perhaps the graptolitic facies of the Ordovician and Silurian, provides a more complete record of continuous evolution in slowly changing organisms.[41]

The commonest fossils of these Cretaceous strata, apart from the micro-organisms, are echinoids and crinoids, ammonites and belemnites, lamellibranchs, brachiopods and fish-teeth (see fig. 3). What have we of evolutionary significance in this most complete of all records? There is the curious case of the echinoid *Micraster* and the devious story of the lamellibranch *Inoceramus* (both of which will be referred to later); but little beyond that, from what the published accounts tell us. We have the usual thing: different genera and species of the several groups, some of them characteristic of horizon, all placeable on a time-chart, but practically no visible, recognizable, morphologically graded sequences. This, then, is about the best that any stratigraphical series, viewed objectively and as a whole, can do for us in our inquiries as to the palaeontological evidence of evolution. Here is the paradox of stratigraphical palaeontology; it persistently appears to withhold sweeping evidence in favour of evolution, but gives away a little here and there, at no point denying that evolution has occurred.

One might perhaps suggest that the accounts of the stratigraphical succession of fossils tend to divide the matter overmuch into compartments. The faunas of successive systems, of the series within the systems, and of smaller subdivisions, are treated as separate units. Changes from bottom to top may be noted, but transitional changes between the somewhat arbitrary divisions are comparatively neglected. Yet it is just these that are the most critical from the evolutionary point of view, especially if it be contended that the faunas of the divisions are distinctive. Let us have more accounts of the transitional faunas, the world over, between, for instance, Devonian and Carboniferous, Cretaceous and Tertiary.

STRATA AND TIME

It is important to bear in mind the limitations to our ability to interpret strata in terms of a general time-scale, to carry over the observable realities of superposition, lateral extension and character of make-up and content, into the intangible dimension of time whose onward flow has been so wonderfully, yet so fortuitously, recorded in the rocks. We may classify the strata as elaborately as we please and locally draw sharp lines to separate our divisions; but we can only approximately correlate the strata, as regards time, from region to region. True 'time-planes', though present in the rocks, can never be detected as such, particularly over a wide area. There is no feature of a rock, structural, lithological or palaeontological, that is any definite indication or test of simultaneity. Fossils lying along one bedding-plane in one exposure may probably be taken, for all intents and purposes, as being those of individuals that lived all at the same time, and an upward succession of fossils must represent an onward succession in time; but beyond that we cannot go with any certainty.

We have to beware, also, of arguing in a circle; for instance,

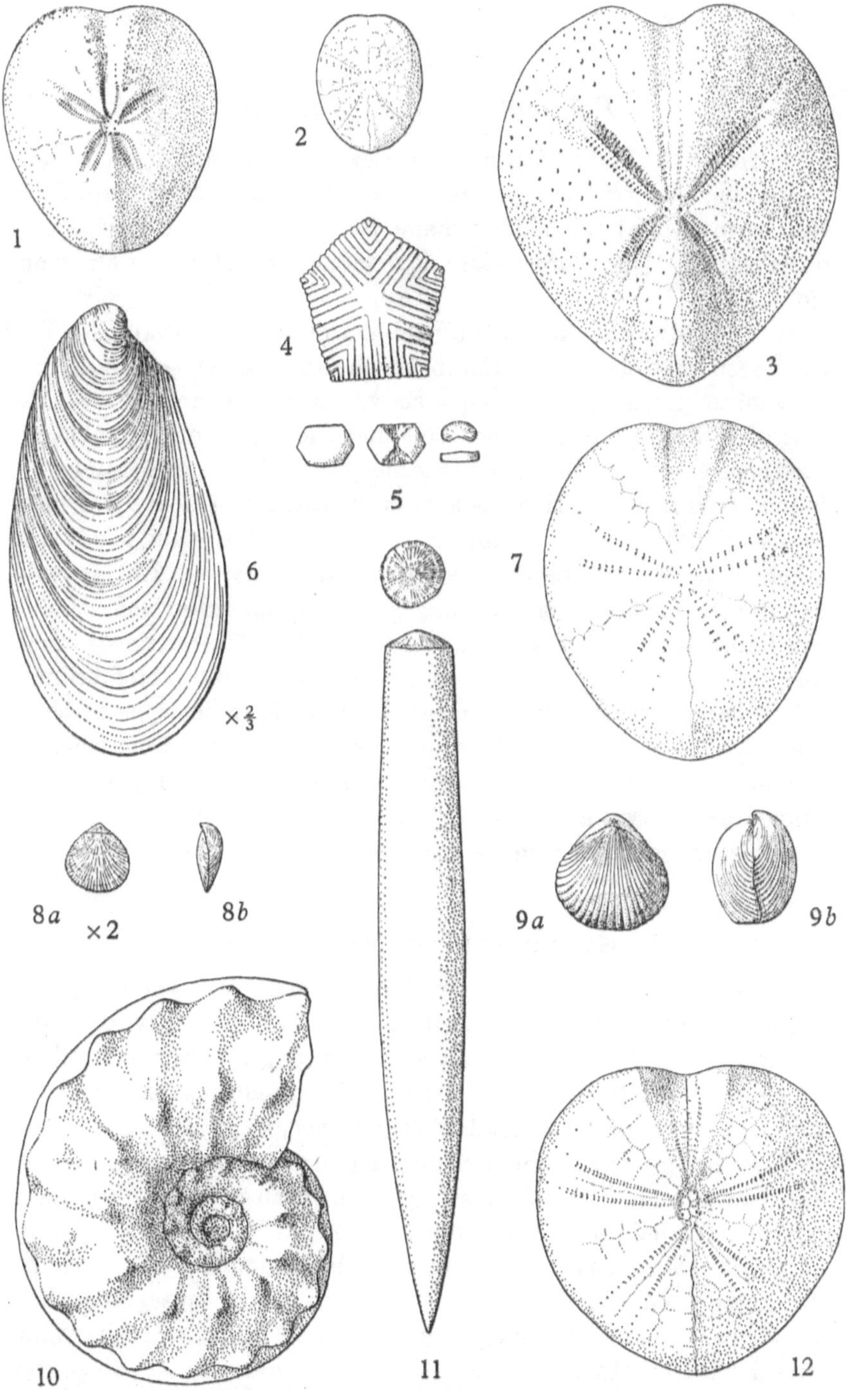

Fig. 3. Fossils from the Chalk. 1, ***Micraster leskei*** (Desmoulins); 2, ***Offaster pilula*** (Lamarck); 3, ***Micraster coranguinum*** (Leske); 4, ***Marsupites testudinarius*** (Schlotheim); 5, ***Uintacrinus westfalicus*** Schlueter; 6, ***Inoceramus labiatus*** (Schlotheim); 7, ***Holaster planus*** (Mantell); 8 *a*, *b*, ***Terebratulina*** Etheridge; 9 *a*, *b*, ***Rhynchonella cuvieri*** d'Orbigny; 10, ***Schloenbachia varians*** (J. Sowerby); 11, ***Actinocamax plenus*** (Blainville); 12, ***Holaster subglobosus*** (Leske). (From C. P. Chatwin, ***The Hampshire Basin and Adjoining Areas***, London, 1948.)

if we define the Middle Cambrian series as comprising those strata containing the trilobite genus *Paradoxides* (and their supposed time equivalents) we cannot make the statement, as a separate observation, that *Paradoxides* is confined to Middle Cambrian times.

There is always with us the problem of Pre-Cambrian life. We have seen that the Lower Cambrian fauna is largely composed of 'advanced' invertebrate groups, so we can only conclude that life existed in Pre-Cambrian times, but that the organisms (animals, at least) were without hard, fossilizable parts. The Lower Cambrian fauna 'is varied, but there is a notable absence of heavily shelled forms'.[42] Whether the evolution of this varied life would need such a long time as might be supposed is another question. In Darwin's day the object called *Eozoon canadense* was supposed to be a definite Pre-Cambrian fossil (a giant foraminifer), but its organic status is now entirely discredited. Up to the present no undoubted animal fossils have been found, but there are reliable indications of algal masses. We can say with Whittard that the absence of evidence of organic remains from Pre-Cambrian rocks, when they may be unmetamorphosed and have all the appearances of proving to be fossiliferous, is a question that still defeats scientists.

Evolution in the Groups

The general plan of evolution must underlie the whole organic world; it is hardly to be supposed that evolution proceeds according to certain principles in one group and according to others in another. But each fossil group provides particular manifestations and problems because of its special biological peculiarities and the geological circumstances under which it is usually fossilized; that is, because of circumstances that are accidental to the evolutionary process. The habitat of a group will determine the amount and kind of strata which may contain its remains, structure will determine the amount and parts of organisms which can be preserved, while habit, particularly crawling, swimming or drifting, will determine dispersal. Nevertheless, although papers on evolutional palaeontology may seem to show that evolution is somewhat capricious in its manner, each group or family behaving according to some special whim, if series or trends are seen in one group, they must be looked for in every other.

We shall now take some of the groups in biological order and discuss very briefly what evolution, in a general way, has been discerned in them, chiefly by referring to an authoritative work or two in each case.

THE FORAMINIFERA

Here we have the simplest of animals, mostly microscopic; but the genera and species as exhibited in the intricate shell-form are as distinct as in the larger and 'higher' animals. The geological record of this group is full and varied. The systematic and economic micropalaeontologists are concerned with comparison and identification of individuals, the general aspect of faunas and the application of these facts to correlation and facies and thus to geological structure. But the question that we have to ask here is: what signs of evolution does all this vast mass of information give us?

If we turn to one of the standard books on the group[43] and see the many attractive diagrams of phylogenetic trees, we might well feel that we need go no further than the Protozoa to find among fossils beautifully detailed examples of evolution. However, when we add the stratigraphical ranges of the forms selected, we by no means always find that these forms come in at successively higher levels. Certainly, it is not impossible for a genus whose range is given as Miocene, for instance, to be derived from one given as Pliocene—we can put the apparent reversal down to the inadequacy of the geological, or at least the palaeontologists', record—but such anomalies should be rare. As the evidence given us stands, the morphological series shown do not always seem to have very strong claims to being evolutionary series.

THE ANTHOZOA

In the corals, to try to solve the difficulty of reconciling the apparently haphazard occurrence of unlike forms, Lang has suggested that there was a fundamental simple type which 'threw off', at various times, and more or less suddenly, forms in which certain evolutionary tendencies in particular combinations were given expression.[44] He applied the idea particularly to the Carboniferous Rugosa; but we do not find even morphological intergrades, except here and there, between the supposed parent form and its (dissimilar) offspring. Thus he postulated the following tendencies: from single to compound (with more and more compaction of the corallites), from conical to cylindrical, from a bilateral to a radial arrangement of the septa, a retreat of the septa from wall and centre, an increase in their number, and the appearance of dissepiments, columella and axial complex.

Much reliance was placed on recapitulation, and the necessarily simple structure of the coral near its tip (the early formed part) was interpreted as an indication of derivation from a simple type. Swinnerton took much the same view independently.[45] However, unsatisfactory as the evidence is, any suggestion is welcome in the attempt to find some evolutionary scheme into which the corals may be fitted.

THE ECHINOIDEA

For the class Echinoidea we have Hawkins's presidential address to the Geological Society.[46] Our knowledge of the evolution of this group is here vividly presented by the leading British authority on it.

The echinoids sort themselves out into groups according to morphological differences, the main division being between those with a (nearly perfect) radial symmetry (subclass Regularia) and those that are bilaterally symmetrical (Irregularia); the latter being grouped particularly according to the degree of bilateral symmetry. When the groups are placed on a range diagram, the usual critical question (for us) arises: what do we find by way of connecting links between them at the right times? As usual, the answer is: very little. One evolutionary trend does emerge, however; a trend towards a higher degree of bilateral symmetry. Though echinoids in the Palaeozoic are peculiar and not common (order Perischoechinoida), they are all Regularia. At the end of the Palaeozoic the peculiarity in the matter of abnormal numbers of columns of plates in the areas disappeared, but the group obviously gave rise to the normally constituted order Cidaroida, which are otherwise rather

PLATE I

Two views of a particularly fine specimen of the Jurassic (Corallian) ammonite *Perisphinctes oxoniensis* Arkell. This is a large species, the specimen measuring 13·5 in. (34·3 cm.) across. The frilled form of the septal suture, characteristic of the ammonites, has been made conspicuous by painting.

similar, and the earliest cidaroid is the most like the Perischoechinoida, thus supplying just the kind of evidence we want in trying to link one group with another. The commonest families of the Regularia, forming the order Centrechinoida, became established in early Jurassic times. The Cidaroida are the only echinoids from which these could have been derived (as there were no others in existence), and here again such fragmentary remains of echinoids as are known from the Trias on the Continent show features somewhat intermediate between the Cidaroida and the Centrechinoida. But almost immediately on the establishment of the latter, several distinct groups of Irregularia appeared. The earliest of these is the least bilateral, which is a very satisfactory evolutionary bridge; but otherwise connecting links are hard to find. The group of most highly bilateral forms does not appear till the Cretaceous, which is again part of the demonstration of the general trend towards bilaterality.

Apart from these facts of morphology and structure, Hawkins considers the environment of the animals and the question of their adaptations and habits. He ends his address with a series of questions arising out of particular facts, questions which come to this, in a general way: what is the explanation of the variety of evolutionary behaviour, of all this ceaseless and apparently pointless restlessness, for 'most of these changes seem to have given no appreciable advantage to their owners'? Such questions arise in the study of every group and 'their number is a measure of our ignorance for they are as yet unanswerable'.

THE BRACHIOPODA

Cooper and Williams[47] have made a very revealing survey of the range in time of practically all known brachiopod genera, classified into their families, superfamilies and larger groups. This was done with the particular object of determining 'the evolutionary pattern of the rise and fall of groups', meaning the shape of the time-distribution strips. The point being examined was whether there were sudden 'bulges' in the strips and, if so, whether these represented 'bursts' of evolution, accelerated proliferations of new genera. The evidence as to this was inconclusive; but a generalization about a more important matter emerged, which may be given in their own words:

> Despite the fact that many thousands of brachiopod species have been described from all parts of the world, and from much of the stratigraphic column, vast gaps still occur in our knowledge. In a generic study of the stropheodontids of North America and Europe it is estimated that the history for at least one-third of the stock is unknown and that the history of a sizable amount of the remainder is inferred from a study of small numbers of specimens. When we consider that this is the story for the two best investigated regions in the world, and for one of the better known brachiopod families, the insufficiency of our data can be appreciated. These gaps are not only temporal and spatial, but genetic as well. We can say by morphological comparison that a genus belongs to a certain family, but precisely when, and from what stock it arose is largely a matter of conjecture. This prompts us to use block diagrams [time-distribution strips] rather than phyletic trees to represent brachiopod history.... In only a handful of genera do we have the preservation of what might be transition stocks from one divergence to another.

Such is the imperfection of the geological record of evolution!

THE MOLLUSCA

These no doubt provide far more fossils (macro-fossils) than any other phylum and most of these belong to the Lamellibranchia and Gasteropoda. But no very coherent story emerges when we trace the lamellibranchs and gasteropods through the stratigraphical systems. This is a very remarkable and highly important generalization about the evidence of the evolution within these two great classes; but, as providing something positive, or at least something for discussion, the class of the Cephalopoda (with their chambered shells mostly coiled in a plane spiral) is far more profitable.

If the text-books tell us in their first editions, as something well-established, that the cephalopods evolved in one way and in their next editions, some years later, they tell us something quite different, we must naturally be curious as to the reason for the change of view. If the later view is a correction of the earlier, as we may suppose, why was the earlier ever put forward, except as something tentatively suggested by the partial evidence then to hand? We are prepared for elaborations and modifications of an evolutionary interpretation as new facts come to light, but hardly for its complete overthrow after it had been generally accepted as, presumably, soundly based on the facts so far known. Facts cannot contradict themselves. These remarks introduce the salutary cautionary tale of the attempts to read the evolutionary story of the cephalopods, a story which (like all the stories of the kind) lies largely hidden, but of which a flash is inevitably given at every point where a fossil appears.

If we call to mind the most familiar British nautiloid fossils, we have the straight forms, especially common in the Ordovician and Silurian, the forms with slightly overlapping whorls from the Carboniferous, and the various species of the completely involute shell of the Jurassic, Cretaceous, Tertiary and, the last lingering survivors of the whole order, to-day. We can add to these the curved and loosely coiled forms from the continental Silurian and Devonian. All this suggests at first sight that there may be a coiling trend from straight to enwrapping.

During the last decade of last century the chief authority on these fossils was Hyatt in America. He elaborated the idea of a coiling-up trend. 'A slowly working tendency is apparent, leading to the production of more and more closely coiled cones, and the elimination of straight and slightly curved forms.'[48] Hyatt largely invoked recapitulation, a proceeding of little, if any, weight in support of an hypothesis as to what has actually occurred in the history of a race; but Hyatt's authoritative pronouncement was for long accepted unquestioningly.

In 1933, Spath[49] went thoroughly into the matter. In a searching review he showed that the hypothesis of a coiling-up trend was quite untenable. 'Hyatt was not unaware of the weak links in his chain of evidence, especially after the exact ages of the various forms became known, and his papers contain many contradictory statements and more or less doubtful admissions; but his followers made light of the difficulties.'

Spath's view of the evolution of the Nautiloidea can be summarized as follows. The earliest nautiloids, from the Upper Cambrian and Tremadocian of North America, are slightly curved cup-shaped forms with a wide *siphuncle* (the tube running through all the chambers). With these in the Tremadocian, and derived from them along two divergent lines, are the straight, elongate, 'orthoceratid' cones with narrow siphuncle and forms coiled in a plane spiral of five slender, slightly overlapping whorls. He thinks it likely that the many loosely coiled and curved forms, and even some of the straight forms, of the Silurian and Devonian were 'uncoiled', being derived from the more completely coiled type represented, very early, in the Tremadocian. By assuming the Mesozoic belemnoids, with a straight internal 'shell', to be derived from the straight-shelled nautiloids, and these to give rise to our commonest cephalopods of to-day (such as the cuttle-fish), he demonstrates the persistence of the straight-shelled type.

This, then, is an entirely different story from that which seemed well established; and the later editions of the text-books have adjusted their remarks.

The Ammonoidea, first appearing as a distinct group in the Devonian, must have been derived from the Nautiloidea, when we have once accepted the obvious inference that since they are so much alike they must be nearly related. The only constant differences between the two groups, when they had become quite differentiated, were that in the ammonoids the siphuncle was always marginal and the septal suture was more intricately folded than in the nautiloids. As we trace the ammonoids back to the Lower Devonian and Silurian we find that they merge morphologically into the nautiloids, so that it becomes difficult to say whether certain forms should be classed with one group or the other. This merging with a parent stock is what we look for in all groups and so seldom find. Where taxonomists are in difficulties, there the seeker after evolutionary evidence will very likely find his way made smooth.

When the coiling-up tendency in the Nautiloidea was accepted, it was naturally thought that the Ammonoidea would be likely to show the same trend. Thus the most primitive ammonoid, in the sense of being nearest the parent stock, was said to be a straight-shelled form, so far not found below the Upper Devonian. But with this now taken to be a 'secondarily uncoiled' ammonoid, there is no need to force on to the material an evolutionary interpretation contradicted by what is known of the stratigraphical occurrence.

Evolution in the Groups

What is perhaps the best-known example among fossils of an evolutionary trend in a separate structural feature is to be seen in following the history of the Ammonoidea. This trend is the progressive elaboration of the pattern of the septal suture. From the Devonian to the Trias this gradually advances from broadly undulating (as we have just seen) through the sharply bent 'goniatitic' stage of the Carboniferous to all degrees of folding and frilling in the Trias. In the Jurassic and Cretaceous the ammonite can go no further in this direction and we have every variety of 'ammonitic' suture line.

It is sometimes claimed that the successive sutures of the individual ammonoid shell provide evidence of recapitulation. The earliest formed septa are very small and could hardly develop highly folded edges, whereas in the later septa there is plenty of room for intricate folding and frilling which may well have been a functional advantage in these larger cross-plates. Thus the simpler early sutures, though in fact recapitulating the known phylogeny of this feature, are no evidence at all of any general principle of recapitulation. Some of the Cretaceous ammonites show a comparatively simple suture, which seems to be a case of evolutionary retrogression. Should the earliest sutures in such forms be highly frilled, on a miniature scale, strong evidence for recapitulation would be apparent, but the writer does not know of any record of such a case.

In the Gault, ammonites are perhaps more abundant than they are in any other stratigraphical series, and they have been intensively studied over a period of 25 years by Spath.[50] After the full description and comparison of over 300 species, illustrated by something like a thousand photographs and very many drawings, we are given a hundred pages of 'palaeontological and stratigraphical results'. These facts are mentioned in order to show that here we have an ideal example of the kind of research, recorded and established in a large monograph, which must form the foundation for our knowledge, hypotheses and speculations about the palaeontological evidence of evolution. The author's own conclusions, after all this work on the fossils themselves, must naturally be of the greatest importance. His discussion and the accompanying tables show that the evolution and relationships within the Ammonoidea are by no means clear, notwithstanding the wealth of material and the prolonged and minute examination of it. Nevertheless, morphological comparison and consideration of the stratigraphical occurrence (and geographical occurrence outside Britain) allow some more or less tentative lines to be drawn showing possible derivations of families and genera. What chiefly emerges is the likelihood of there being a few persistent stocks which gave off from time to time subsidiary stocks in which certain characters, particularly of ornamentation, became elaborated. The persistent stocks seem to have been smooth forms, living in the open seas of the 'Mesozoic Mediterranean', the ribbed forms being evolved on migration to the shelf seas of the 'British area', the ribbing being some sort of response, presumably, to the environment. A perusal of the author's statements and arguments gives us an illuminated view of a representative sample of what one fossil biological group in one stratigraphical formation can be made to show in the way of evolutionary evidence, when all the conditions of the investigation are most favourable.

THE TRILOBITA

We have a good record of that very interesting group of extinct arthropods, the trilobites, with their elaborate dorsal armour. In spite of this, no clear plan of evolution is discernible. Swinnerton has postulated evolutionary tendencies, 'trend lines of structural change',[51] and has taken as the primitive type one group conspicuous in the Lower Cambrian. But there are more ordinary trilobites also in the Lower Cambrian, and it is thus by no means certain that the group to which *Olenellus* and *Callavia* belong is essentially primitive. We do not get gradations of form occurring successively in time between these so-called primitive types and other genera, so that the suggestion remains merely an interesting speculation. We have already seen that the Cambrian record, at least, of the trilobites reveals very little of the evolutionary paths they followed.

THE GRAPTOLITHINA

This group, of uncertain affinities, is divided into the Dendroidea and the Graptoloidea. These latter, which are much the commoner, range from the Tremadocian series of the Cambrian to the Lower

Ludlow series of the Silurian, inclusive. From the evolutionary point of view it is an outstandingly interesting group. Trends are very clearly shown, and these are sufficiently regular to give a stamp to the aspect of each successive graptolite-fauna. Moreover, at first sight, genera or groups of species within genera, when their stratigraphical ranges are set out, may be arranged in a fairly satisfactory (but always necessarily hypothetical) scheme of evolutionary lines. The chief trends are three: reduction in number of branches from many to one, change in direction of growth, relative to the *nema* (the thread of attachment), from pendent to scandent, and increase in the elaboration of the *thecae* (the individual cups). The evolution of the graptolites was first fully expounded by Elles,[52] her exposition being based on the monograph of British graptolites of which she was part author. Since then, however, it has been found that the links in the supposed evolutionary chains are not so secure as was thought, owing chiefly to the more exact knowledge now available of stratigraphical 'dates' of first appearances. Recent advances in our knowledge of graptolites are very largely due to Bulman's work.[53]

Although the origin of the Graptolithina as a whole is particularly obscure, the derivation of the (simpler) Graptoloidea from the Dendroidea, the group which appeared first (and died out last), is neatly confirmed by the Anisograptidae, 'a family so completely transitional between the Dendroidea and Graptoloidea that it could be included in either', and one which is also a link in time, appearing just before the typical graptoloids became established.

It should be noted that the latest Graptoloidea are altogether the simplest (even the thecae have become simple again), while the earliest are the most complicated; a warning that the simpler forms in a group are not necessarily the earlier.

THE VERTEBRATES

The geological history of the vertebrates is well known and has been summarized in many works, in more or less detail.[54] Thus it is not proposed to treat it here other than to quote two passages, eminently pertinent to our present purpose, from recent authorities in this field.

Palaeontology and Evolution

To-day we are close to the centenary of the publication of the *Origin of Species*, and it is interesting to reflect on the controversy to which this book gave rise against the background of our present knowledge of the fossil record of evolution. It is, indeed, almost startling to observe how completely this record [of the vertebrates] has provided the verification for prognostications based primarily on the comparative morphological study of living types. We see the graded succession through time of jawless ostracoderms to jawed fishes, of lobe-finned crossopterygians to the first land vertebrates, of labyrinthodont amphibia to primitive reptiles, of mammal-like reptiles to the first true mammals, and so forth. The temporal sequence is as impressive as the structural sequence, and, as the story is unfolded age by age, and phase by phase, we recognize that it is the combination of the two which provides the final proof of the evolutionary hypotheses.[55]

The reptile-birds *Archaeopteryx* and *Archaeornis* remain outstanding examples of a link of this kind between two major organic groups. That this link should be dependent upon two or three specimens only is an eloquent reminder of the frailty of palaeontological evidence.

We shall soon be celebrating the centenary of the first publication of the paper by Darwin and Wallace which was one of the most epoch-making events in the history of science and humanity. In view of the great day shortly to come, it will not be expected of me that I should devote myself to a treatment of the whole subject of evolution, and I have in consequence selected a more restricted field. I have a certain apprehension in speaking on any scientific subject, particularly that of evolution, without having some solid object to talk about so as to anchor my words to a firm sea-bed of evidence, and the material which I have selected for this purpose is the specimen of the fossil *Archaeopteryx lithographica* preserved in the British Museum (Natural History); probably the most precious, the most beautiful, and the most interesting fossil hitherto discovered in the world. The fossil is preserved in a slab and corresponding counter-slab of a block of limestone from the Solnhofen deposit of the Jurassic, about 150 million years old. At the very first glance its most important features become apparent, for while some of them are thoroughly characteristic of reptiles, others are no less completely characteristic of birds. It is this intermediate position which *Archaeopteryx* occupies that makes it an object of such enormous interest.[56]

The invertebrate animals must always supply the bulk of the evidence regarding evolution because they occur so much more profusely as fossils than do the vertebrates and plants (and in so much wider variety than the vertebrate animals); so it is rather curious, perhaps, that the vertebrate phylum should provide a better evolutionary story, with more connecting links between its main groups, than does any other animal phylum. Nevertheless, the origin of the vertebrates is no more clearly revealed than the origin of any other phylum, and this also is curious, as this phylum is probably the only one that arose after the progress of life had begun to be recorded by fossils.

PLANT LIFE

Although the no less interesting history of plant life lies outside the scope of this essay, reference may be made to several authoritative works. There is the excellent British Museum guide,[57] Seward's well-known book,[58] so richly packed in text and illustration, and Scott's short, masterly review.[59] The meagre evidence about some of the main groups of plants, the algae (sea-weeds and similar fresh-water plants) and the bryophytes (mosses and liverworts) at one end of the scale and the angiosperms (flowering plants) at the other, does not prevent the outline being clear, and the extinct groups, especially the seed-ferns and 'flowering' cycads, though surprising, fall into the general evolutionary picture.*

Evolutionary Series

Let us look now at some of those instances of evolutionary series that have been detected, especially those of the invertebrates.

THE CORALS

In four parts of the Lower and Middle Carboniferous of the Midland Valley of Scotland, where limestone bands are found, occur species of the coral genus *Zaphrentis*. These species form a distinct group within the genus as they have certain special features in common (e.g. external longitudinal ribbing).[61] Moreover, the species of the group form a closely graded morphological series. These forms are

* One of the most remarkable evolutionary series yet described in the plant kingdom, based upon excellent morphological evidence and extending from the Carboniferous into the Mesozoic, is that of the conifers.[60]

arbitrarily grouped into four (or by some, six) named species. Any specimen can thus be placed in one of the four (or six) restricted species within the species-group, which bears the name of one of them, *Zaphrentis delanouei, sensu lato*. The variation is in the length and thickness of the *septa* (the vertical partitions) and the shape of the *cardinal fossula* (break in the ring of septa), as seen in cross-section. When the frequencies of the specimens referable to each species, conveniently reduced to percentages, are tabulated for each of the four parts of the geological succession, the highest frequency shows a continuous lateral shift as we read the table, in the usual geological way, from below upwards. A regular morphological progression with time, and so, presumably, evolution, is thereby expressed.

Much has been made by Carruthers and his followers of the supposed evidence of recapitulation within the group (well-preserved corals are excellent fossils for studying skeletal ontogeny). Hill goes so far as to say, 'Carruthers' classic studies on this group remain unassailable as evidence of slow evolution...and of the repetition of the phylogeny during ontogeny',[62] and Swinnerton uses the material as the chief example to illustrate his address to the British Association.[63] But none of these authorities seems to have adequately considered what is left over of evidence in favour of recapitulation, as a law, after the facts inseparable from normal growth have been eliminated.

Slow and more or less gradual evolution has also been recorded for the coral species-group *Caninia cylindrica*.[64] There are four fairly well marked forms, constituting a series of morphological grades, each of which, in order, occurs commonly at a successive particular stratigraphical level in the Carboniferous Limestone series. An interesting feature here is that the forms characterizing the lower two levels have been found only in Belgium, the third form is found also in Britain, while the latest is common in Britain and rare in Belgium. Thus we have here a record of the migration of an evolving stock.

Studies of two genera of Silurian corals have produced instances where the morphology and occurrence suggest a particular evolutionary drift and at the same time give precision, because of the amount of detail described, to the limited degree of drift revealed.

These genera are *Acanthocyclus*[65] and *Cymatelasma*,[66] from the Welsh Borderland. The interpretation of the evolutionary evidence is, however, obscure, firstly because of the small proportion of the species which fall into a morphological series, and secondly because of the lack of strictness in the agreement between morphological state and stratigraphical level.

These instances of slow, gradual evolution are exceptional. What are the rare conditions, biological or geological, that have caused these particular corals to evolve in this way?

It need hardly be said that studies that reveal exactly and fully the facts of the history of a genus or family, or some such small group, and thus give the precise degree to which graded change is evidenced, if at all, are always valuable, whatever the particular outcome. There is, perhaps, a tendency for investigations of the kind to be pursued and published only where positive evolutionary evidence is seen to be emerging. But we cannot have too much detail of the precise level, place and manner of occurrence of carefully discriminated forms.

Lang has put forward an interesting evolutionary idea exemplified, as he suggests, in certain genera of Mesozoic corals.[67] Certain well-known long-ranging genera are distinguished chiefly by the degree of compaction of the component corallites; *Montlivaltia* is single, *Thecosmilia* branching, *Isastraea* has polygonal corallites everywhere in contact and *Thamnasteria* is similarly massive, but with indistinct walls and confluent septa. It is supposed that, instead of each of these genera constituting a separate line of forms nearly related by descent and persisting for a long time, it is the forms occurring at any one time (approximately) and covering all degrees of compaction, that are the more nearly related. Stratigraphical occurrence gives no evidence as to the direction of evolution, either among the genera (each of the four appears in the Trias, outside Britain) or among the species in a particular formation; the idea is solely that the simpler type is the more primitive and 'throws off' strings of other genera in much the same way as was considered in connection with the Palaeozoic rugose corals as a whole. The inference that particular species are very nearly related must rest primarily on the possession of features common to those species and distinguishing them morphologically from other groups of species. Thus for Lang's idea to have substance, his species at each horizon, distributed among the several genera (defined by the degree of compaction), should have features characteristic of that horizon; but he does not seem to have applied this test very rigorously, judging by what he gives us and what one can find out about the precise characters of the species he mentions.

THE ECHINOIDS

The echinoid genus *Micraster* is exceptional among the many Chalk echinoids in showing gradual, though apparently not quite even-paced, evolution throughout its range from the middle of the Middle to the middle of the Upper Chalk. The forms within the genus do not fall naturally into separate and distinctive species, but rather make a morphological continuum, various parts of which may be given specific names for convenience, the species merging one into another. The echinoid test lends itself to analysis into a large number of features and, within the narrow limits of the genus, most of these are found to vary rather strictly with time. A very detailed study, probably the first of its kind in Britain, was made by Rowe and published in 1899.[68] About 2000 specimens were collected, from precisely recorded horizons in the Chalk, and minutely examined. The conclusions are well summarized by Swinnerton,[69] the main varying, and evolving, features being proportion of length to breadth, depth of anterior notch, positions of apical disc and mouth, and (particularly) the character of the inner ends of the *ambulacral plates* (the double rows of plates forming, in this genus, petal-like areas radiating from the apical disc).

THE BRACHIOPODS

Some of the well-known brachiopod genera are common throughout a system and a careful scrutiny of these, or of species-groups within them, has revealed, here and there, some indications of evolution. Thus the species-group which includes the well-known *Sowerbyella sericea* has been traced through a large part of the Ordovician and Silurian in Britain.[70] Some changes, particularly in details of ornamentation, having a serial time-distribution, have been noted.

Perhaps the best instance is the change with time in the Silurian *Atrypa reticularis*, one of the most familiar of all species of fossils.[71] On the whole, as we go up the succession through the Llandovery, Wenlock and Ludlow series, the smaller dorsal valve becomes more and more convex and the larger ventral correspondingly flatter, and the shell itself becomes larger. Detailed studies of brachiopod species and genera, as of the species and genera of any group, are almost certain to suggest some possible evolutionary derivations and connections. One example may be mentioned, the study made by Prentice of certain forms lying within the Carboniferous genus *Gigantroproductus*.[72] This is primarily a variation study, of a kind that necessarily precedes most kinds of evolutionary study.

THE LAMELLIBRANCHS

Among the lamellibranchs there is one family that provides interesting material, the Ostreidae.[73] Here there is a wide diversity of form, but a diversity in which the more extreme members are bound by intergrades to a central type, *Ostrea*. The morphological continuum may be visualized as three-rayed, the forms towards the further ends of the rays being sufficiently distinct from *Ostrea* as to warrant their being given separate generic names. There are four variable structural features, varying in certain combinations: (1) valve-thickness, (2) incurvature in the region of the beak of one valve (the left), (3) lateral twisting in the region of the beak, and (4) degree of development of coarse angular folds. The genera are: *Gryphaea*, characterized by (2), together with (1); *Exogyra*, characterized by (3), together with (1) and something of (2); and *Lopha*, characterized by (4) with a little of (3). *Ostrea*, the oyster, itself has a long range, from the Trias to the present day and, apart from its simplicity, its position at the centre of the continuum of variation strongly suggests that, if there has been evolution in the family, this has been 'outwards' from *Ostrea* leading to one or other of the three distinctive genera by various combinations of the four structural trends. It would, presumably, be theoretically possible for the three 'outside' genera each to be the parent of flat-valved forms, characteristic of *Ostrea*, by the morphological convergence of their derivatives; but this is hardly likely. Also, the three 'outside' genera are not so persistent throughout the strata as is *Ostrea*, and there is at least one clear instance of a particular evolutionary series, to be described in the next paragraph, of an *Ostrea* giving rise to a *Gryphaea*. Thus we are justified in inferring that evolution has probably occurred in this family by a persistent parent stock producing the same kinds of derivatives several or many times. No doubt such evolutionary offshoots would be adaptations either to changing conditions on the spot or to new environments sought. *Gryphaea*, for instance, seems to be an adaptation to muddy surroundings; but why is there no *Gryphaea* to-day? These derivatives, each kind rightly called by the one generic name, but usually sufficiently distinct at the several times for their representatives at those times to be given separate specific names, will thus not be directly related, each along one straight persistent line. We have, then, to be prepared, everywhere in the fossil record, to find evolution working in this way; indeed it seems a very natural way, and perhaps it is rather surprising that we do not find it more often.

Gryphaeate forms are well known from the Lower Lias. Large numbers were collected from the cliffs on the Glamorgan coast, particularly at five stratigraphical horizons, and were the subject of a study by Trueman.[74] The essence of this study has often been summarized, and is particularly clearly shown in the annotated diagram given by George[75] (fig. 4). The main facts are that the degree of curvature varies, from flat to an enrolment of one-and-a-half coils, at each horizon the range of variation is about three-quarters to one coil, and the average degree of enrolment steadily increases from about one-quarter of a whorl at the lowest horizon to about one-and-a-quarter coils at the highest. These variations are arbitrarily grouped round five named species, the first two being placed in the genus *Ostrea*, the other three in *Gryphaea*.

Inoceramus is a very common genus in the Gault and Chalk. There is great variation in shape and ornamentation, and the lines of variation are found to correspond to some degree with time.[76]

The non-marine lamellibranchs of the Coal Measures are distinctive of particular zones. The ranges and abundance of the genera are shown in Trueman and Weir's diagram,[77] but there is little to be discerned in the way of general evolution among them. George has shown how communities of several species of *Anthra-*

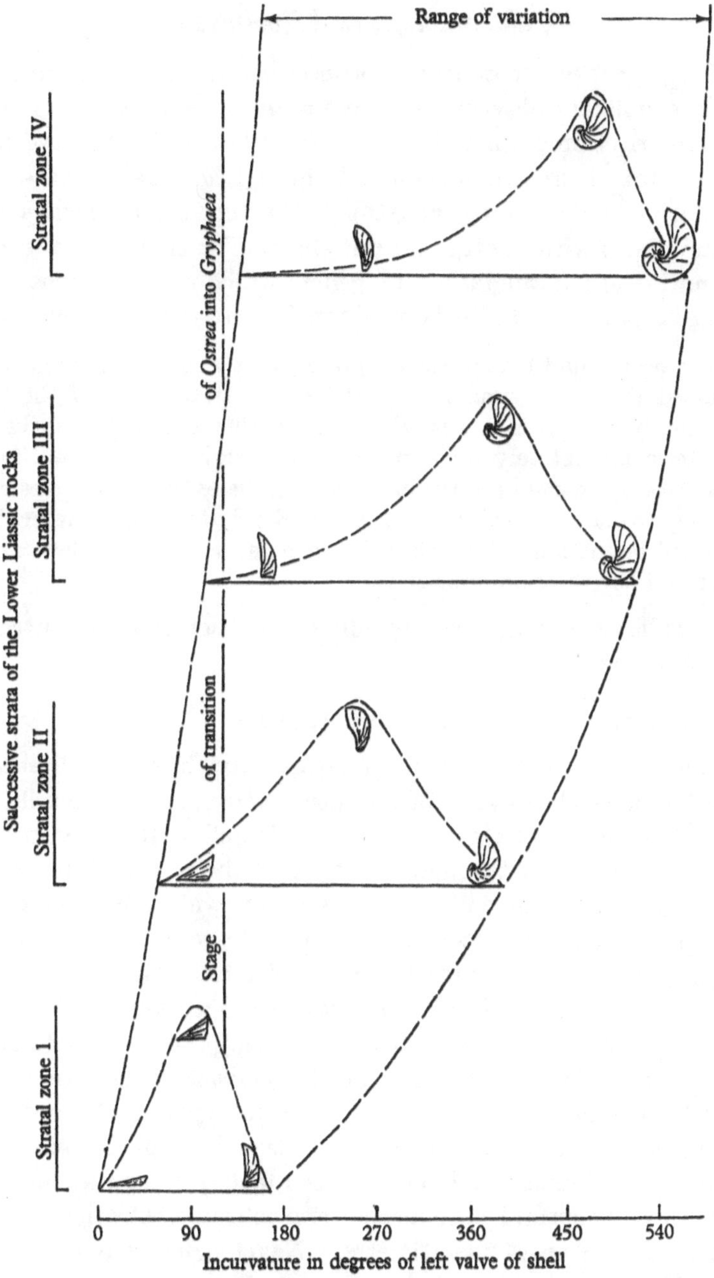

Fig. 4. Diagram illustrating the evolution of gryphaeate oysters in the Lower Lias. Each of the four graphs shows, at successive levels, the proportionate abundance of forms with varying amount of incurvature. (From T. N. George, *Evolution in Outline*, London, 1951.)

conaia merge into one another morphologically and also in time.[78] Highly significant observations have been made by Eagar. He has detected 'rhythmic units' in which, from bottom to top, we tend to pass from a marine shale-band through gradually coarsening non-marine shales to a coal-seam at the top, after which a new rhythm starts with another marine shale. (The shales described as non-marine are distinguished by genera which resemble those containing certain non-marine lamellibranchs of to-day.) He remarks:

> It has been found that in each rhythmic unit where shell horizons are developed there is a similar shell sequence. At the base of the unit, immediately overlying a *Lingula*-band (marine shale), the shells are elongate, comparatively small and mostly referable on their outline to *Anthraconaia*. Traced upwards in coarsening shales the fauna gives place gradually to one of *Carbonicola*. Higher in the rhythmic unit larger shells are added to the fauna, and there is a general increase in relative shell height and length of anterior end.[79]

Thus we have here, apparently, micro-evolution strictly controlled by environment.

THE AMMONOIDEA

Ammonoids evolve and disperse so quickly, it seems, that it is almost impossible to stay in one place and see in the fossils any gradual change, and almost as difficult to follow them by keeping track of their career through the strata of the world. A few cases, however, of evolution within a genus are known. Thus Bisat's work shows that the genus *Reticuloceras*, one of the commonest goniatites of the Millstone Grit, forms a morphological series as its representatives are traced through a part of the succession.[80]

In 1929, Brinkmann published in Germany the results of a study he had made of the ammonite genus *Kosmoceras* in the Oxford Clay at Peterborough. This work was later brought to the notice of British geologists by Arkell.[81] Three clay-pits provided exposures to give a continuous section of thirteen metres of strata, and more than three thousand specimens were collected, stratigraphically 'dated' to the nearest centimetre. 'Never before has a group of ammonites been subjected to such a searching statistical analysis based on such large quantities of individuals.' Brinkmann concluded that within the genus there were four separate lines along which evolution proceeded more or less gradually except across certain levels where the lithology would in any case suggest breaks in the continuity of sedimentation. It is thus shown that very intensive study of exceptionally rich material will, as is to be expected, break down, to some extent at least, the intractability of fossils in providing evidence of graded change in time.

THE TRILOBITES

Little can be found in palaeontological works as to evolutionary series among the trilobites. However, if we take a glance at the matter by examining the figures of some of the genera in a few families, as set out in Swinnerton's text-book, adding the ranges from his text, there appear to be some indications of evolutionary series here and there. For instance: *Cheirurus* through *Sphaerocoryphe* to *Deiphon* (Ordovician to Silurian); *Acaste* through *Phacops* to *Trimerocephalus* (Ordovician to Devonian).[82]

THE GRAPTOLITES

The Graptolites often occur abundantly usually to the virtual exclusion of other kinds of fossils, in the mudstones of the Ordovician and Silurian systems, yet it is rare to find one species merging into another in the course of time, in spite of the fact that the graptolites show clearly defined evolutionary trends. One case of such a merging has, however, been briefly described by the present writer.[83] There is a particularly fossiliferous locality in the gorge of the River Rheidol, three-quarters of a mile S.S.E. of the village of Pont-erwyd in Cardiganshire. The particular species of graptolite show that the rocks at this spot belong to the *Monograptus triangulatus* zone of the Llandovery series of the Silurian system. Many species are found, and among the more abundant are forms which belong to the species *M. fimbriatus*, *M. raitzhainiensis* and *M. triangulatus*; moreover—and here is the exceptional condition—there are specimens that fill in the gaps between the typical forms of these species, so that there is something of a morphological continuum (fig. 5). They occur through a thickness of 25 feet, which here represents the whole zone. To see what evidence the morphological series might provide of evolution, it was obviously desirable to collect as copiously as possible from as many stratigraphical levels as possible. The specimens are common in five bands, each about a foot thick; but they are mostly fragmentary and it was found that the only satisfactory way to deal with them was to have two morphological

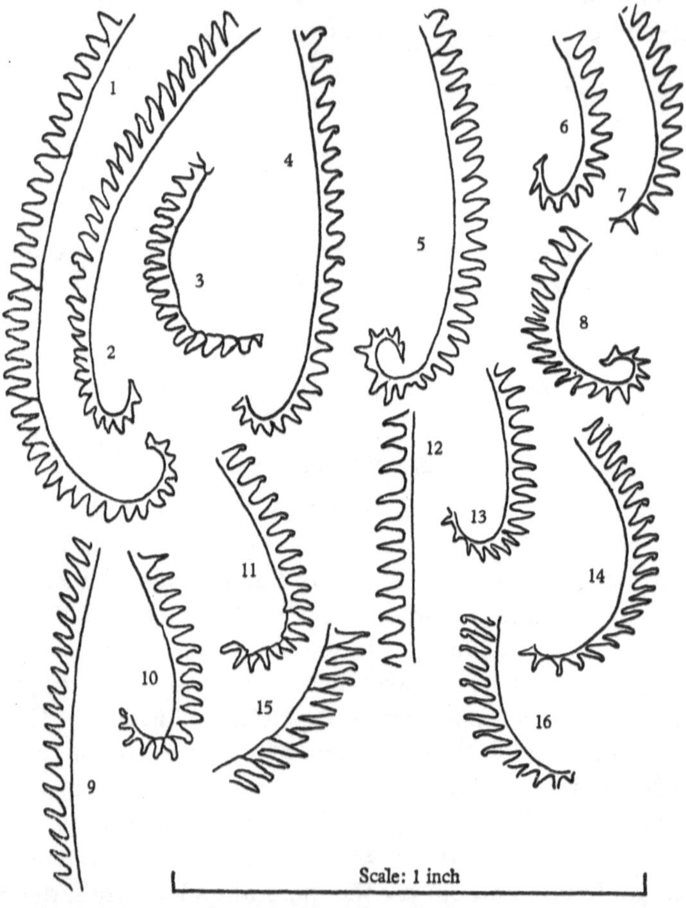

Fig. 5. Outline sketches of a morphological series of specimens of graptolites from *Monograptus fimbriatus* (1–5) through *M. raitzhainiensis* (6–10) to *M. triangulatus* (11–16).

grades only and to place each specimen in either a *fimbriatus* group or a *triangulatus* group. This was necessary partly because the shape of the thecae is not the same along all parts of the branch, particularly in the intermediate species *M. raitzhainiensis*, and partly because the shape of the thecae varies with some distortion in fossilization. Thus a too refined grouping of the fragments, on the shape of the thecae, would be deceptive. The essential facts are shown in Table I.

TABLE I. *The frequencies of species of* Monograptus *in relation to stratigraphy in an exposure of Silurian rocks in Wales*

Height above base of lowest band (in feet)	*Fimbriatus* group (per cent total)	*Triangulatus* group (per cent total)
24·0–25·0	77	23
11·5–12·5	78	22
8·0– 9·0	82	18
2·5– 3·5	94	6
0·0– 1·0	100	0

It will be seen that although the *fimbriatus* group of forms was everywhere commoner than the *triangulatus* group (the zone might here be called more appropriately the zone of *M. fimbriatus*) there was a fairly steady rise in the proportion of the latter. A tendency for evolution to occur from *M. fimbriatus* to *M. triangulatus* thus seems to be demonstrated.

THE VERTEBRATES

Perhaps the most familiar example put forward to illustrate an evolutionary series is that of the horse family, where an increase in size, in height of the teeth and especially the acquirement of the 'one-toed' hoof is described. (The fossils do not occur in Britain.) However, it is salutary to read the remarks of a specialist and to realize that the evolution is far from the simple matter that might be supposed.

> The most famous of all equid trends, 'gradual reduction of the side toes', is flatly fictitious. There was no such trend in any line of Equidae. Instead there was a sequence of rather rapid transitions from one adaptive type of foot mechanism to another. Once established, each type fluctuated in the various lines or showed certain changes of proportion related to the sizes of the animals, but had no defined trend.[84]

However, it seems that the term 'trend' is being used here in a rather restricted sense and that we might legitimately refer to a 'sequence of rather rapid transitions' as a 'trend'.

Appendix

THE GEOLOGICAL TABLE

(Time-scale: the 'B' scale of Holmes) [85]

Era	Age (in millions of years)	Duration	Period		
Cainozoic	0–1	1	Quarternary		Recent Pleistocene
	1–26	25	Tertiary	Upper	Pliocene Miocene
	26–58	32		Lower	Oligocene Eocene Palaeocene
Mesozoic	58–127	69	Cretaceous	Upper Lower	
	127–152	25	Jurassic	Upper Middle Lower (Lias)	
	152–182	30	Triassic	Upper (Rhaetic and Keuper) Middle Lower	
Palaeozoic	182–203	21	Permian	Upper Lower	
	203–255	52	Carboniferous	Upper (Coal Measures) Lower	
	255–313	58	Devonian	Upper Middle Lower	
	313–350	37	Silurian		
	350–430	80	Ordovician		
	430–510	80	Cambrian	Upper Middle Lower	
Pre-Cambrian	510–				

3

J. B. S. HALDANE

NATURAL SELECTION

A Sketch of Darwin's Opinions

The full title of the book by which Darwin is best known is *The Origin of Species by Means of Natural Selection, or the Preservation of Favoured Races in the Struggle for Life.* In his chapter headings he used two other phrases, *Struggle for Existence* and *The Survival of the Fittest.* Of the latter phrase he wrote (chapter III) 'I have called this principle, by which each slight variation, if useful, is preserved, by the term Natural Selection, in order to mark its relation to man's power of selection. But the expression often used by Mr Herbert Spencer of the Survival of the Fittest is more accurate, and is sometimes equally convenient.' In the same chapter he explained that he used the term *Struggle for Existence* 'in a large and metaphorical sense', 'including dependence of one being on another, and including (which is more important) not only the life of the individual, but success in leaving progeny'. In *The Origin of Species* he regarded sexual selection, which 'depends, not on a struggle for existence in relation to other organic beings or to external conditions, but on a struggle between the individuals of one sex, generally the males, for the possession of the other sex', as a form of natural selection. Later, he was inclined to give it independent rank.

It is worth mentioning two other phrases which Darwin did not use. Herbert Spencer in his *Principles of Biology* stated that organisms became adapted to their environment by direct and indirect equilibration. Direct equilibration is the response of an organism to the environment in which it lives, for example the growth of muscles which are constantly used, or the acquisition of a skill or habit. He believed that such changed structures and habits would be transmitted to descendants. Indirect equilibration was the result of the survival of the fittest, those beings which were pre-adapted to the environment or adapted to it most quickly. Romanes, in his *Darwin and after Darwin*, used the phrase 'winnowing' for natural selection. He thought that this process was more like the mechanical selection of heavy seeds by a current of air which blew away chaff and lighter seeds than conscious human selection. This at least serves as an antidote to the notion of Nature as a stern but benificent female who weeds out the 'unfit' in each generation, a notion which lies at the back of some eugenic thinking. It would make for clarity of thought if the name of Siva or Durga were used in this context.

The fact of natural selection was quite well recognized before Darwin's time. I quote two passages.

Lucretius (*De rerum natura*, v, 855) wrote:

> Multaque tum interiisse animantum saecla necessest
> Nec potuisse propagando procudere prolem.
> Nam quaecumque vides vesci vitalibus auris
> Aut dolus aut virtus aut denique mobilitas est
> Ex ineunte aevo genus id tutata reservans.*

Dante (*Paradiso*, VIII, 139) wrote:

> Sempre natura, se fortuna trova
> discorde a se, com'ogne altra semente
> fuor di sua region, fa mala prova.†

Lucretius recognized natural selection as acting between species, Dante as acting between ecotypes. Neither saw it as an agent which could gradually transform a species. Aristotle, in the remarkable passage quoted by Darwin in his *Historical Sketch* which opens the *Origin*, came near to such an idea, but never developed it.

Lamarck, as Darwin points out in the same sketch, was Darwin's main precursor in calling attention to evolution as a historical fact, and, in Darwin's words: 'He first did the eminent service of arousing attention to the probability of all change in the organic, as well as

* And many races of living beings must then have died out and been unable to reproduce themselves. For whatever you see absorbing oxygen, it is cunning, valour, or finally nimbleness, which has protected and preserved that race from its origin.

† Nature, if it meet with unfavourable conditions, like any other seed out of its own region, always fails to thrive.

in the inorganic world, being the result of law, and not of miraculous interposition.' Darwin, basing himself on half a century's progress in palaeontology, geographical distribution, comparative anatomy, and embryology and other branches of biology, was able to give stronger arguments for evolution, and a more accurate account of it, than Lamarck. He also gave a different account of its causation. As he states, Wells, Matthews, and others, had anticipated him, but had not worked out the theory in any detail.

The fact of evolution was accepted largely because Darwin gave an intellectually satisfying account of how it might have happened. To-day the situation is different. The vast majority of biologists believe that evolution has occurred, even if some of them postulate supernatural intervention in connection with certain great changes. But it is quite possible to accept evolution as a historical fact, and reject Darwin's theory of how it happened wholly or in part. In the same way one can accept the fact that human societies have changed, while wholly or partially rejecting Marx's theories as to why and how they did so. In what follows I shall take evolution for granted as a historical fact, and discuss the questions whether natural selection occurs, and if so whether it can account, wholly or in part, for evolution.

In the sixth edition of *The Origin of Species* Darwin did not take up the question of evolution as a historical fact till the tenth chapter, and then devoted five chapters to it. His first chapter deals with variation in domestic animals, its inheritance, and the effects of artificial selection, the second with variation in nature. The third and fourth deal with natural selection, and the fifth with laws of variation. Then follow four chapters on various difficulties in the theory, on instinct and hybridism. The first chapter was later expanded into *The Variation of Animals and Plants under Domestication*, a book in two volumes, which Bateson regarded as Darwin's greatest work, and which is still indispensable to geneticists.

Darwin was convinced that variation was constantly occurring, and that variations are generally inherited. 'Perhaps the correct way of viewing the subject would be', he wrote in *The Origin of Species*, 'to look at the inheritance of any character whatever as the rule, and non-inheritance the anomaly.' He showed that artificial selection was effective in domestic animals, and (though he knew less about them) in domestic plants, and proceeded to argue that the analogous process of natural selection could not merely give rise to new species, but could account for the very great transformations, such as the origin of mammals, through many intermediates, from fish, in which the palaeontological record compels us to believe.

His argument was very simple; it was sometimes deceptively simple. He supported it by a mass of examples. Most of the contemporary arguments against him were worthless, but he devoted many pages to refuting them. I shall try to consider those criticisms which, in my opinion, have most weight to-day.

The most important advance in knowledge bearing on Darwinism in the last century has been the elucidation of the fundamental laws of genetics. In this field the greatest name is that of Mendel, but Mendel would be as lost in modern genetics as Dalton in modern atomic theory. It is a sufficiently exact branch of science to permit of a considerable mathematical superstructure. So we can now answer questions exactly, where Darwin could only deal with tendencies. Darwin stated truthfully in *The Origin of Species*: 'The laws governing inheritance are for the most part unknown.' They are now for the most part known, even when they are unexplained. We know that genes are usually copied accurately when a nucleus divides. We do not know how this is done, but the laws of inheritance do not depend on how it is done.

If Darwin could be resurrected for a week, I should recommend him to read Mayr's *Systematics and the Origin of Species*, and Simpson's *The Major features of Evolution*. He would find most of the facts new and thrilling. But he would understand most of the arguments. I would not recommend him to read Dobzhansky's *Genetics and the Origin of Species*. He could not profitably do so without a grounding in elementary genetics.

A Sketch of Genetics

This sketch is simplified and dogmatic, and, therefore, inexact, but is a necessary prolegomenon to what follows. Readers who are not familiar with genetics should consult a less condensed account.

A Sketch of Genetics

If two members of a species can be distinguished by methods other than breeding from them, they are said to possess, or to belong to, different *phenotypes*. This word is used both for a quality and the class possessing it. The differences may be visible, as those between a brown-eyed man and a blue-eyed woman, or only detectable by other means, as, for example, between members of different blood groups. If two organisms cannot be distinguished as regards some character they are said to possess the same phenotype with regard to it. For example, two black cats may possess the same phenotype for hair colour, but different phenotypes for tail length. If two organisms can be distinguished by their breeding behaviour they are said to belong to, or possess, different *genotypes*. For some characters in some organisms there may be a full correspondence between genotype and phenotype. Thus human beings carry one or both of two antigens called *M* and *N* on their blood corpuscles, and this may be important if they are given repeated blood transfusions. The three phenotypes are called *MM*, *MN* and *NN*. An *MM* man or woman has only *MN* children by an *NN* spouse, an *MN* has both *MN* and *NN* children (if enough are born), and an *NN* has only *NN* children. But two or more genotypes can correspond to one phenotype. Thus there are two sorts of tabby cat. One sort, called *homozygotes*, have all tabby kittens if mated to a black; the other sort, called *heterozygotes*, have about equal numbers of tabby and black kittens. Similarly, several phenotypes can correspond to one genotype. Thus a stock of 'Himalayan' rabbits may breed true. They are all born white, and later develop black noses, ears, feet and tails. If, however, such a rabbit is kept warm enough, the black pigment does not develop. It has the phenotype of an albino rabbit. However, it breeds as a Himalayan. The children of two heat-whitened Himalayans develop dark extremities when kept cool.

Often we do not know why animals of the same genotype have different phenotypes (for example, mice of the same pure line may or may not have an interfrontal bone). We only know that the two sorts of mice have the same genotype. The children of a pair with interfrontals are no more likely to have an interfrontal than are the children of a pair from the same pure line without interfrontals. But interfrontals are much commoner in some pure lines than in others.

Differences of phenotype may be qualitative, as in the examples given, or quantitative, if, for example, we are concerned with lengths of body or life, milk production per year, or intelligence quotients. These seem to vary continuously. It is not possible to get a set of quite similar organisms. The heights of pea seedlings of identical genotype, kept under identical conditions, have a coefficient of variation of about 4 per cent. That is to say about two-thirds of the heights are within 4 per cent of the average, and twenty-one out of twenty-two within 8 per cent of the average. It is possible by asexual propagation, continued inbreeding, and other methods, to get a set of organisms of identical (or at least indistinguishable) genotype. Within such a set selection is ineffective, as with the Himalayan rabbits. The differences between its members are believed to be wholly due to environment, and are not inherited. However, differences of this sort, especially maternal ones, are sometimes inherited for one generation in higher plants and animals, and in bacteria and yeasts they may be inherited for several hundred generations.

While some phenotypes, at least—for example *metrical* characters such as height and milk yield—vary continuously, it is believed that genotypes always vary discontinuously, by 'steps' which may be large or small. These steps are due to differences between units called genes, which were discovered by Mendel, though the word 'gene', with 'phenotype' and 'genotype', is due to Johanssen. My account of them is only true for *diploid* organisms, that is to say organisms whose nuclei contain two like sets of chromosomes, one set descended from a set contributed by the mother to a fertilized egg, the other from a set contributed by the father, in a spermatozoon or a pollen grain. It is not quite true when the maternal and paternal sets differ, as they do in mammals, birds, and many insects, in respect of one pair of 'sex' chromosomes. It is quite untrue in haploid organisms, such as many fungi, whose nuclei contain only one set.

Consider the blood types *MM*, *MN* and *NN*. All the spermatozoa of an *MM* man carry a gene called $\mathbf{L}^{M}$. So do all the female pronuclei of the eggs of *MM* women. These are the nuclei in the eggs with which the nucleus of a spermatozoon fuses. Similarly, the *gametes* (spermatozoa or unfertilized eggs) of *NN* people all carry a gene called $\mathbf{L}^{N}$. Such people are *homozygous*. But *MN* people are *heterozygous*. Half their gametes carry $\mathbf{L^M}$ and half $\mathbf{L^N}$. We thus distinguish:

Genotypes	$\mathbf{L^M L^M}$	$\mathbf{L^M L^N}$	$\mathbf{L^N L^N}$
Phenotypes	*MM*	*MN*	*NN*

There are six types of possible mating, as shown in the table:

Mating type	Progeny *MM*	*MN*	*NN*
MM × *MM*	All	0	0
MM × *MN* or *MN* × *MM*	½	½	0
MM × *NN* or *NN* × *MM*	0	All	0
MN × *MN*	¼	½	¼
MN × *NN* or *NN* × *MN*	0	½	½
NN × *NN*	0	0	All

Very often we can only distinguish two of the phenotypes. Let us call our two genes **A** and **a**. If the homozygote **AA** and the heterozygote **Aa** have the same phenotype, **A** is said to be *dominant* over **a**, and **a** to be *recessive* to **A**. Thus in cats, **AA** and **Aa** are tabbies, and **aa** is black. So a homozygous (**AA**) tabby mated with a black gives heterozygous (**Aa**) tabbies. These give half tabby and half black if mated with blacks (**Aa** × **aa** → ½**Aa** + ½**aa**), and three-quarters tabby and one-quarter black if mated together (**Aa** × **Aa** → ¼**AA** + ½**Aa** + ¼**aa**). Blacks mated together breed true (**aa** × **aa** → all **aa**). Sometimes a biochemical test will enable us to distinguish homozygotes from heterozygotes. Thus normal rabbits possess an enzyme called atropine-esterase in their plasma, which destroys atropine and allows them to eat moderate amounts of deadly nightshade with safety. The absence of this enzyme is recessive. Using the symbols *Ae* and *ae*, *AeAe* and *Aeae* rabbits have the enzyme, *aeae* lack it. But careful work shows that a given amount of serum from an *AeAe* rabbit destroys atropine at about twice the rate as it is destroyed by the same amount from an *Aeae* rabbit. Thus, if a mixed population of such rabbits were given moderate amounts of nightshade in their food, most of the *aeae* rabbits would die, and rather more of the *Aeae* than of the *AeAe*. *Ae* is not completely dominant over *ae*. This incomplete dominance is usual for pairs of genes controlling quantitative characters such as size.

If we cross breeds which differ by two different genes, for example, dominant rose-combed white fowls with recessive normal-combed black fowls, the genes usually segregate independently. On crossing the mongrels to the double recessive (here normal-combed black) we get about equal numbers of rose-combed white, rose-combed black, normal-combed white and normal-combed black. If, however, the genes are carried on the same chromosome they tend to go into the same gamete, and most of the 'back-cross' from mongrel to double recessive resemble one or other grandparent.

Genes are believed to be small regions in a chromosome controlling the synthesis of larger molecules, usually enzymes, which are proteins that in their turn control other metabolic processes. Genes vary sharply because they are chemical molecules or parts of them. For example, adenine and guanine are thought to be constituents of genes. There is no intermediate between adenine and guanine, and the substitution of adenine for guanine in a gene will presumably alter its function. It will synthesize a slightly different protein, which may have no enzymatic power, or may act more quickly or slowly than the original, or more quickly on some substrate but more slowly on others. A great many different genes (called *multiple allelomorphs*) have been found at the same place (or *locus*) on a chromosome in different individuals of a species. The maximum number yet found is about fifty.

As a general rule a gene reproduces its like when a nucleus divides, or we may say that a gene is generally copied accurately. But sometimes a 'mistake' occurs. Either a gene is copied incorrectly, or it is altered in such a way that a new type of gene is produced when it is copied. Both these events are called *mutation*, and the new type of gene is called a *mutant*. Mutation is a rare event. The most mutable human genes known mutate about once in 50,000 human generations, or, since each human generation involves 50 or 100 nuclear divisions, about once in three million nuclear divisions. The figure for bacterial gene mutation is not very different. The least mutable human genes known probably mutate about once in a thousand million generations, or about once per generation in the entire human species. Mutation can be speeded up several hundred times by X-rays and other agencies such as alpha particles and neutrons which produce ionization in tissues, and by a variety of chemical substances. The effects are specific. X-rays speed up mutation of some genes, one chemical mutagen another set, a second chemical mutagen yet another set of somewhat different genes. Besides these molecular mutations, others occur with much larger effects on the chromosomes. Sections of chromosomes may be lost, duplicated, or inverted, two chromosomes can exchange segments, and so on.

Most mutations are disadvantageous, for the very simple reason that organisms are sufficiently well adapted to their environments to ensure that most changes are for the worse. But this does not prove that mutation is mere damage to the hereditary material. On the contrary, a mutation may be provoked by X-rays. A further dose of X-rays to the descendants may then induce a gene descended from the mutated gene to mutate back again and resume its original function. Both changes cannot be for the worse.

The better an organism is adapted to its environment, the greater will be the fraction of mutations which are harmful. When it is placed in a new environment, many previously harmful mutations will be beneficial. In particular, domestication has this effect. To take a simple example, whiteness makes a wild hen conspicuous to predators, and shortens its expectation of life. It makes a domesticated one more easily caught in the evening by its human owners, and, therefore, more likely to be shut up safe from nocturnal predators.

Mutation is described as a random process. It is no more and no less so than any other chemical process. One cannot predict which gene will mutate in a given nucleus. One can predict the frequency with which a particular gene will mutate, and sometimes at least choose a mutagenic agent which will affect this particular gene rather than some others. From the biological point of view it is random, as is, for example, predation of flying insects by birds or the finding of a suitable site for attachment by small floating larvae. But as we shall see, a function of natural selection is to negate the randomness of mutation and of the events which determine the death or survival of individuals.

The genetical analysis of metrical characters is much harder than that of discontinuous ones. Some of the variation is always due to environmental differences. In man we can only say that the statistical facts about inheritance of stature and similar characters are compatible with determination by genes at many loci. In the small fly *Drosophila*, however, it is possible to say that so much of the difference in average wing length between the two stocks is due to genes on one chromosome, so much to genes on another, so much to interaction between these sets. Some differences between races are due to self-reproducing (or accurately copied) particles outside the nucleus, and handed down by mothers only.

Darwin believed that the differences between related species were due to the same kind of causes as those between breeds of the same species. This hypothesis can only be verified when the hybrids between two species are at least partly fertile. For example, the primrose *Primula veris* and the Caucasian *P. juliae* differ by two genes for colour, one being responsible for yellow pigment in the primrose, the other for purple pigment in *P. juliae*. By crossing hybrids together one can get plants which lack both, and have white flowers. The species also differ in respect of genes for leaf shape, hairiness, and many other characters. The taxonomic differences between the skins of related newt species and subspecies, such as *Triturus marmoratus* and *T. cristatus*, or *T. cristatus cristatus* and *T. cristatus carnifex*, similarly depend on quite a lot of gene pairs. Immunology has opened up new possibilities. The blood corpuscles of related species of Columbidae (pigeons and doves) differ in the antigens which they carry. These differences have in some cases been completely analysed as due to a dozen or so genes, each responsible for making a different kind of antigen.

Sometimes the hybrids resulting from reciprocal crosses are indistinguishable. Sometimes they differ in one sex only, the difference being explicable by sex-linked genes. They may also differ because the female gamete contains more chromosomes than the male gamete, as in some roses. Sometimes, however, they differ considerably, and the difference has been shown to be due to cytoplasmic particles transmitted by the mother only. This is probably commoner in plants than in animals. But there is no evidence whatever for the theory, which was common a generation ago, that specific differences are mainly determined by the cytoplasm, while genes in the nucleus are responsible for the more 'superficial' characters distinguishing members of the same species. It has also been possible, in several different ways, to produce races of the same species which give very few hybrids, sterile hybrids, or both.

Some readers will suppose that recent discoveries in the Soviet Union have overthrown Mendelism. Exceptions to it have been claimed, probably sometimes correctly and sometimes incorrectly, However, many Soviet geneticists who make such claims fully realize that they are dealing with exceptions. I quote two phrases from a recent paper of Gluschenko on such exceptions. 'But this segregation is unusual and greatly diverges from normal rules of dominance.' 'In control crosses...we found the usual segregation into the intermediate and parent forms in the F_2.' I think it has yet to be shown that the results obtained from such processes as grafting are likely to have been of any importance in evolution. For example, Michurin's claim that grafting may facilitate hybridization of species has been verified. And hybridization has played a part in evolution. But grafting must be very rare in nature, particularly between animals, and between herbaceous plants. On the other hand, in bacteria a process very like Michurinism is normal.

Definition and Measurement of Selection: Effects of Genotypic Selection

Selection is most easily considered where generations do not overlap, and particularly in annual plants and animals. Matters are still simpler when each organism has only one parent, as in apogamous plants like the dandelion (*Taraxacum officinale*), or self-fertilized hermaphrodites like the pea (*Pisum sativum*). Suppose we count all plants of such a species at a definite stage of their life-cycle (say when the cotyledons first appear), and can distinguish two phenotypes (say hairy and glabrous). We then follow up the seeds of each plant, and find how many seeds from each plant reach the same stage in the life-cycle at which the parents were counted, in the previous year. We then make up a fictitious parental population in which the parents of no seedlings are not represented at all, those of one seedling once, of two seedlings twice, and so on. We then ask if this fictitious population contains the same fractions of hairy and glabrous plants as the population from which it is drawn. If not, we ask whether the difference could be due to random sampling, or whether it is statistically significant. If it is statistically significant, we say that there has been selection in favour of one phenotype and against the other.

For example, if the population in the first year consisted of 500 glabrous and 500 hairy, while of the parents of 900 seedlings in the second year 380 were glabrous and 520 were hairy, we should say that this parental population differed from equality by 4·67 standard errors. This could only occur once in about 327,000 experiments by chance, and we should neglect this possibility.

In the case of biparental inheritance it would be best to do the calculation separately for each sex. For a character may increase the probability that the female carrying it should become a mother, and diminish the chance that a male should become a father, or conversely. If generations overlap, there is no difficulty provided the population is not, on the whole, increasing or decreasing. If it is increasing, an offspring produced early in its parents' life must count for more than one born later, and conversely. However, populations cannot increase or decrease steadily over long periods. A steady increase of 1 per cent per generation for 10,000 generations would increase the number of members of a species by $1 \cdot 5 \times 10^{43}$, that is to say the progeny of a small bacterium would occupy a volume much larger than the earth. We can, therefore, neglect this complication in considering anything but very rapid evolution.

It is at once obvious that the kind of counting suggested is usually impossible in practice, like the 'ideal experiments' of physicists, except in a few cases. It can be done in human populations. We can follow up 100 dwarfs from birth, and find that they produce about a quarter as many children as 100 normal people. It can be done in domestic animals, in which, of course, artificial selection is occurring. It can be done in some laboratory animals. It cannot be done in most 'wild' animals and plant species, because although it may be possible to estimate the success of a particular type as mothers, it is very much harder to estimate their success as fathers, particularly as pollen parents. However, we can often estimate components of selection; for example, selective resistance to disease or escape from predators.

It is obvious that we can extend the method to as many phenotypes as we can score, provided we can score them early enough, and in particular we can grade a metrical character. For example, we can compare the survivals of human babies weighing 2·5 to 3·5 lb., 3·5 to 4·5 lb., 4·5 to 5·5 lb., and so on. Perhaps posterity will follow up a set of several thousand babies weighed at birth and determine their fertilities. We can then estimate the intensity of selection in several different ways.

The over-all intensity of selection (I) can be measured as follows. Suppose we have n phenotypes whose frequencies in the original population are $p_1, p_2, p_3, \ldots, p_i, \ldots, p_n$, and in the fictitious parental population $P_1, P_2, P_3, \ldots, P_i, \ldots, P_n$. The ratio $(P_i/p_i) \div (P_j/p_j)$ can be called the relative fitness of the ith phenotype compared with the jth. For one of these phenotypes P_i/p_i is a maximum. Call this the optimal phenotype, with frequencies p_o and P_o. If all the population had had the mortality and fertility of this phenotype there would have been P_o/p_o times as many parents as there were in fact. If selection was by juvenile deaths or 'survival of the fittest', then the average survival was p_o/P_o times that of the optimal genotype; the population was reduced to a fraction p_o/P_o of what it would have been had all belonged to the optimal phenotype. I take $I = \ln P_o - \ln p_o = \ln (P_o/p_o)$ as my measure of the intensity of selection.* This measure has the following advantages. When P_o and p_o do not differ greatly, I is nearly $(P_o - p_o)/P_o$, or the fraction of deaths which were selective. And if I is measured in several parts of the life-cycle we can add the different values together to get an over-all value.

If we are dealing with a metrical character x (say weight) we can find the distribution of x in the original population, and also among the parents. The means may differ. Lush[1] calls the difference between parental mean and population mean the *selective differential.* Thus, if pigs are being selected for weight at the age of 100 days, and the average weight of parents (counting a parent of 9 pigs 9 times over, and so on) is 10 kg. greater than the average weight of all pigs at 100 days, the selective differential is +10 kg. We can also compare the other moments of the distribution of x, and in particular the variance, which is the simplest measure of spread round the mean. Most forms of selection reduce the variance.

I have, so far, dealt only with *phenotypic selection*, selection based on phenotypes. This may have no effect whatever on the next

* $\ln x$ means $\log_e x$ or $2{\cdot}3026 \log_{10} x$.

generation. It will certainly have none if the population is genetically homogeneous, as de Vilmorin found a century ago with wheat. Small plants usually produce fewer seeds than large ones. This is phenotypic selection in favour of large size. But small plants are often small because of poor soil or competition. Hence the next generation is not usually larger. It may even be smaller in the same conditions if plants with a genetically determined capacity for slow growth produce more seeds than those with greater growth capacity.

Genotypic selection can be measured in just the same way as phenotypic if the genotypes can be determined. And now we can also calculate its effect on later generations. First suppose that there are two competing genotypes A and B, which do not interbreed, as in self-fertilized or clonally propagated plants, or which are due to a maternally inherited cytoplasmic difference. Suppose that in the next generation the frequency of A is increased by a factor $(1+k)$ relative to B and that the population is so large that we can neglect the effects of chance survival. If to begin with there are u_0A for every B, then after n generations there will be u_nA for every B, where $u_n = (1+k)^n u_0$. The frequencies of A and B will thus be $u_n/(1+u_n)$ and $1/(1+u_n)$, or $1/[1+(1+k)^n u_0]$. Thus the frequency of B will diminish indefinitely so long as k is positive. For example, if k is 0·001, giving A an advantage of one-thousandth, and $u_0 = 1$, that is to say, half the original population consists of B, then after 1000 generations the frequency of B will be reduced to 26·93 per cent, after 3000 to 4·75 per cent, after 10,000 to 0·0046 per cent, and so on. It will disappear in a time which is very short on the geological scale.

Things are quite different in a small population. If, for example, there is only room for two plants in each generation, chance survival is far more important than selection, and the probability that only B will survive is $1/(2+2k+k^2)$ or 49·95 per cent.

Now consider a pair of allelomorphic genes in a diploid. I shall call them **A** and **a**, but make no assumption as to dominance. Let the relative fitnesses of **AA**, **Aa**, and **aa** be $(1-K):1:(1-k)$, where K and k may be positive or negative, but cannot be larger than 1. I assume that they are constant, and in particular do not depend on the frequencies of the three genotypes. The fate of the population depends in part on the mating system. There may be a good deal of inbreeding or assortative mating, that is to say mating of like with like. I shall only consider large populations where mating is at random, that is to say inbreeding and assortative mating are negligible. In such a population, if the frequencies of **A** and **a** in the gametes forming the nth generation are p_n and q_n, the frequencies of genotypes in this generation are p_n^2 **AA**, $2p_nq_n$ **Aa**, q_n^2 **aa**.

Among the parents of the next generation the frequencies are

$$\frac{(1-K)\,p_n^2}{1-Kp_n^2-kq_n^2}\ \mathbf{AA}, \qquad \frac{2p_nq_n}{1-Kp_n^2-kq_n^2}\ \mathbf{Aa},$$

and

$$\frac{(1-k)\,q_n^2}{1-Kp_n^2-kq_n^2}\ \mathbf{aa}.$$

It follows that

$$p_{n+1} = \frac{p_n-Kp_n^2}{1-Kp_n^2-kq_n^2},$$

so

$$\Delta p_n = p_{n+1}-p_n = \frac{p_nq_n(kq_n-Kp_n)}{1-Kp_n^2-kq_n^2}. \tag{1}$$

This fundamental equation is due to Wright.[2] There are five possibilities.

(1) If K and k are zero, the population will not alter. More accurately it will change by chance. The rate of change of p_n may be of evolutionary importance in populations of 100 or so, but not in populations of many millions.

(2) If K and k are both positive, so that heterozygotes are fitter than either homozygote, the population will come to an equilibrium with gene frequencies of p and q given by

$$p = \frac{k}{K+k}, \quad q = \frac{K}{K+k}. \tag{2}$$

Genetic variation is permanently conserved.

(3) If K and k are both negative, so that heterozygotes are less fit than homozygotes, then the equilibrium given above is unstable. If p_0 is larger than p, p_1 will be larger still, and so on. The population will finally consist of all **AA** or all **aa**, apart from effects of mutation.

(4) If K is negative and k positive, or if one has the above sign and the other is zero, p_n will increase with n, and the population will finally consist of **AA** only.

(5) If K is positive and k negative, or if one has the above sign

and the other is zero, p_n diminishes, and the population finally consists of **aa** only.

Only the last two cases are of evolutionary importance, for the main unit process of evolution is the replacement of a gene present in almost all members of a species by one of its allelomorphs. If K and k remain constant, we can calculate how long this takes provided they are small. In this case we may write (1) approximately as

$$dp/dt = pq(kq - Kp). \tag{3}$$

If K is negative and k positive, p_n increases, and

$$\begin{aligned} t &= \int_{p_o}^{p_n} \left[\frac{1}{kp} + \frac{1}{-K(1-p)} + \frac{(K+k)^2}{Kk\{k-(K+k)\,p\}}\right] dp \\ &= k^{-1}\ln\left(\frac{p_n}{p_o}\right) - K^{-1}\ln\left(\frac{q_o}{q_n}\right) + (k^{-1}+K^{-1})\ln\left(\frac{kq_o - Kp_o}{kq_n - Kp_n}\right). \end{aligned}$$

If p increases from a small quantity p_o to a value p_n such that $1-p_n$, or q_n is small, the number of generations needed is very nearly

$$n = k^{-1}\ln(p_o^{-1}) + K^{-1}\ln q_n + (k^{-1}+K^{-1})\ln(-k/K). \tag{4}$$

For example, to change the value of p from 10^{-4} to $1-10^{-4}$ would take about 18,400 generations if $k = -K = 0{\cdot}001$, and about 9760 if $k = 0{\cdot}001$, $K = -0{\cdot}005$.

These times are short on a geological scale, even if a generation lasts 25 years as with man. And a difference of fitness of one-thousandth could only be detected with fair certainty by counting about thirty million individuals!

If, however, either of K or k is zero, so that the gene in question is completely dominant or recessive, selection is much slower. Suppose $K = 0$, so that **A** is completely dominant as regards fitness, (3) becomes

$$dp/dt = kpq^2$$

and (4) becomes

$$n = k^{-1}(q_n^{-1} - \ln p_o - \ln q_n). \tag{5}$$

Thus if $p_o = 10^4$, $p_n = 1-10^{-4}$ as before, and $k = 0{\cdot}001$, n is just over ten million generations. Selection for or against fully recessive genes is a very slow process. For almost all the recessive genes are in heterozygotes, and shielded from selection. When I first obtained the equivalent of equation (5)[3] I thought that this might account for the slowness of evolution. However, I now think that complete dominance is very unusual, and equation (4) is generally valid. The reasons why evolution is so slow are given on pp. 135–41.

Elsewhere I have given solutions of equation (1) which are valid when K and k are not small. However, the solutions (4) and (5) are not very inaccurate in such cases. If, say, $k = \frac{1}{3}$, the value of n will be out by something like 30 per cent. Since no values of k are accurately known, this makes little difference. Inbreeding and assortative mating have no great effect unless K or k are zero (or one is very much nearer zero than the other), in which case one allelomorph is wholly or nearly recessive. In this case they speed up selection very greatly when the recessive is rare. Thus if **a** is recessive and $q = 10^{-4}$, the frequency of recessives would be 10^{-8}. But a coefficient of inbreeding of $\frac{1}{1600}$, which is the value due to 1 per cent of first-cousin marriages (as in France to-day), would increase it by a factor of 7·25, and speed up selection correspondingly. Inbreeding is much more intense than this in many plant and animal populations.

The above equations also hold for the selection of more complicated mutants such as inversions. K and k may, of course, change their values with time, or with the frequencies of the genes concerned. And in small populations a slightly disadvantageous gene may oust a more advantageous one by chance. The full consideration of such cases has occupied much of the lifetimes of Wright, Fisher, Kimura and myself, while other workers have contributed to it to a less extent.

If there is genotypic selection of types (2), (3), (4) or (5) and the phenotypes corresponding to **AA**, **Aa** and **aa** fall into that order as regards a metrical character, its mean will be altered. Lush calls the change in the mean resulting from selection the *selective advance*. It is never as great as the selective differential. In practice it is never more than half of it.

Centripetal Selection

The biometric school, associated with the journal *Biometrika*, founded in 1901, set out in the hope of measuring evolutionary rates and the intensity of natural selection in living populations.

They succeeded in the second aim, but not in the first, though as we shall see, this has since been done. Instead, Weldon[4] discovered a case where selection was occurring, probably with no evolutionary effect. Such cases are, in fact, quite common. It is easy to see why. We now know that selection may be balanced by other agencies, particularly mutation, segregation, and migration. Genes, and, more rarely, chromosomal rearrangements, responsible for unfit phenotypes are constantly appearing as the result of mutation, and being destroyed by natural selection. Where homozygotes are less fit than heterozygotes the fact of segregation renders it impossible for a whole population to belong to the fittest genotype. This is only possible if reproduction is clonal, that is to say asexual. And immigration of organisms (including pollen-grains) adapted to a different environment keeps up a supply of genes which must be eliminated by natural selection.

Now in these cases selection may be very intense, and it is quite easy to measure it. But directed natural selection, which alters gene frequencies and has evolutionary consequences, can only be intense under the following circumstances. A population is placed in a new environment where most of its members are far less fit than a small minority which is favoured by natural selection. The change of environment must be sudden, and the majority, though relatively unfit, must be fit enough to propagate themselves. This is a rare combination of circumstances except in the case of domestication, and here artificial selection is usually being practised.

I take one example of selection against a mutant. Achondroplasic dwarfs are a rather rare variant of the human species. In 1940 there were 86 of them among 3,793,000 Danes.[5] The condition appears to be due to a dominant gene, for about half the children of such dwarfs are themselves similar dwarfs. However, most dwarfs are not the children of other dwarfs, but of normal parents in one of whom a gene (or possibly a section of a chromosome) has mutated. Many dwarfs die in their first year; those who survive are infertile, for dwarf women can usually only have a child by Caesarean section. On Mørch's data the fitness of dwarfs is about a fifth of that of normal people, and the mutation rate about 4×10^{-5} per generation. If some of the dwarfs who died in infancy were due to physiological accidents and not to the action of the gene in question (that is to say if some babies with achondroplasic dwarf phenotype were of normal genotype) the fitness is greater and the mutation rate less. But the fitness is certainly less than half the normal.

A number of other rare human dominant and sex-linked recessive genes are kept rare by the same process, which was first described simultaneously by Gunther and Penrose[6] for epiloia, a condition characterized by tumours and fairly often by mental defect, and by Haldane[7] for haemophilia, a disease in which the blood takes a long time to clot. Rare and harmful autosomal recessives such as microcephaly, phenylketonuria, and foetal ichthyosis are probably kept in being by a similar process, but for various reasons we cannot measure the mutation rates so accurately. Similar unfit mutants occur in animal populations.

Now when a rare gene of this kind affects a metrical character, natural selection has a much larger effect on the variance of this character (or its standard deviation) than on its mean. Consider haemophilia. Suppose that the mean coagulation time of normal bloods is 5 minutes with a standard deviation of 2 minutes, that of haemophilics 60 minutes with a standard deviation of 24 minutes (the figures vary greatly with the method used) and that the frequency of haemophilics is one in 30,000 at birth and one in 90,000 among parents. Then selection has reduced the mean coagulation time from 5·0020 to 5·0007 minutes, that is to say by 0·027 per cent, and the variance from 4·140 to 4·047 minutes, that is to say by 2·3 per cent, or the standard deviation by 1·1 per cent. This is characteristic of any selection which weeds out extremes. Human genes may well be found which cause the blood to coagulate too quickly, and cause people possessing them to die of thrombosis. So we cannot be sure that natural selection lowers the mean coagulation time. We can be quite sure that it lowers its variance.

The second process which weeds out extremes is selection in favour of heterozygotes. A good example is a human gene pair which we may call **Si** and **si**. (The correct designation is $\mathbf{Hb^A}$ and $\mathbf{Hb^S}$.) Most people are **Si Si** and have a homogeneous, or nearly homogeneous haemoglobin, which is soluble in the high concentration found in corpuscles, even when reduced. **sisi** people are not rare in tropical Africa. They have an abnormal haemoglobin which is insoluble when reduced. The corpuscles in their venous blood assume abnormal forms and are called sickle cells. They are rapidly destroyed, and such people usually die of anaemia in childhood. The heterozygotes **Si si** have a mixture of the two haemoglobins, which is still soluble when it has lost most of its oxygen. They appear to be quite healthy, though their fitness may exceed those of **Si Si** or fall short of it by 1 per cent or more. But in many parts of tropical Africa they have a considerable advantage. They are somewhat resistant to infection with *Plasmodium falciparum*, which causes malignant tertian malaria.[8] If they are infected, they are, on an average, infected later in life and less severely. Let us suppose that, in a malarious region the relative fitnesses are:

Si Si 0·9 : **Si si** 1·0 : **si si** 0·2.

That is to say, **Si Si** have 90 per cent of the chance of **Si si** of becoming parents, **si si** only 20 per cent, through premature death, and, perhaps, impaired fertility of those who survive. In equation (2) $K = 0{\cdot}1$, $k = 0{\cdot}8$, so there is a stable equilibrium with gene frequencies $\frac{8}{9}$ **Si**, $\frac{1}{9}$ **si**, giving $\frac{64}{81}$ **Si Si**, $\frac{16}{81}$ **Si si**, and $\frac{1}{81}$ **si si**. We find that before selection the mean and variance, if the three phenotypes are scored as $+1$, 0, and -1, are 0·7 and 0·19753, and among the parents 0·7 and 0·17802. Here both extremes are eliminated, and the mean unaltered but the variance reduced. Usually the heterozygous phenotype is not exactly intermediate, and there is a slight change in the parental mean, but no selective advance.

The most striking cases of equilibrium of this sort are found in *Drosophila* species such as *D. pseudoobscura* where one chromosome can exist in several different orders, and flies with two different orders are fitter than those with two like orders. Formally two such orders behave like two allelomorphic genes. These are discussed in detail by Dobzhansky.[9] K and k are positive and large. They can reach values as high as $\frac{1}{2}$. These values depend to some extent on the temperature, one order (presumably because of the genes which it carries) being favoured by heat, the other by cold.

When a metrical variate is acted on by selection either against homozygotes or against several kinds of rare mutants, the phenotypic selection found will be against extremes. The mean is little altered, and the variance reduced. Unfortunately this kind of selection has not been followed through entire life-cycles, but only measured during parts of them. Karn and Penrose[10] recorded the weights of 13,730 babies born in London between 1935 and 1946. Of these 614 were still-born or died before 28 days. The weights were grouped in half-pounds, and fig. 1 shows the percentage surviving for birth weights between 3½ and 11½ lb. Rather more girls

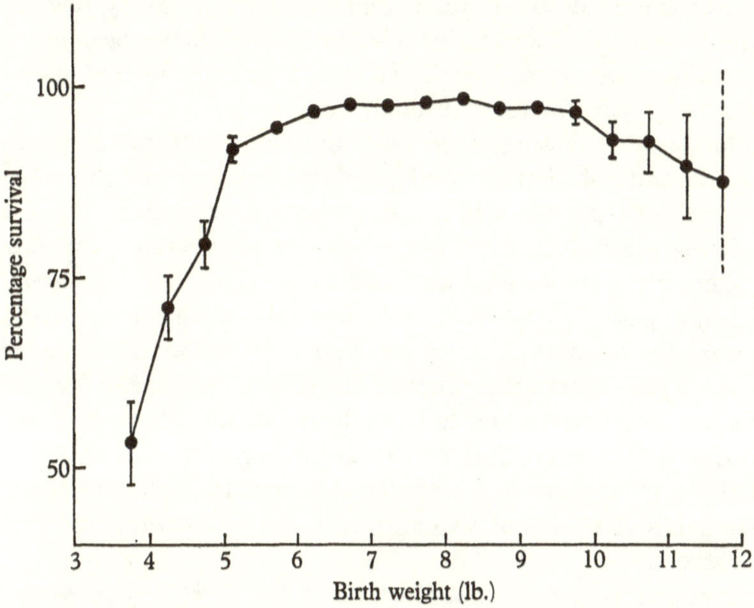

Fig. 1. The relation between weight at birth and mortality. The vertical lines represent standard errors. For explanation, see text.

than boys survived, but the graph is smoother when the sexes are pooled. Most of the very light babies were premature, but light babies born at full term have a slight chance of survival. 180 babies weighed under 3½ lb., but only 151 of them survived, and none survived below 2 lb. One of the four babies weighing over 11½ lb. died. Clearly there is an optimum weight, namely 7·97 lb. Only about 1·8 per cent of babies of this weight died as compared with 4·47 per cent of all babies. Thus 2·7 per cent or three-fifths of all deaths at birth and in the first month of life are due to natural selection for weight. That is to say $I = 0{\cdot}027$. Of course, the babies do not die *directly* because they are too light or too heavy. Many probably die because the same abnormalities of gestation which caused them to have an abnormal weight handicapped them in other ways.

This is an example of *centripetal* phenotypic selection, a weeding out of extremes. The word is due to Simpson.[11] In this case the mean weight of survivors was slightly higher than that of all babies, in fact higher by 0·08 lb., or 1 per cent of the mean. But the standard deviation of weights was reduced from 1·32 to 1·17 lb. in males, and from 1·22 to 1·10 lb. in females, that is to say by 10·6 per cent on an average. Though this selection has doubtless been going on for millions of years, there is no reason to suppose that it has led to a progressive increase in birth weight.

Waddington[12] distinguished two kinds of genotypic selection (though he had not distinguished genotypic from phenotypic selection) corresponding to centripetal phenotypic selection. In the *normalizing* selection, genotypes which, on an average, generate high phenotypes are weeded out, and so are those which generate low phenotypes. In *stabilizing* selection genotypes which give a more variable phenotypic response than the average are weeded out, even if they have little effect on the mean. The genetic basis of human birth-weight is not fully understood, but Robson[13] has found that it depends mainly on the mother's genotype, and not on the baby's. For there is an appreciable correlation between the birth-weights of babies of two sisters, and none between those of two brothers.

It must be emphasized that Karn and Penrose[10] only measured selection during a short but critical phase of the human cycle. We know that adult weight is not highly correlated with birth weight. It is conceivable, but not at all likely, that the selection is later reversed, that, for example, people who weighed 10 or 6 lb. at birth are on the average more fertile than those who weighed 8 lb.

A number of other cases of centripetal selection are known. Adult human beings who are much lighter or heavier than the average value for their height have a short expectation of life. Rendel[14] found figures for the hatchability of ducks' eggs like those of Karn and Penrose, but the selection was more intense. In one group of ducks 70 per cent of the eggs weighing between 70 and 75 g. hatched, but only 64 per cent of all eggs. The average weight of all eggs was 73·9 g., of those which hatched 73·8 g. So there was no effect on the mean. But the standard deviation, or 'spread', was reduced by 10 per cent. The weight of an egg depends entirely on the mother's physiology, and is well known to be in part genetically determined, for races of domesticated birds can be artificially selected for high or low average egg weight.

Weldon[4] invented a most ingenious method of measuring natural selection. In molluscs, and some other invertebrate phyla, the parts of the shell laid down in early life are not subsequently altered, whereas vertebrate bones and echinoderm plates are reshaped. So an adult snail carries about in its smallest whorls a record of its juvenile performance. Weldon collected 100 young snails (*Clausilia laminata*) whose length was under 9 mm. and another 100 adults whose mean length was about 15 mm. He ground them on a stone so as to prepare sections, and carefully measured several metrical characters, of which the most important is (roughly) the number of turns of the spiral in a given length of axis. He compared this number in the young snail shells, and in that part of the adult snail shells which had been laid down in youth. He found no differences in the means, but the standard deviations were consistently reduced by about 9 per cent. This means that the snails which, when young, had made too many or too few turns for a given increased length of the shell axis were less likely to survive to full size than those whose performance was near the average. Weldon found no such effect in another snail population, but his pupil di Cesnola found it in a third. Haldane[3] pointed out that if, in the population exposed to natural selection, the mean and standard duration are m_1 and σ_1, while after selection they are m_2 and σ_2 where $\sigma_2 < \sigma_1$, then if the character measured is normally distributed, the intensity of selection is

$$I = \ln \sigma_1 - \ln \sigma_2 + \frac{(m_1 - m_2)^2}{\sigma_1^2 + \sigma_2^2}.$$

Weldon's I was about 0·08, since m_1 and m_2 did not differ significantly.

There are enough similar examples to show that centripetal selection is of very general occurrence. If this fact had been discovered about 1860 we can imagine some of Darwin's opponents writing as follows:

> In view of Weldon and Rendel's findings Mr Darwin's absurd speculations may now be relegated to the obscurity from which they should never have emerged. He postulated the existence of natural selection

to account for the evolution in whose existence he believed. He doubtless deserved some credit for stimulating others to carry out accurate measurements. Natural selection was in fact found to occur. But so far from causing species to change, it actually prevents such change. Not only does it preserve the type of a species by eliminating deviants. It eliminates hybrids between species, which, if they are not too weak to be capable of development, are sterile.

Centrifugal selection occurs, but rarely except after the crossing of two species. Here it is sometimes found that the survivors arising from breeding hybrids together tend to resemble one or other parent species. Within a species centrifugal selection occurs with respect to inversions. The heterozygote for an inversion usually produces some inviable gametes, and is, therefore, less fit than either homozygote, though in *Drosophila* and some other insects this is not so. Again, in mammals, heterozygotes for certain antigens may be killed by provoking immunity to these antigens in their mothers. This is the case in our own species with several antigens responsible for neonatal jaundice. Selection proceeds nearly according to case (3) of equation (1) (p. 115). One or other of the chromosomal and antigenic types will spread through the whole population.

We may make the following tentative generalization. Genetically determined differences between members of a species belonging to a gamodeme (population whose members mate together) are usually due to pairs of allelomorphic genes where the heterozygote is fitter than either homozygote. Genetically determined differences between related species are at least frequently due to genes or larger differences such that the heterozygote is less fit than either homozygote.

Directed Selection

We have now to consider whether selection of the kind postulated by Darwin does in fact occur. Let us begin with artificial selection. Fig. 2 shows the results obtained by Falconer[15] in selecting mice for high and low weight. In each generation he picked out the heaviest mice in the plus line at the age of 42 days, and bred from them. Similarly, in the minus line he bred from the lightest mice. Some of the progress in one direction or the other could have been due to agencies other than selection. If his methods of keeping mice had improved, all the mice might have got a little heavier. And inbreeding, which is inevitable unless the parental population in each generation numbers many thousands, might have been responsible for a drop in weight. The progressively increasing difference between the two lines is convincing evidence of the efficiency of artificial selection. Improvements in yields of milk, eggs, wool, corn, fruit, cotton, and so on, are usually at least partly due to better conditions. In fact Falconer tested the descendants

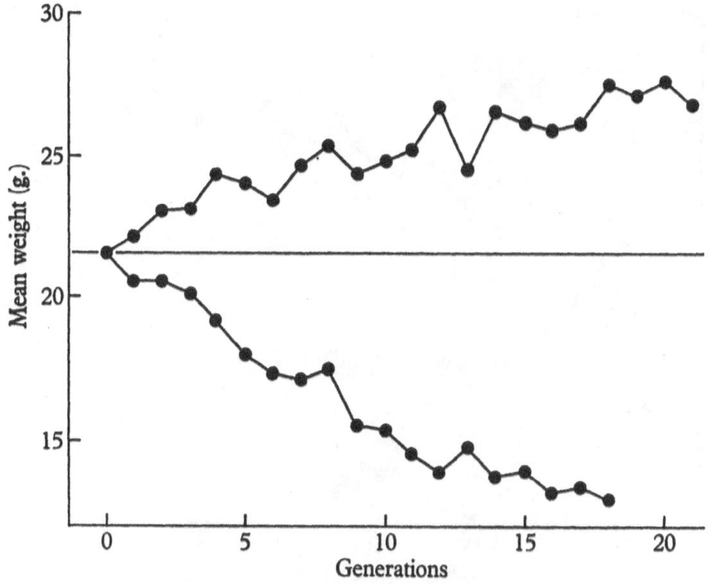

Fig. 2. The results of selecting mice for high and low weight; the plus line above, the minus below. For explanation, see text.

of his original stock to see if there was any change, and found none. He also tested both selected lines to see if they reverted towards normal when selection was relaxed. There was only a slight effect of this kind.

It will be seen that the progress per generation in each direction slowed down with time. This could have been because the stocks had both become nearly homozygous, and no more gene differences were available for selection. This often occurs when inbreeding is intense. But it had not happened with Falconer's mice. In each generation he measured the selective differential. Thus, in the first generation the over-all mean weight was 21·6 g., those of the parents of the high and low lines being 24·0 and 19·6 g. So the selective differentials were +2·4 and −2·0 g. The ordinates of fig. 2 are the sums of the selective advances in successive generations, sometimes advances backwards! The abscissa is the number of generations. In fig. 3 the mean weights of the selected lines are plotted against the sums of the selective differentials. We can see that the selective

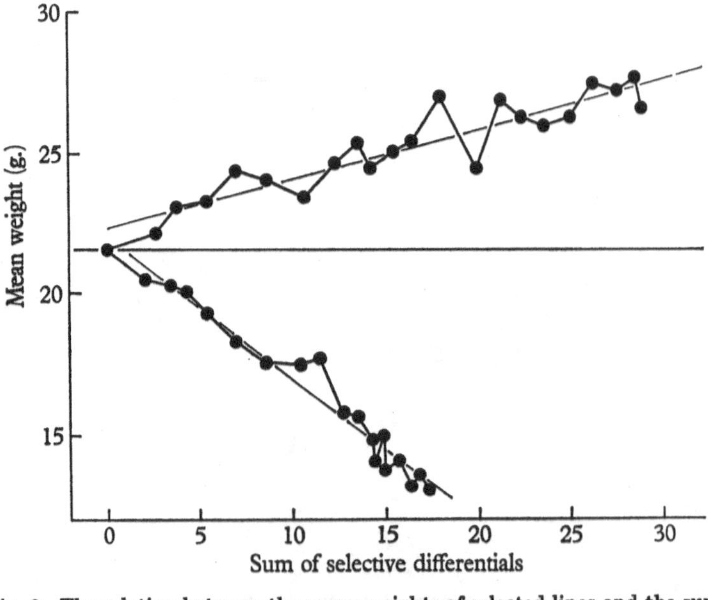

Fig. 3. The relation between the mean weights of selected lines and the sums of the selective differentials in mice selected for weight. For explanation, see text.

differentials became progressively smaller. This was because the most 'advanced' mice were often weak or sterile, and those which finally qualified as parents were little different from the average, particularly in the light line. This result is unpleasantly familiar to animal and plant breeders. It can often be overcome by selecting for vigour and fertility for a few generations. The over-all efficiencies of selection, that is to say selective advances divided by selective differentials, were 0·175 for the heavy, and 0·518 for the light line.

The final result of artificial selection may be a stock which is fairly homogeneous genetically. Or it may be obviously heterogeneous, either because the ideal animal is heterozygous, the homozygotes being viable but less valuable, as with blue breeds of poultry, cross foxes, and so on; or one homozygote being inviable, as with crested canaries, short-legged poultry such as Scots Dumpy, and Dexter cattle. Such stocks are only kept in being by culling at least half of each generation, or by special breeding systems. They are almost irrelevant to evolution.

Now let us study some well-authenticated cases of directed natural selection. Many species of Lepidoptera have become dark in areas polluted by industrial smoke. The change is extremely rapid. About sixty years after the melanics were first reported as great rarities they had replaced the original type almost completely. The best studied case is that of the peppered moth, *Biston betularia.*[16] The almost black variety *carbonaria* was first found near Manchester about 1848. It is now found with frequencies up to about 99 per cent, though not often over 90 per cent, over a wide area of England where the lichens on tree trunks have been killed by smoke, and the bark more or less blackened. It is due to a single dominant gene C. Another less dark variety, *insularia,* also due to a dominant gene, is found in partly polluted areas (fig. 4). Haldane[3] showed that this rapid change could not be due to mutation. The mutation rate would have to be over 10 per cent; and in fact the recessive type has always bred true, so the rate must be much less than 1 per thousand, and may well be 1 per million. He concluded that the observed evolution could be explained if the recessives had an average disadvantage of about 30 per cent, that is to say in equation (1), K was nearly zero and k about 0·3.

Kettlewell released about equal numbers of dark *carbonaria* and light recessive moths in a highly blackened wood near Birmingham and an unpolluted wood in Dorset. He and Tinbergen obtained films of predation by small birds which ate the more conspicuous type preferentially. In each wood he released about equal numbers of dark and light moths at daybreak. They were marked with paint under their wings, the mark being invisible when they were resting. The next night males were caught with light traps, and assembling to virgin females. In each wood the number of the more conspicuous phenotype caught was about half that of the phenotype

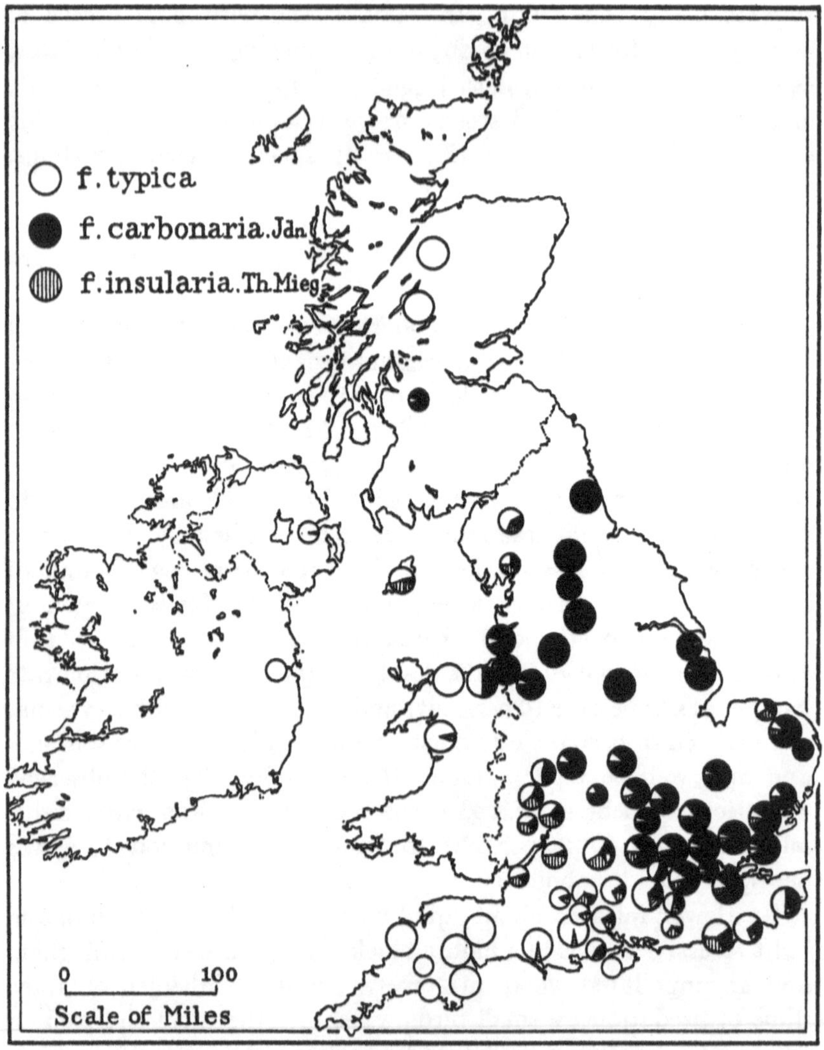

Fig. 4. *Biston betularia.* A frequency map of the peppered moth and its two melanic forms in Great Britain. Compiled by H. B. D. Kettlewell from 20,000 records received from over seventy observers in the past four years.[17]

PLATE II

(i)

(ii)

Biston betularia L. (peppered moth). (i) The normal form and its melanic (f. *carbonaria*) on a lichened tree trunk. Dorset. (ii) The same on a polluted tree trunk. Birmingham.

PLATE III

Ancient and modern forms of heterozygotes of *B. betularia* f. *carbonaria*. Left-hand column: modern heterozygotes; right-hand column: specimens extracted from the oldest collections made during the last century showing incomplete dominance.

which is hidden when at rest in the wood in question. We do not know the intensity of selection, as we do not know how quickly these moths mate and lay their eggs in nature. It is possible that in these woods k might be as high as 0·5. Kettlewell doubtless chose them to give as strong selection as possible. At the present time the frequency of *carbonaria* is increasing in some parts of Southern England, at rates between 1 and 3 per cent per year, in other areas it is stationary. We do not know whether it is still increasing in the Black Country. The latest figures for Manchester were 346 out of 350, or 98·9 per cent, and for Birmingham 1403 out of 1611, or 87·1 per cent. If another sample of the same size were taken we should not be sure of a change even if a sample from Manchester contained 100 per cent *carbonaria* and one from Birmingham 91 per cent. We can only say that natural selection has largely done its work.

But Kettlewell has shown something more, and something equally important for evolutionary theory. The *carbonaria* moths collected in the nineteenth century are not so black as those collected in the twentieth century. They often show little white patches which a bird might notice. And when heterozygous *carbonaria* (**Cc**) was crossed with the original type (**cc**) from 1900 to 1905, only 47 per cent of the moths which emerged were **Cc**. But in similar families bred from 1953 to 1956 the frequency was 67 per cent. The difference is significant. It was almost certainly due to the selection of genes which improve the health of **Cc** larvae relative to **cc**. These larvae must be biochemically different, even though no colour difference is visible. Similar results have been obtained with cultivated plants under artificial selection.

Besides **C** a number of other genes must have had their frequencies increased by natural selection. Such genes have been detected in organisms like *Drosophila* which can be bred by the hundred thousand. It may not be possible to detect them in *Biston*, but probably the number of 'modifiers' selected is at least four, and may quite well be more. Besides the gene **C** for *carbonaria*, another dominant, **I**, has spread. This produces a grey form, *insularia*, which is found in about 40 per cent of moths in some moderately polluted areas such as Oxford and Severn Estuary.

About seventy other species of Macrolepidoptera appear to be evolving in a similar way in Britain. In no other case are there so many data, but the trends are similar. It is quite likely that in some or even most of these cases centripetal as well as directed selection is going on. **CC** is very possibly not as fit as **Cc**. If so a state of balanced polymorphism has perhaps been reached in which the gene **c** persists in small numbers even where **cc** is conspicuous. It might take several more centuries before **CC** became as fit as **Cc** by the selection of modifiers. Before this time we may hope that smoke pollution may be ended in Britain, and the evolutionary trend will then be reversed.

TABLE I. *Numbers of snails of different colours killed by thrushes (after Sheppard)*

Date	Pink and brown	Yellow	Total	Per cent yellow
11 April	4	3	7	42·9
23 April	10	7	17	41·2
30 April	21	11	32	34·4
7 May	25	9	34	26·5
19 May	16	3	19	15·8
22 May	6	1	7	14·3
26 May	12	2	14	14·3

In another case studied at Oxford it has been possible to measure natural selection of very weak intensity. The snail *Cepea nemoralis* is polymorphic for several visible characters, of which I shall only consider one. Yellow is recessive to various shades of brown and pink. It is less conspicuous to human eyes on a green background such as grass, and more so on a dark background such as fallen beech leaves. Yellow snails are rare (often 5 per cent or less) in beechwoods, but are in a majority in permanently green grass.[18] One of their predators, the thrush, *Turdus ericetorum*, leaves a record of its predations. For it breaks snail shells on stones or stumps, especially during the breeding season. Sheppard[19] obtained several records of this predation, of which I give one. He released marked snails in a small wood. They were marked underneath, the mark being invisible to predators. Table I shows the numbers killed by thrushes in successive periods during the thrushes' breeding season. The wood was at first brown, and yellow snails were picked out. As it got greener, fewer and fewer were killed. Probably other birds behaved in a similar way. The selective advantage of one type over another was clearly an average, and in fact one type had little advantage over the other. But if the background had been permanently brown or green, there would have been a considerable advantage. The few yellow snails which survive in beech woods may be on their way to extinction. Or they may be protected because, to birds, they look so unlike brown snails that each bird must learn separately that they are edible.

It would not be possible to detect the selection of Table I by counting survivors. This is only possible because one can count those killed by a particular agent. The same is true for the human blood group genes. They have no detectable effect on mortality from all causes, but an appreciable one on death from gastric ulcers and cancer. We do not yet know whether this effect is directed or centripetal.

Most supposed examples of natural selection are very hard to detect. Thus Darwin pointed out how frequently insects on oceanic islands are wingless. He attributed this to the winged forms being blown out into the ocean. He was probably right. But no one has yet deliberately introduced a winged species on to such an island to see if it passes through a period when some members are wingless; much less to count the numbers blown out to sea. Again many economically harmful insects have developed resistance to poison. Enough laboratory research has been done to show that this is almost certainly a Darwinian rather than a Lamarckian effect. But this is not easy to prove in the field.

Sexual Selection

Darwin considered the selection of characters which increase the probability of members of one sex which possess them mating with the other sex. He considered that sexual selection acted mainly on males, and accounted to a large extent for the bright plumage of many male birds, the horns of male deer, and so on. The characters selected may be roughly divided into those which primarily attract the female, and those which conquer or intimidate other males. The distinction is in fact difficult, because the songs of male birds certainly, and their bright colours probably, drive other males away, as well as sometimes attracting females.

The clearest measurements of the intensity of sexual selection have been made in my laboratory on *Drosophila subobscura*. A sex-linked recessive gene, y, gives a yellow body colour in place of the black of the wild type. Spurway found difficulty in mating yellow males with black females. Rendel[20] investigated this fact. He found that whereas about 86 per cent of black females confined with a black male, and of yellow females confined with a black or a yellow male for a week laid fertile eggs, only about 3 per cent of black females confined with a yellow male did so. The black females actively repelled yellow males, by dodging, kicking, or extruding the ovipositor. However, by breeding for ten generations from those black females which accepted yellow males, he was able to produce a race of black females which accepted 82 per cent of yellow males. Thus, as Darwin had predicted, the choice depends on hereditary characters of *both* sexes.

Maynard Smith[21] showed that virgin females accepted only 45 per cent of males of two inbred stocks within 1 hour, but 90 per cent of outbred flies. There is a fairly elaborate courtship in which the female 'dances' from side to side, while the male faces her and does his best to keep time with her movements. If he can do so for a second or two he is permitted to mate. He concluded that the hybrid males are superior in 'athletic ability'. As they also fertilize a larger fraction of the eggs, this selection is definitely advantageous to the species. Whereas the selection against yellow males may not be so. In some environments yellow flies might be fitter than black.

A good deal of work on similar lines has been devoted to the question of how far, and why, members of one species choose a mate of their own species rather than of a closely related one. Other workers have altered secondary sexual characters artificially. For example, they have painted the ceres (coloured facial skin) of budgerigars, which lessened their mating success. But it is too early to say whether sexual selection is as important as Darwin believed, or perhaps even more important. There is at least no doubt that it is a reality.

Speciation by Allopolyploidy

We now come to a fact which would have surprised Darwin considerably, and which is perhaps the most important correction which must be made to his theory of the origin of species. When two related species of flowering plant are crossed, the progeny is often vigorous, but more or less completely sterile both with itself and with the parent species. Such a sterile hybrid may, however, produce a fertile branch. The progeny of self-fertilized flowers on such a branch are fairly uniform, and fertile with one another; but they are usually sterile on crossing with either parent species, and if they can be crossed the progeny are usually sterile. In fact they behave as a new species.

Suppose the parent species each have seven pairs of chromosomes. We may denote them by

$$\begin{matrix} A\,B\,C\,D\,E\,F\,G \\ A\,B\,C\,D\,E\,F\,G \end{matrix} \quad \text{and} \quad \begin{matrix} A'\,B'\,C'\,D'\,E'\,F'\,G' \\ A'\,B'\,C'\,D'\,E'\,F'\,G' \end{matrix}$$

The sterile hybrid is

$$\begin{matrix} A & B & C & D & E & F & G \\ A' & B' & C' & D' & E' & F' & G' \end{matrix}$$

It is sterile, either because A will not pair regularly with A′ in meiosis, or because if it does so this gives rise to inviable rearrangements, or for both these reasons. The fertile hybrid is found to have 28 chromosomes. In its meiosis like usually pairs with like, A with A, A′ with A′, and so on. Its gametes are therefore uniform, containing a set

$$A\,B\,C\,D\,E\,F\,G\,A'\,B'\,C'\,D'\,E'\,F'\,G'$$

of chromosomes. Such a plant is called an allotetraploid or an amphidiploid.

Several such 'species' have been made artificially. Müntzing[22] was the first to repeat, under controlled conditions, a major evolutionary step, when he crossed *Galeopsis speciosa* and *G. pubescens* (hemp-nettles). A triploid raised from this hybrid back-crossed with *G. pubescens* yielded an allotetraploid very similar to the existing *G. tetrahit*, and crossed with it freely. There can be little doubt that *Galeopsis tetrahit* originated in this way, though it then evolved to some extent, but not enough to make a sterility barrier with newly constituted members of its species. However, allopolyploids can evolve quickly, particularly by losing one or more pairs of chromosomes, becoming, to take a possible example,

$$\left.\begin{matrix} A\,B\,C\,D\,E\,F \quad A'\,B'\,C'\,D'\,E'\,G' \\ A\,B\,C\,D\,E\,F \quad A'\,B'\,C'\,D'\,E'\,G' \end{matrix}\right\} \text{(which may give a fitter phenotype)}$$

The majority of species of flowering plants may well be allopolyploids. Many economically important plants, including all the cultivated wheats and potatoes, the plum, and the New-World cottons, are so. Some have six, eight, or even more chromosome sets. But the condition is quite unusual in animals; probably because it interferes with the mechanism of sex determination. There is, however, good reason to think that many of the Salmonidae have evolved in this way.

Autopolyploidy is also found. Some races have three sets (triploids) or four sets (tetraploids) derived from the same species. These are usually sterile. Autopolyploid races reproducing parthenogenetically are not very rare in insects and isopods. But they are probably of little importance in evolution, and are doubtfully described even as new species.

A species originating by allopolyploidy is of course subject to natural selection. Unless in some habitats at least it is fitter than either parent, it will die out, and I have little doubt that most allopolyploid species do so. Further it must evolve as the result of natural selection, and probably does so rather quickly, as it is likely to be less perfectly adapted to its environment than either parent. However, such species have not arisen as the result of natural selection, and therefore constitute a genuine exception to Darwin's conclusion. An excellent account is given by Stebbins.[23]

The Rate of Evolution, and the Argument from Artificial to Natural Selection

Darwin was unique in his time in giving the rate of evolution approximately correctly. In the first edition of the *Origin*, he estimated the age of the Wealden (Upper Jurassic) as 300 million years, on the basis of a very simple calculation about the rate of erosion. The modern estimate, based on radioactive minerals, is about 130 million years. Some of his contemporaries estimated the age of the earth at under 6000 years, a value based on ancient Hebrew documents. Ancient Sanskrit documents allow a much wider choice, which includes the correct value of about 4000 million years, but no European took them seriously. On the basis of theories as to the cooling of an originally molten planet they reached figures of 10 or 20 million years for the age of the earth, which would have corresponded to about half a million for the Wealden. In the sixth edition of the *Origin* Darwin withdrew his estimate, but did not accept a shorter one.

Simpson[11] gives two types of estimate of the rate of evolution. The average duration of a genus of carnivores is about 8 million years, and this is fairly representative for mammals. That of a species is less than a million. Zeuner[24] gives half a million, but his time scale for the Pleistocene may be too long, and perhaps 300,000 years would be a better figure. Some insects seem to have evolved at about the same rate. But the molluscs were much slower, and a few molluscan genera are over 300 million years old. Another sort of estimate is based on the rate of evolution of quantitative characters. Simpson corrects some calculations of my own, though never by a factor of as much as two. The teeth of the ancestors of horses evolved fairly quickly from short structures like those of modern pigs to the remarkable ever-growing teeth of modern horses. But if we measure linear dimensions we find that the average increase in mean measurements was only about 3½ per cent per million years. The rate of change of shape parameters, that is to say ratios of lengths, was about half this. But in some extinct groups such as Toxodonta it must have been a good deal quicker.

I used the term, a *darwin*, for a rate of increase or decrease by a factor of e (2·718) per million years, or in other words by a factor of 1·001 per 1000 years. The average rate for horse teeth, and some other structures, was about 30 or 40 millidarwins. Only very rarely were rates of 1 darwin reached, whereas the rate of change of a character under consistent artificial selection, such as the butter content of the milk of Danish cattle, may be about 10 kilodarwins, or 300,000 times the natural rates.

We can now see, at least roughly, why evolution under natural selection must be slow. First of all provided that no other agency is opposing selection, the rate of evolution is proportional to the intensity of selection, as long as this is fairly small, say less than a 10 per cent advantage to the favoured phenotype. But we saw earlier that centripetal selection is generally acting to weed out extreme phenotypes. Weak selection will seldom overcome centripetal selection. The values of K and k in equation (1) will alter a little, but neither will change its sign. A new equilibrium will be reached. For example if k is increased by 1 per cent and K diminished by 1 per cent, the ratio of **A** to **a** will increase by 2 per cent. It may be that modifiers of fitness of the genotypes may be selected, and the population may change further, but this is a very slow process.

Darwin was justified in arguing from artificial to natural selection. But the argument is not so strong as he believed. There are stabilizing agencies of which he knew nothing, and these may prevent evolution occurring, and lead to stagnation or extinction.

Suppose, however, that evolution occurs, it cannot occur at the rate which would have been necessary had the geological time scale been a matter of a few million years. We can see roughly why. Evolution occurs by natural selection because some members of a species are better adapted to their environment than others. As Fisher[25] pointed out, though it is implicit in Darwin's writings, the environment of most species is deteriorating, largely because other species are evolving too. This is a stimulus to evolution, but if the deterioration is too great, it leads to extinction. We saw that, about 1800, the woods round Manchester were so blackened that about half the Peppered Moths in some of them were eaten on the first day after emergence. I do not think, however, that even in a somewhat blackened beech wood the light moths would be so conspicuous as they are on sooty oak or elm trunks. Fortunately a few black mutants were available, but even with this intensive selection the evolutionary episode of the blackening of the species lasted about a century. If selection had been ten times as intense the species would presumably have become extinct. And it is very likely that even if a few black mutants had been present they would not have found mates.

Now we can make calculations on this basis which, though rather rough, are quite as accurate as the numerical data yet available. Suppose that after n years of selection the three genotypes of moth hatched out with frequencies

$$p_n^2\,\mathbf{CC},\ 2p_n q_n\,\mathbf{Cc},\ q_n^2\,\mathbf{cc}\quad (\text{where } p_n+q_n = 1).$$

The **cc** moths were light, the others dark. A fraction k of the light moths died before leaving offspring, a fraction which would not have died had they been black. That is to say the parents of the next generation were in the ratios p_n^2 CC: $2p_n q_n$ Cc: $(1-k)q_n^2$ cc. So $p_{n+1} = p_n/(1-kq_n^2)$, so the value of p_n increases. If there were N moths, kq_n^2N died which would not have died had the light-coloured ones not been at a disadvantage. We shall suppose that selection began suddenly, and remained at a constant level, and that p_o, the frequency of the gene C, was quite small, say 10^{-5} or 1 in 100,000 when selection began, and finally reached a value very near unity, that is to say light moths became very rare. We can then ask what is the sum of the values of kq_n^2N over all generations. It turns out that it is equal to $-N \ln p_o$, or $-2{\cdot}30N \log_{10} p_o$, regardless of the value of k, when k is small, and a little less when k is moderately large. Thus if $p_o = 10^{-5}$, the total number of deaths is about $11{\cdot}5N$. We can see why the number is almost independent of k. Suppose we compare selections where $k = 0{\cdot}01$ and $0{\cdot}001$, the latter kills one-tenth as many moths in each generation, but it takes 10 times as long to produce the same change in the value of p_n, and these two influences cancel out. If C is not completely dominant, the number of deaths is increased. And to allow for this I have suggested that the average number of deaths for each gene substitution is about $30N$. Now this is so for a trivial gene substitution altering the average number of bristles on the abdomen just as much as for one with an important effect, for example one giving protective coloration or absence of wings. The calculation can be extended to cover cytoplasmically determined genetical differences, the effects of partial inbreeding, and many other possibilities. The total of deaths due to slow selection is never below $-\mathrm{N} \ln p_o$, though it may exceed it considerably.

If we suppose that on an average, through a million years or so, the fraction of individuals killed selectively by natural selection was 10 per cent this would mean that on an average about one gene substitution occurred every 300 generations. This would mean 100 years for insects with three annual broods, 1000 years or so for animals about the size of deer, and 8000 years or so for man. I think the figure of 10 per cent is a reasonable average over geological time. It may have been a good deal higher during the Pleistocene, and it is high at present because man is altering the environments of animals and plants over most of the world's land surface and in some areas of the sea.

Now a pair of species A and B derived from an ancestral species X living a million years ago may each differ from X by g genes (say 500). But some of their evolution will have been in parallel, so they differ from each other by less than $2g$, but probably not much less than g. How large is g for a pair of species as different as, say, the horse and donkey? We have very little idea, but my own guess would be as follows. The taxonomic differences such as size, bone shape, hair form, etc., might be due to changes at about 30 loci. There are also physiological differences, congenital differences in behaviour, immunity, and so on, which might be due to another 70 genes. And each of these genes might require four others to restore physiological balance and perhaps to secure dominance, as Kettlewell found. The total would be about 500. My own guess would be nearer 1000. These 500 changes would take 300×500 or 150,000 generations, that is about half a million years for moderate-sized mammals. This is a figure of the right order of magnitude. The time taken to increase the frequency of a dominant gene from 1 in 100,000 to 99 per cent is about $23{\cdot}8/k$, for example 2380 generations if dominants have a 1 per cent advantage. The last few recessives are eliminated very slowly. But the process is greatly speeded up by a little inbreeding, or incomplete dominance.

I think Darwin may have made calculations on these lines. If he had seen, even not quite consciously, the point that it would take about twice as many deaths to increase the frequency of a character from 1 in 10,000 to 1 in 100 as to increase it from 1 in 10,000 to 1 in 1000, and that the intensity of selection had little effect on the total, he must have arrived at figures like my own, though he would not necessarily have been willing to publish them.

But now we must deal with a fundamental question on which I think Darwin was misleading, though not inaccurate. He gave examples of the enormous mortality in many species between the beginning of life and reproduction. He pointed out that there were great possibilities of selective death. This is true. But I doubt if there is any correlation between the intensities of juvenile death-rate and of natural selection. A rough idea of the juvenile death-rate can be obtained from the number of eggs in a female. Suppose each female mayfly which emerges from the water contains 500 eggs, but that one-third of the females are eaten by birds before they can lay, then the 500 eggs must give about 1½ females and 1½ males in the next generation, or 3 on an average. So larval deaths kill about 166 out of 167.

Let us compare an oyster, where about one in a million of the eggs is destined to become an adult, with a zebra before the advent of man, where the average number of offspring was probably 4 to 6, so that a half or a third of the foals survived the hazards of infancy. The oyster's eggs are shed into the water, and give rise to larvae which swim with cilia and eat single-celled algae. But their survival is almost wholly a matter of chance. If at the right stage they happen to collide with a hard object on the sea bottom, they will attach themselves to it, but most get no chance to do so, and it is hard to suggest any improvements in oysters which would give the larvae a better chance of surviving the animals which eat them, or of finding a place to settle. Of course if they were larger they would not have to grow for so long before settling, and they might be able to produce organs which enabled them to find the right sort of site to settle down. But if so there would be fewer of them. The baby zebra, on the other hand, must be able to run within an hour or so of birth. It must recognize its mother, and be able to call to her in the correct way, to answer her calls, and so on. I think it is quite likely that a larger fraction of deaths is selective in zebras than in oysters. By this I mean that if all baby zebras could be replaced by animals of the fittest genotype available, the juvenile deaths would be reduced by a larger fraction than if the same were done in oysters. I do not suppose that a 5 per cent difference in the proportions of two of the lobes on a larval oyster would appreciably alter its chance of surviving till old enough to settle. A difference of 5 per cent in the lengths of the front and hind legs of the zebra, let alone of a left and right leg, would make a great deal. We have got to recognize the fact that we do not yet know the over-all intensity of natural selection for any species. We know that it is not zero, and occasionally it exceeds 10 per cent. This is, however, usually due to centripetal selection, with which we are not concerned at the moment, as it does not cause evolutionary change. We also know that in the zebra it could not possibly be 90 per cent, for if it were, the average female would have to bear over twenty foals. It could be as high as this in the oyster, though I doubt it.

One of the stock arguments against natural selection is that if a complicated organ such as the eye is to change for the better, a number of characters must change simultaneously; if one only changes, fitness is lowered. To take a concrete example, the jaws of a dachshund are lengthened by genes, some of which act independently on the two jaws. So hybrids between dogs with long and short jaws may give offspring with lower jaws much too short for the upper one. This is, I think, the only anti-Darwinian argument which has been strengthened by genetics. Suppose, for example, that most of a population is **aa bb cc**. Three genes **A**, **B** and **C** occasionally turn up by mutation. Each, by itself, lowers the fitness by 1 per cent. If a pair, such as **A** and **B**, are present, it is unaltered. If all three are present, it is raised by 5 per cent. If the mutation rate is 4×10^{-5}, as for the most mutable human genes, the frequency in the population of individuals carrying **A** will be about 0·8 per cent. So only about one in two million will carry all of **A**, **B** and **C**. The slight increase in the frequency of the gene **A** caused by the increased fitness of such individuals is quite negligible compared with the diminution caused by the lesser frequency of those carrying **A**, **B** or **C** only. In fact it can be shown that, if the frequencies of the three genes are equal (which is of course very improbable), natural selection would not begin to increase their frequencies until each was present in over 30·9 per cent of the population.

Animal breeders attempt to get over this difficulty by 'nicking'. They have an ideal for a given breed of horse or sheep. They mate animals each of which approaches it in a different respect, though each may be decidedly imperfect, and indeed ill-balanced. They hope that some of the offspring will combine the desirable qualities of each parent. I know of no adequate tests of the efficiency of this method in animal breeding, but it is certainly in accordance with genetical theory, provided that if necessary, the progeny is inbred for a further generation. It is certainly efficient in plant breeding. It is most unlikely that anything of this kind occurs in nature.

There are at least two possible ways out of this impasse. Wright considered a species divided up into a number of small endogamous tribes. Such a tribe might easily, by chance, become homozygous for several slightly unfavourable genes. If it happened to hit on a favourable combination, its numbers would increase. To my mind the weakest point in Wright's argument is that he has not adequately considered what happens when this tribe starts hybridizing with others. It is also possible to suppose, with Fisher, that the transformation takes place by very small steps. Each step would give an improvement in some structure or function, partly, but not wholly, cancelled by misfitting with other structures. If so each step, by itself, might be favourable. But if, say, gene **a** changes to **A** through six intermediate alleles $\mathbf{a}^1$, $\mathbf{a}^2$, $\mathbf{a}^3$, $\mathbf{a}^4$, $\mathbf{a}^5$, $\mathbf{a}^6$, each change involves the death of about $30N$ individuals, and the selective process is extremely slow. If each gene substitution conferred an advantage of one ten thousandth, it would take several million years. And a thorough-going improvement in such a structure as the vertebrate eye might take a hundred million or so. There is certainly no difficulty in explaining the slowness of evolution.

Evolution and Geography

Darwin was converted to a belief in evolution by what he saw in his voyage round the world in the *Beagle*. The fact that groups of animals were confined, or nearly confined, to certain areas, for example marsupials to Australia, sloths and armadillos to South America, though they are well adapted to live elsewhere, is a powerful argument for evolution.

In late years, Mayr[26] has taken up the problem in great detail. Suppose it could be shown that a species X living in the Pliocene has given rise to two species now living, and giving no hybrids or sterile hybrids where their habitats overlap, one could hold either of two hypotheses. The two races gradually (or suddenly) diverged in the same country. This is called sympatric speciation. Or the original species was separated by a geographical barrier, and the groups on the two sides of it evolved in different ways. When the barrier broke down hybridization was impossible. This is called allopatric speciation. The barrier may arise for geological causes. For example a peninsula may become an island, a belt of desert or forest may divide a habitat, or a range of mountains may be formed. Or the barrier may have been there all the time, but only a few individuals may have passed it. This is particularly true of islands in the Pacific which Mayr has studied.

Now there is no doubt at all that allopatric speciation has occurred, and every stage between local races and distinct species has been studied. Thus there are several subspecies of *Triturus cristatus*, the great crested newt, but only one in one region. If one crosses individuals from London and Naples, 80 per cent of the young larvae grow up. If one crosses these hybrids together only 4 per cent do so. This is partly because their chromosomes differ so much that pairing breaks down to a considerable extent. They might be (and have been) described as distinct species. But they do not overlap geographically. On the other hand, the marbled newt (*Triturus marmoratus*) rarely hybridizes with the great crested newt. The male hybrids are quite sterile, the females nearly so. Their habitats overlap in western France, and there is little evidence that genes are transferred from one to the other in nature, and no suggestion of a fusion. The marbled newt seems to have evolved in Spain, and crossed the Pyrenees rather recently. They were probably an impassable barrier during ice ages. Perhaps if the newts on the two sides of the Alps evolve in their own ways for another 100,000 years they may become so distinct that they can live side by side without appreciable mixing.

One way in which species descended from the same ancestor can come to differ was discovered by Harland.[27] Homologous structures in two species may develop under the influence of different genes in them. In the F_2 from their hybrids some individuals lack both genes, and are consequently abnormal. For example, radially symmetrical (actinomorphic) flowers are found in the F_2 between two bilaterally symmetrical (zygomorphic) species of snapdragon. Thus there may be sharp genetical differences between very similar-looking species. One can quite understand that it may be advantageous to a species in a new environment to do a physiological job in a new way, as grain was ground by water-mills in hilly country and windmills in flat. But any such change is a step towards speciation.

Mayr doubts whether sympatric speciation has ever occurred in animals. It certainly has in plants by polyploidy. I think it is very hard to see how the wealth of species swimming and floating in the open sea developed by allopatric speciation. I have no doubt that allopatric speciation has been much commoner than sympatric among most land and freshwater animals. But I think there is a case for sympatric speciation in stenophagous animals (that is to say those restricted to a small range of food) and in parasites. Thorpe[28] has shown that many adult insects seek plants with the smell to which they were accustomed as larvae on which to lay their eggs. Thus the difference between races eating different plants is at first sight a matter of tradition. If the males seek females only in the neighbourhood of the food plant this difference of tradition can become a barrier to crossing. Again a difference of a month or so in times of emergence may be as effective a barrier as a mountain range. The whole question is under vigorous discussion. And in particular it may be that chromosomal changes are most readily established when a single fertilized female crosses a barrier.

My guess is that, as Mayr's views are now orthodox, they will be violently criticized in twenty years or so, and that most, but not quite all, of these criticisms will prove to be invalid.

Instinct

Darwin devoted a chapter to instinct. Instincts are hard to explain by natural selection, because we know almost nothing about their variation and inheritance, as we do know something about variation and inheritance of shape, size and colour. And they are easily explained as inherited memories. One objection to this explanation is that in the only species where we are certain that memories exist, they are not inherited, though minute details of form, and probably of character, are inherited. As Darwin pointed out, the most remarkable of all instincts are those of the sterile worker 'caste' of social insects such as bees. As their ancestors, the fertile 'queens' and 'drones', have never displayed these instincts (at least not for some millions of years) they cannot be inherited memories, and would be lost by disuse if disuse necessarily led to the loss of faculties. Finally, attempts to prove inheritance of memory in animals have failed.

An instinct may be defined as a mode of behaviour which develops in an animal without specific external stimuli which could reasonably determine one particular mode of behaviour rather than another. No sharp line can be drawn between a reflex and an instinct, but the latter is usually much more adjustable than the former, and involves many more parts of the body. A typical instinctive pattern of behaviour involves three stages. There is a stage of appetitive behaviour characterized by restlessness and 'searching' which, however, need not involve consciousness of the goal, even in men. Then a fairly specific sensory stimulus arouses a goal-seeking behaviour. For example, an animal wandering about, in part at least under the influence of its sex hormones, smells the odour characterizing a member of the opposite sex, and directs itself towards the source of this odour. It may run, jump, climb, or swim, as the case may be. The final stage is a consummatory act such as swallowing, copulating, breathing air at the end of a dive, or fixing a twig into a nest. This may be a reflex action, and is commonly accompanied by satisfaction (in man by pleasure). The drive builds up again and the process is repeated. Learning may greatly facilitate instinctive activity. An animal learns where and how to get food. But it only does so if it has an instinct to eat, which some adult insects have not. The specific stimulus varies from one species to another, and the 'innate releasing mechanism' postulated in the brain, which responds to it, must have evolved. So must the structure responsible for the nearly fixed pattern of the consummatory act.

There is a sense in which we can say that instincts are inherited. Many breeds of fowls lack the instinct to brood on eggs. This may be because they do not make enough of the pituitary hormone which releases brooding, or because they require more hormone than they produce to release the instinctive behaviour. But this tells us little about how instincts evolved, though it does show that they can easily be lost. A piece of instinctive behaviour depends on many physiological processes, any one of which can be interrupted by a gene substitution.

However, we know that selection can rapidly change instinctive behaviour. Rendel[20] abolished an instinctive mating preference in one species of *Drosophila*, and Koopman[29] greatly accentuated another by keeping two species of this genus together and killing the hybrids in each generation, breeding only from the progeny of flies which had mated with their own species.

Again Tryon[30] selected rats for speed and slowness of learning to find their way through a dark maze. After seven generations a congenitally 'bright' and 'dull' race were established. Krechevsky[31] found that when searching a maze the 'dull' rats tended to use visual cues such as light and dark, black and white, the 'bright' rats using right and left (for example, usually trying a left turning before a right or conversely or trying left and right alternately). The 'dull' rats were naturally at a disadvantage in darkness. These two types of behaviour, which were hereditary, and did not depend either on a rat's own experience or its ancestors', might well be called instinctive. And they can be changed by selection. We shall understand better how an instinct can evolve as if memory were inherited by discussing genetic assimilation.

Genetic Assimilation

Waddington[12] and his colleagues have studied a number of cases of which this is typical. Larvae of *Drosophila melanogaster* are exposed to heat for a short time. Among the flies which develop from them several kinds of morphological abnormality are found. One is the absence of cross-veins on the wings, an apparently harmless defect. This occurred in about 34 per cent of his flies. Waddington selected the flies which showed it, bred from them and those of the progeny which showed it, and so on for many generations. The fraction with no cross-veins increased. After fourteen generations a few cross-veinless flies appeared even from larvae which had not been heated. By breeding from these he ultimately obtained a stock, most of whose members had no cross-veins. It was shown to differ from the original type by genes in three chromosomes.

This might very well be interpreted as follows. The heat has caused changes in the developing wings of these flies. It has caused comparable changes in their ovaries and testes, so that the changed character is inherited. Waddington disproved this explanation by consistently breeding from the flies whose development had not been affected by heat. He found that the fraction of heat-changed flies in their progeny became successively smaller in successive generations. This is quite inexplicable on theories of the type favoured by Lamarck and Lysenko. Let us see what has presumably happened in these flies. Some substance is needed for the development of cross-veins. The amount of this in the relevant place (probably the wing-buds) in a pupa is x. The amount formed can be altered by selection, like the amount of milk, hair, or any other quantitative character. The substance is partly destroyed by an amount of heating which kills very few larvae. If x is between 0 and 1, no cross-veins are ever formed; if x is between 1 and 2, cross-veins are usually formed, but not if the larva is heated in the standard way; if x is greater than 2, cross-veins are formed even if the larva is heated.

Waddington started with flies where x was generally greater than 2, and never less than 1. For some generations he selected flies where it was less than 2, until the value of x was below 2 in most flies, and below 1 in a few. He was then able to select flies where it was below 1, that is to say where absence of cross-veins was hereditary. Similarly, he could obtain a stock in which x almost always exceeded 2. This anti-Lamarckian effect is quite useful. Bateson was able to eliminate bolting (that is to say flowering in the first year, which uses up the food stored in the root) in biennial root crops by submitting them to 'temptation' by sowing earlier than usual, and not breeding from those which flowered prematurely.

If we substitute learning for 'not developing a cross-vein' we have a possible parallel with the development of an instinct. I take a hypothetical example. In area A a particular volatile substance is produced by a nutritious plant, in area B by a poisonous plant. In area A those insects of a certain species which learn most readily to recognize this odour and associate it with food are at an advantage. As the features in the nervous system which favour such learning are accentuated, a few insects appear to whom the odour is attractive without learning, as the odour of sheep appears to be attractive to sheep-dog puppies. They are at a double advantage, and after some time all members of the insect species are attracted by the odour without any learning. Similarly, in area B a race evolves which finds the odour repulsive. We know that there is in fact 'raw material' on which selection can act from a study of our own species, where there are considerable differences in the capacity for detecting smells and tastes, and in judgment as to whether they are attractive or repulsive. Some at least of these differences are genetically determined.

Other Difficulties

I think that Kettlewell's work[17] on *Biston betularia* and my own on the rate of evolution give us a roughly quantitative theory of natural selection. It will take centuries to fill in the fine details. Supposing it to be accepted, the next question to be asked[32] will perhaps be why some kinds of mutation occur frequently in one species and never in another; or at least appear to do so. If white bristles on the eyes are never found in one species of fly, while in a closely related species, mutations giving this character have occurred at three different loci, the reason is probably that in the first species mutations interrupting the synthesis of pigment in these bristles also interrupt some important developmental process, and cause death. If this is correct, the possibilities of evolution open to a species depend not so much on its genes and their mutability, as on its developmental processes. Too great an integration of these processes makes for smooth development, but makes evolution impossible. Which, if any, of the evolutionary paths open to it a species will take is determined by natural selection. I can give no hint as to why some paths appear to be open to one species and not to another. This may be the most important of all questions. Men have expressed various ideals in their religion, art and literature. We should like our descendants to attain or approach these ideals. We neither know how to attain them nor whether they are attainable. There may be some ineluctable barrier, original sin, *avidya*, or what you will. Whereas no such barrier, perhaps, stands between mankind and extinction. *Patet isti ianua leti.*

However, many people refuse to accept Darwinism on some such ground as that the design, organization, or, as Darwin called it, purpose, embodied in living organisms cannot be the product of 'blind chance'. The answer is twofold. On the one hand, natural selection acts so as to eliminate chance effects. Variation is in some sense random, but natural selection picks out variations in one direction, and not in another. On the other hand, all natural processes appear to depend on 'blind chance'. This is obviously so at the level of quantum mechanics. We cannot predict just which atoms or molecules in an assembly will pick up a quantum of radiation in the next second; but in a sufficiently large assembly, say a gram, we can often predict the fraction which will do so with very great accuracy. This holds for human voluntary actions also. When I strike a match I contract various muscles rather weakly. Just which fibres will contract, and for how long, is probably unpredictable. But the average force exerted is regulated with great precision. The match head rubs on projections and is locally heated. When a large enough volume is heated to a sufficient temperature it bursts into flame. One cannot predict the details of the process; one can predict its outcome.

Again it is urged that mind cannot be a product of evolution. This difficulty does not arise for a consistent idealist, nor for a consistent materialist, whether he adopts a materialism based on modern science, or on the Hindu notion of *prakriti*, which is the basis not only of matter but of individual mental phenomena, or on the Chinese notion that *li* (meaning something like substance) and *chih* (meaning something like organization) are omnipresent, the latter in various degrees.

It is possible, however, that Darwin may have started a revolution in philosophy as great as that initiated by Aristotle, whose logic is largely based on his efforts at the classification of animals. Darwin, in the last chapter of *The Origin of Species*, wrote that in studying any two forms 'we shall be led to weigh more carefully and to value higher the actual amount of difference between them'. In ordinary thought we sometimes know more about the difference than the experiences which differ. For example, the interval between two tones is more easily recognized than their absolute pitch. A distance is a much less sophisticated notion than a point. The 'observables' of modern physics, such as wavelengths, are regarded as indicating differences between unobservable atomic or molecular configurations. It may be that some future philosophy will be based on differences as our common-sense philosophy of to-day is based on similarities, even if we are not Aristotelians or Thomists. If so we do not know how these problems will appear to our descendants.

The history of science makes it almost certain that facts will be discovered which show that the theory of natural selection is not fully adequate to account for evolution. But the same history makes it extremely improbable that these facts will be in any way related to the criticisms at present made of it. The physics of

Newton and Galileo have proved inadequate in several respects, and are being replaced by relativistic and quantum mechanics. These, however, are even further from the medieval physical theories than were the theories of Galileo and Newton. They were discovered because when the consequences of Newtonian physics were fully worked out, certain facts disagreed with them. It was not possible in Newton's time to guess at these discrepancies, though Newton's own guesses, in his *Opticks*, were far more pregnant than those of his contemporaries. Darwinism will, I do not doubt, be modified. Like any other successful theory it will ultimately develop its own internal contradictions. I think that a major task of biologists, perhaps for a century or more, will be to work out the full consequences of Darwinism. My own life has been largely devoted to this task. I have not got far with it; but if, as I believe, it is worth pursuing, others will pursue it.[33]

4

P. MARLER

DEVELOPMENTS IN THE STUDY OF ANIMAL COMMUNICATION

It is a significant reflection, both on scientific method and on the genius of Darwin, that little real attempt was made to take up the study of animal communication where he left off. Instead, as so often happens, the subject lay fallow for half a century, until others rediscovered it. With the stimulus of new methods, the same problems were approached from a new angle, and it is in fact difficult to find any major work on animal communication which is directly and obviously inspired by *The Expression of the Emotions in Man and Animals* (London, 1872), unless it be that of J. S. Huxley. Nevertheless, perhaps von Uexküll unconsciously served as a bridge between the two generations, for several of the later workers came under his direct influence.

The change of emphasis in the new approach to animal communication has been towards the direct study of animals in natural or semi-natural environments, which we owe mainly to J. S. Huxley and K. Lorenz. With captive animals, on the other hand, H. Lissman and N. Tinbergen were pioneers in the complementary approach, exploiting the fact that some animals can be induced to respond to artificial models as they would to the social partner. By use of a wide range of models, it has often proved possible to break down apparently highly complex behaviour into surprisingly simple basic elements.

These two methods of approach have stimulated an enormous variety of work on a great many animals, the most important contributions coming from Holland, Germany and Switzerland. It is the aim of this essay to review some of these developments in order to assess how our understanding of animal communications has developed since Darwin's time, and to point out that there are still certain fields where we may look to him for further inspiration.

Perhaps Darwin's greatest limitation was his inevitably rather fragmentary knowledge of how animals actually communicate among themselves in their natural state. It is no accident that his main subjects, apart from man, were the domestic cat and dog. The need to be thoroughly intimate with the animal's life before its social communication can be really understood has since been made evident by K. Lorenz. As a result we now have a better grasp of how widespread the phenomenon of animal language is, and how complex some of the signals are which individuals exchange.

The term *communication*, which can be applied to all kinds of interaction between cells, organs and organisms, is restricted here to exchanges of information between individuals. There are various methods available to animals for signalling to each other. One of the most widespread is by chemical signals which others can either taste or smell. The substance may be carried either by diffusion or by a current to the other animal, or it may be deposited on some object where others are likely to encounter it. As we shall see, this method has many disadvantages, and more complex languages usually make use of the senses of sight and hearing. We know most about visual communication since it is the easiest for us to investigate. With the development of new techniques we are now learning more about vocal language. But the sense of smell is still rather a mystery, and the possibilities of communication by touch have scarcely been recognized, although it provides the basis for the most complex animal language so far known, discovered by K. von Frisch in the honey bee.[1]

Communication by Taste and Smell

Because most of us neglect our chemical senses, we are apt to overlook their importance in animals. They play an important role in the life of most, with the exception of birds. A dog's life, the visual aspect of which had such careful attention from Darwin, is probably more dominated by the sense of smell, and dogs' powers of discriminating between slightly different odours, recently confirmed in a series of experiments by Neuhaus,[2] are already familiar to

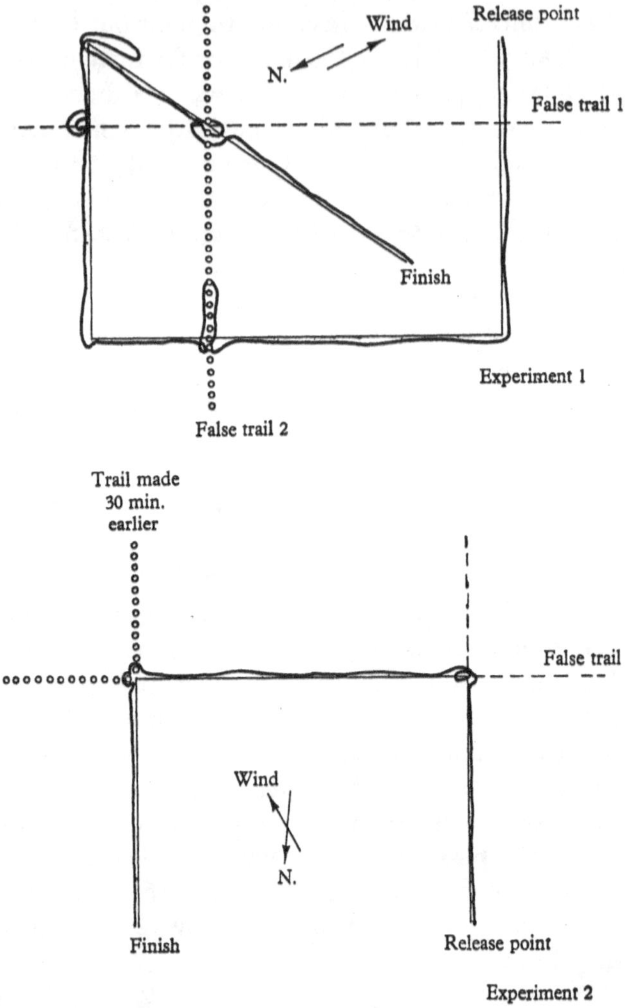

Fig. 1. Tests of the ability of a police dog to follow scent trails of other dogs. In the first experiment it successfully followed a trail without being more then momentarily confused by the false trails of two other dogs. In the second experiment it again avoided a false trail, and distinguished the fresh track from another laid by the same dog 30 minutes earlier. (After Schmid[3].)

anyone who has kept them. German police dogs, for example, can readily follow the fresh trail of a horse or another dog, and recognize it within a network of false tracks. They can even distinguish two tracks of the same dog, one 30 minutes older than the other[3] (fig. 1).

This ability to distinguish other individuals by smell is by no means confined to mammals. Goz[4] found that a minnow could distinguish the water passing over one member of its shoal from that passing over another. The powers of discrimination extend to other species and genera, even to other families, though they become blunted with the more distant relatives. There is one exception to this in the remarkable responses to water passing over a predatory fish, such as a pike. Exposed to this, a minnow either keeps quite still, or edges gently away, and occasionally goes into a curious trance-like state.

So fish can get a surprising amount of information about other fish by the chemical sense. In these examples it appears that the chemical signal which passes from one fish to another is an accidental by-product, probably produced more or less continuously. This is no disadvantage for communication purposes, the only requirement being that there should be little chance of two individuals producing exactly the same combination of chemicals. It probably suffices in some cases to rely on the varied diet of animals. Any two individuals are likely to eat slightly different foods, or will metabolize it in slightly different ways, so producing different excretory products.

Apparently the colony odour of bees arises in this way, where it is combined with the habit of free exchange of food between all members of the hive.[5] The repulsion of strange bees from the hive is known to be based on their different smell, as well as behaviour. Kalmus and Ribbands[6] divided a colony into three parts and periodically examined them to see if marked bees returned to the wrong part. Two groups were kept unfed and the third was given heather honey and black treacle. Within eight days, the fed group had only a small proportion of intruders, while the other two were still freely exchanging individuals. So it appears that the change in diet had already enabled the fed group to distinguish between the smells of their own comrades and those of the unfed bees. However, the actual nature of the substance secreted is still in doubt, for Renner[7] has recently shown that the bee's scent organ plays no part in the identification of the colony. Bees did not distinguish between the substances squeezed from the glands of members of different hives, so some other secretion must be involved.

Even in their simplest form, chemical signals make possible the identification of other animals down to species, colony or individual. And they can also communicate the presence of danger. Some fish will flee from a wounded companion. Von Frisch[8] and his pupil Schutz[9] demonstrated that this was a response to something released from the damaged skin. It is most common in species which form schools, though it is not confined to them, and the most effective warning signals come from the fish's own species. So this *Shreckstoff* or *fright-substance* is a quite effective method of communicating danger to other members of the species. It even helps to protect newly hatched fry from being eaten by their parents, since the adults are frightened by the substance produced if they should snap at and wound young of their own species.

Though first described in fish, the effects of these fright-substances may be much more widespread. Toad tadpoles behave in the same way[10] and there is perhaps something similar in *Paramecium*, *Drosophila* and mice, though this is not yet properly established.[11]

The problem of identifying a chemical signal is rather different from that of discovering the direction from which it has come. Perception of the presence of the stimulus only tells the animal that the source lies within a certain distance, depending on the concentration and characteristics of the substance, the forces carrying it from the source, and the sensitivity of the animal's receptors.

There are several methods to attract or repel animals from a particular point by chemical means. The substance can be deposited on some object, which subsequently serves as the signal source. A honey-bee leaves its odour behind when it visits a flower, which helps other bees to find it.[12] The same method can be used to establish contact between individuals if the marking animal remains near the signal points, or visits them regularly. A solitary male bumble bee marks signposts along its track of the day by biting at leaves and twigs, so depositing scent from special glands in the mouth. The track is laid in a zone which varies with the species, either in the tree canopy, in small trees or shrubs, or near the ground. He then patrols the posted area for the rest of the day. If a queen bee looking for a mate should strike this trail, she will follow the scent signals, and so eventually meet the male.[13]

Attraction to a point can also be achieved with a chemical signal from the animal's own body, as long as it keeps still for periods of time, so that the partner can track the smell down. Some of the most elementary organisms, the slime moulds, seem to form their cellular aggregations in this way; they illustrate how a complex communication system can be built up from a simple basis. Individual cells are attracted to the organizing centre by a chemical substance, acrasin, which diffuses from it. Shaffer[14] demonstrated that, although the same substance will attract the cells of more than one species, there is nevertheless a specific response. This is ensured by the secretion of a second substance which destroys acrasin. By secreting these two chemicals alternately, waves of acrasin are produced, spreading outwards from the centre with a particular interval between them. Apparently because of variations in such factors as cellular adhesiveness, each species will only aggregate in response to a particular rhythm of acrasin production, and so intraspecific communication is ensured.

Fig. 2. A male newt directing a water current over his cloaca towards the female. (After Tinbergen[15].)

For these slime moulds, simple diffusion suffices to carry the chemical, but for some purposes this is too slow, or the concentration at a distance may be very small indeed. Some improvement can be achieved if the animal creates a current over the scent gland. This may simply serve to get the scent off the gland more quickly, as in the honey-bee, or may send a concentrated signal in a particular direction. A male newt, courting his mate under water, turns to face her and curls his tail forward, so sending a current over his cloaca and towards her, apparently carrying chemical stimuli[15] (fig. 2).

A directed current will have limited value in chemical communication, unless the animal knows where to direct it. When this is not so, as in the scent signalling by which some female moths attract a mate, a natural current can be used instead, in this case air movements. Here we see some of the most remarkable achievements of this kind of communication, and also some of its inevitable limitations.

In a series of careful studies, Schwinck[16] found that the male silkworm moth can only locate the female at a distance if there is a wind blowing. He can perceive the female odour at very low concentrations, at which direct location would be inconceivable. However, he can respond to the air movement and he starts searching upwind, zig-zagging to and fro. If he loses the scent, he goes downwind and starts again. In this way he can approach close enough to change over to direct orientation, guided by the diffusion gradient around the female. The details of this second stage vary with the species.[17] One species will wheel back down-wind and return two or three times, gradually homing to the female. In another species, the male alights, and makes the last part of the search on foot. Obviously there are disadvantages in this method of communication, remarkable as its achievements are. It only works when there is a wind blowing, and the male spends a lot of time searching. Many animals have exploited visual and auditory signals which lack many of these drawbacks.

Perhaps the greatest restriction on the development of chemical language is the limited repertoire of possible signals. An animal may have several glands which produce different substances.[18] Solitary bees may have at least two distinct smells, and a tropical butterfly, *Heliconius*, probably has at least four.[19] This is a difficult subject to study and no doubt more elaborate chemical languages have still to be discovered. But no external gland is yet known to produce, as the larynx does with sound, a whole spectrum of different signals. The difficulties of rapid exchange of chemical signals between individuals are also great, though this might be partly overcome with shortlived compounds, or by destroying one substance with another,[20] as in slime moulds. Whatever further work may reveal, it seems probable that we shall always have to look elsewhere for the most elaborate forms of social communication.

Visual Communication

In contrast with chemical stimuli, an animal cannot perceive a visual stimulus without also getting some idea of whence it has come. A creature with the simplest eyes may only know that the signal source is in the approximate direction in which the eye is pointing. With more elaborate eyes, each sensitive cell receives light from a particular direction, so that very accurate location of the source becomes possible. Vision is, in fact, the most efficient means many higher organisms have of locating each other. This is not even impossible for animals which are active in darkness, for some, such as many deep-sea fishes, can produce light themselves.

Fireflies, too, are well-known examples of this, making use of what must be the simplest type of visual signal, a flash of light. From this basis, they have built up a remarkably efficient system of signals for mating purposes.[21] As many as a dozen species may be active at the same time of night, each readily distinguished by its manner of flashing. Some have long flashes, others shorter ones. Some twinkle and others are steady, and some have a pattern of successive flashes like a lighthouse lamp (fig. 3). One species, *Photinus pyralis* has been particularly carefully studied.[22] At mating time the male flies around giving a short flash at regular intervals. The female climbs a blade of grass and remains there, flashing in reply to any male who signals nearby. How she recognizes his flashes is still not known, but it may be because of the regular interval of about 5·8 seconds. Her response is to flash back at the male just 2 seconds afterwards, and this time interval is critical. A male has no response to other males' signals, but will approach lights of all sizes, colours and durations that flash 2 seconds after he does. These flashes are exchanged between the male and female as he approaches, until he is near enough to mate with her. In the crucial experiment, males were made to flash 2 seconds after another male by pinching them, and they were tracked down just as though they were females.

Besides being easy to locate, visual signals have many other advantages. They are easy to produce, they can easily be turned on and off, and under the right conditions they can be received at considerable distances. However, some animals can only see detail at close range, if at all, and this has inevitable effects on their social communication.

Most spiders are short-sighted, and have to rely mainly on the senses of smell, taste and touch to identify each other. When the

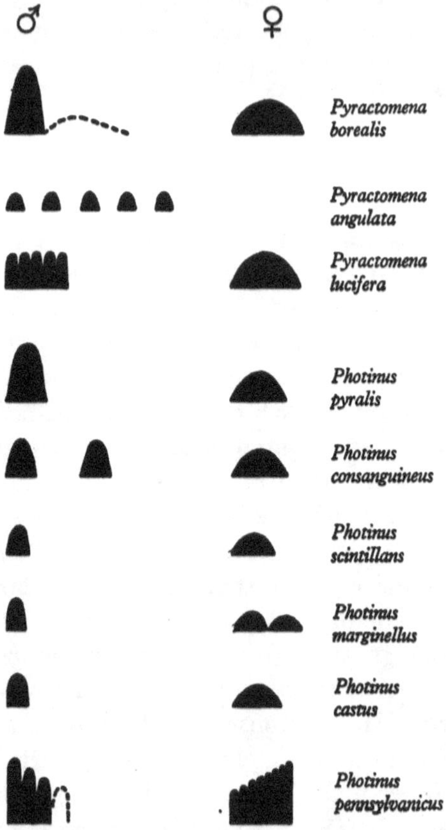

Fig. 3. Diagrams of the flashing patterns of various species of firefly. The height of each mark is roughly proportional to light intensity, the length to duration. (After McDermott[21].)

male approaches the female there is considerable danger that she will treat him as prey. He has to go to considerable lengths to prevent this by advancing very carefully and announcing his approach in various ways. He may jerk on her web with a particular rhythm,[23] he may stroke her, or even rush in and interlock her jaws with his own. In another species the male goes so far as to tie the female down with web, once he has subdued her (fig. 4). All of these methods of mollifying the female tend to take a long time and are not a little dangerous for the male. They seem to be made necessary by the difficulties of identification among these short-sighted spiders.

Some hunting spiders have much better vision, and their behaviour differs in many ways. Instead of waiting for their prey to

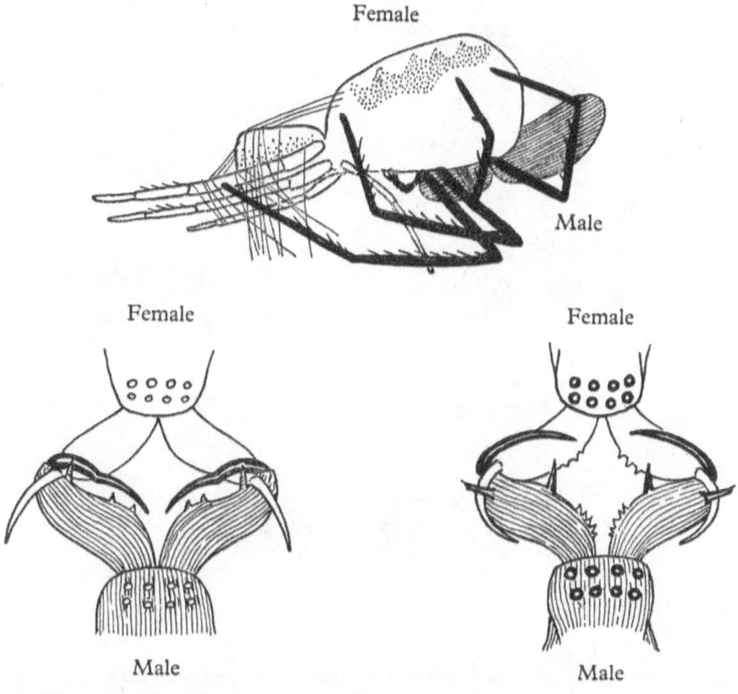

Fig. 4. The courtship of spiders. Above, the male *Xysticus lanio* copulates with his female, after fastening her down with threads. Below, the methods by which males of species of *Pathygnatha* (left) and *Tetragnatha* (right) interlock their jaws with those of the female. (After Bristowe [24].)

enter a web, they go in search of it, running to within jumping distance and then pouncing. They can identify prey quite well by vision alone, and can be seen scanning over an object with their eyes, assessing its size. Similarly, during courtship, they have little difficulty in identifying each other at a distance, for they have evolved all kinds of elaborate movements and colours by which the males announce their species and sex (fig. 5). They avoid the risks that short-sighted spiders have to face.[24]

Again, among the insects, many have to rely on communication by smell and taste in finding mates, often because their eyes are not very good or because they are nocturnal. Nevertheless, some day-flying insects, such as some butterflies, have an elaborate type of colour vision,[25] and quite elaborate visual signals may occur.

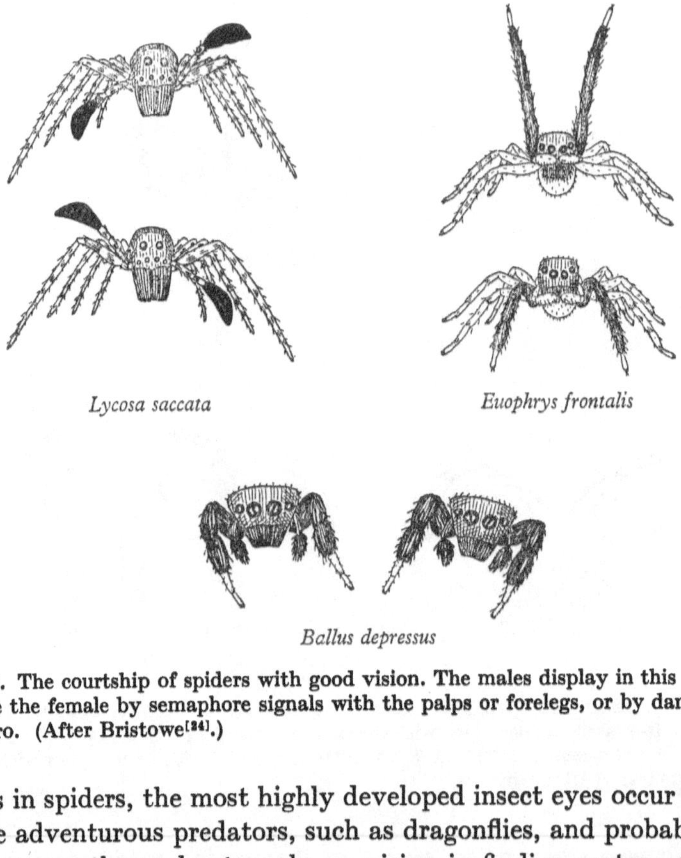

Fig. 5. The courtship of spiders with good vision. The males display in this fashion before the female by semaphore signals with the palps or forelegs, or by dancing to and fro. (After Bristowe[24].)

As in spiders, the most highly developed insect eyes occur in the more adventurous predators, such as dragonflies, and probably for this reason they rely strongly on vision in finding mates. At the other extreme we find such insects as the pond skater, *Gerris*, which uses chemical signals in finding a mate, and the Colorado beetle, *Leptinotarsa*, which seems to have no direct means of finding a mate at all, the sexes meeting fortuitously. The male even has to find which are the female's front and rear ends by trial and error.[26]

Visual Communication

Although the eyes of many lower animals are not well-suited to form vision, they can readily detect movement. This is fully exploited in the communication of spiders, butterflies and dragonflies, and we find similar developments in some Crustacea. Best known are the little fiddler crabs which occur on the sea-shore in many parts of the world. One claw is larger than the other, and the males of each species have a characteristic way of waving it in the air which attracts females and in some cases repels males.[27] It is striking that species living together all wave in a different way, immediately revealing their identity to an experienced observer (fig. 6). Some are distinguished by the direction in which the claw

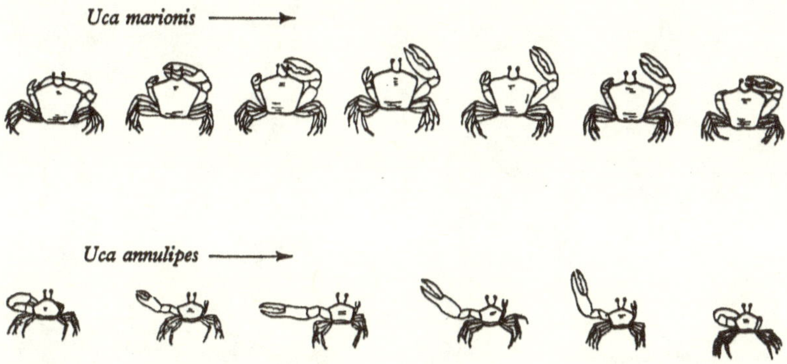

Fig. 6. Two methods of signalling by male fiddler crabs. (After Altevogt[27].)

is moved, some by the speed of movement and others by tilting or rocking of the body. There is also a relation between the pattern of movement and the colours of the body and claws, so that the former is made as conspicuous as possible.

In the study of visual communication in higher animals, there are many difficulties in imitating natural forms and movements well enough to induce animals to react to models. One solution is to disguise living animals. Both methods have been tried with lizards and fish. Some lizards will respond simply to coloured Plasticene models. In species with a dull female and a blue-throated male, disguising the female with blue paint will immediately cause her mate to attack her as though she were a male.[28] Similarly among fish, the male three-spined stickleback will attack very simple models of a fish, provided that they carry the male's red underside.[29] With live jewel fish, Noble and Curtis[30] exploited the fact that the red colour of a reproductive male can be induced in a female by the injection of a drug. This red colour serves as a signal to the female; if given a choice between two males at spawning time, she selects the one with the brighter colour. Similarly, she will also respond to a red-induced female.

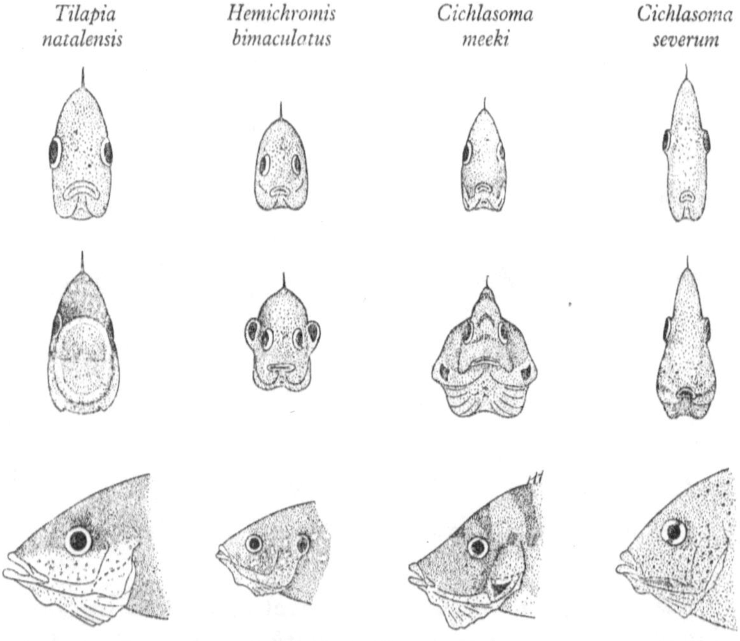

Fig. 7. Displays of four species of cichlid fishes, seen from the front and the side. Note how the visual effects are enhanced by the markings on the gill covers. (After Baerends and Baerends-van Roon[31].)

Sometimes the mode of behaviour of the partner is more important than colour in the visual communication of fish. Males may distinguish between the sexes solely by the way in which they respond to attack or aggressive display, other males replying with aggression, and females with a different 'inferiority' response.[31] In these exchanges between the sexes, a complete range of movements and postures is used, each playing a different role in courtship or defence of the territory (fig. 7).

Birds generally respond to models much less readily because their usually incessant movement is impossible to imitate. However, when the females of many species are ready to copulate they take up a special posture and keep quite still, while the male approaches and mounts. This is a signal we can easily represent with models, and many male birds are so completely deceived that they will mount and even deposit semen (fig. 8).

When movement is involved, we can only experiment with living animals in disguise. This method has hardly been explored yet, but has great possibilities. Another pioneer, Cinat Tomson,[32] applied it to the budgerigar. The female seems to choose a mate at least

Fig. 8. The posture of a female chaffinch soliciting for copulation. On the right is a male, mating with a stuffed model of a female. (From photographs, after Marler[59].)

partly by the spots on his collar. Given a choice between one with small spots and another with artificially large spots, applied with paint, she selects the latter. Another important signal is the cere above the bill, which is blue in males and brown in females. If a female's cere is painted blue, her mate immediately attacks her as though she were a rival male.

In another sexually dimorphic species, the chaffinch, the female is greenish brown, while the male has, among other things, a reddish breast. Just as in budgerigars, females disguised as males, in this case simply by staining the breast feathers with red ink, are treated as though they were males. In the winter flock, male chaffinches dominate females in the social hierarchy. Red-painted females, on the other hand, dominate other females, and may even win some fights with males.[33]

In both of these examples the colours are on display more or less permanently. The most that can be done to remove the signal is to turn in another direction, so that the colour is hidden. If the colours fall on the wings or tail or other mobile feathers, more subtle methods of concealment and display are possible. The chaffinch's wings carry two white bars, which can be concealed by the flank feathers, or exposed by raising the wing. With intermediate stages

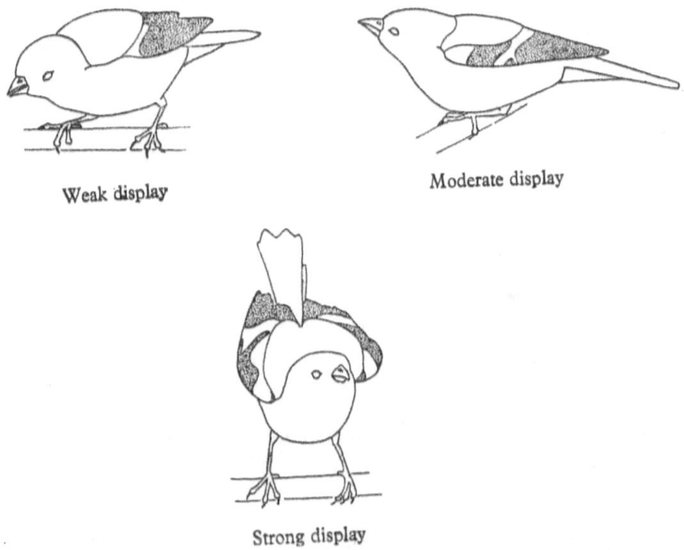

Fig. 9. Three chaffinch threat postures of increasing intensity. Note the continuous gradations of the movements of the feathers and wings. (From photographs, after Marler[59].)

between complete exposure and concealment, a continuously varied gradation of display is possible, a development of great importance in communication (fig. 9).

Many visual signals appear in an all-or-nothing way. That is to say, each is usually true to one type and seldom grades into other signals.[34] Chaffinches have about eight special visual displays. Some of these, like the courtship displays, are either shown completely or not at all. And if we consider their communicatory function, this fits with the requirements, for a bird is either ready for courtship and mating or not.

In fighting the circumstances are different. Here each opponent is repeatedly faced with the question whether to withdraw or to press the attack. Rather than a single constant display, it is more useful to have two, grading into each other, one associated with fear, the other with aggression. In this way the slight changes in a bird's confidence as it advances and retreats are immediately manifest in its signals, and so communicated to its opponent. We can see this exchange in the close-range fighting of chaffinches in winter, and also more noticeably in the 'head-up' display with which males defend their territory. In early spring a male is commonly seen driving an opponent back with vigorous, confident display. As he advances across the boundary, elements gradually drop out of his display, to be replaced by others. When finally he gets too far into his neighbour's territory, he hesitates, his rival takes his chance and drives him back again. In this fashion the two males may spend half an hour going to and fro across the boundary.

The effectiveness with which these graded displays communicate subtle changes in mood depends on how many different changeable elements there are. The function required of the aggressive displays of chaffinches is relatively simple, and the number of transitional changes through which they pass is small. The more complex the function becomes, the more elaborate we may expect the transitions to be.

In the visual displays of the black-headed gull, Moynihan[35] discovered transitions between not just two, but several types. Here there may be several complete changes from one type to another as the relative degrees of fear and aggression change (fig. 10). As a result, slight changes of the bird's mood are registered with extraordinary accuracy. Yet even this is not the sum of the black-headed gull's visual signals. There are probably at least seven more, giving an approximate total of at least twelve, with many transitions. An equally elaborate visual language occurs in some ducks. The common mallard has at least twelve visual signals and probably more, with again many transitions which can be correlated with mood.[36]

The head and face seem to be the most important of all the parts of the body used in visual communication, perhaps partly because

Fig. 10. Some displays of the black-headed gull, all involving varying degrees of conflict between aggressiveness and fear. Aggressiveness is relatively stronger in the postures on the left. The 'upright' and the 'anxiety upright' are associated with moderately strong aggression and fear, increasing in the 'oblique' and 'forward' postures, and reaching a maximum in 'choking'. Several of the phases grade into each other, providing continuous series of visual signals. (After Moynihan[85].)

they can be fully exposed to the opponent while still keeping him in full view. Not only can the way in which animals look at each other be very expressive, but also the head often carries the main fighting weapons. In many species the face plays an important role in another aspect of social communication, namely individual recognition. Disguise of either a chicken or a jewel fish on the head is much more likely to prevent familiar individuals from recognizing each other than alteration of any other part of the body,[37] and Nice[38] has listed numerous other examples. This is of course especially true of ourselves.

Darwin's interest in communication largely centred on mammals, and facial expressions received much attention, especially those of dogs, cats, apes and man. The tendency to nocturnal habits in this group discourages any great development of visual communication. As Schloeth[39] has recently pointed out, the first act of many of them on meeting for the first time is to smell or lick at the nose, or the anal and genital regions (fig. 11). Nevertheless, visual signals do occur. The hamster has at least six displays,[40] and in the wolf there are many more.

Fig. 11. Two stages in the meeting of male and female Grant zebras. First the naso-nasal contact, then the naso-genital contact. (After Schloeth[39].)

The special communicatory movements of many mammals, such as ungulates, are largely confined to the ears and tail.[41] The general body posture often plays a part as well,[42] but only in animals with a more highly developed social behaviour, such as some carnivores,[43] does the face come to play a dominant role (fig. 12).

As in birds, the greatest development of communication signals occurs in mammals with a complex society. Wolves, for example,

Fig. 12. Variations in the bodily and facial expressions of cats. In each section, aggressiveness increases to the right and fear increases downward, to give a continuously graded series of signals which is ideally suited to communicate subtle changes in mood. (After Leyhausen[43].)

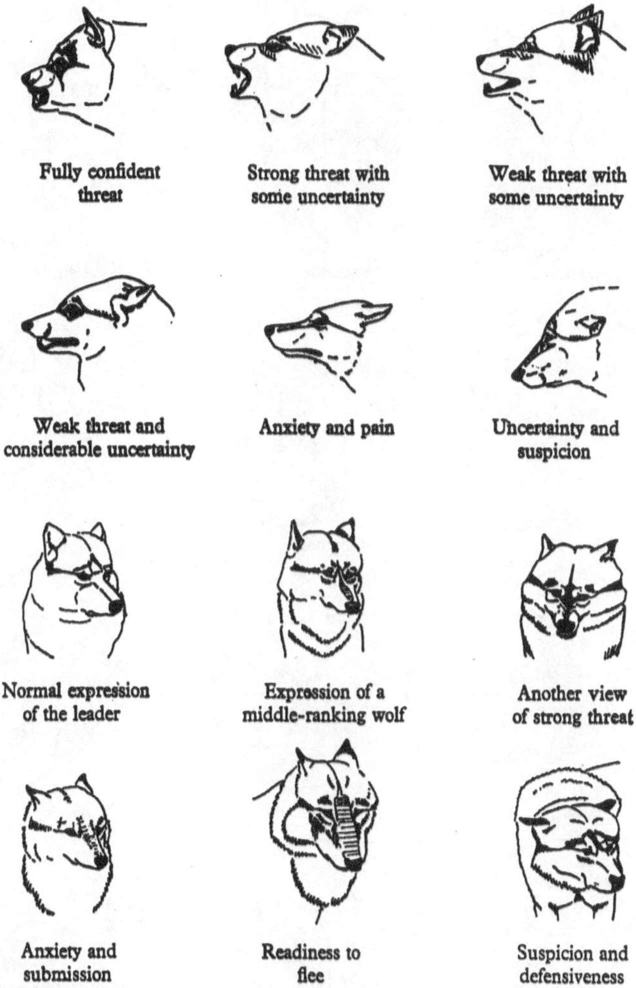

Fig. 13. Facial expressions of wolves. Slight changes in mood are accompanied by many variations, especially in the position of the ears, mouth and eyes, the changes emphasized by marks on the face. (After Schenkel[44].)

exist in highly organized groups with many communal activities. From the careful studies of Schenkel[44] we know that they use at least twenty-one communicatory signals, of which fifteen probably involve some visual elements, the others being olfactory and tactile. They take many forms, and the face is particularly expressive. Eyes, ears, mouth and the creases of the face all co-operate to give

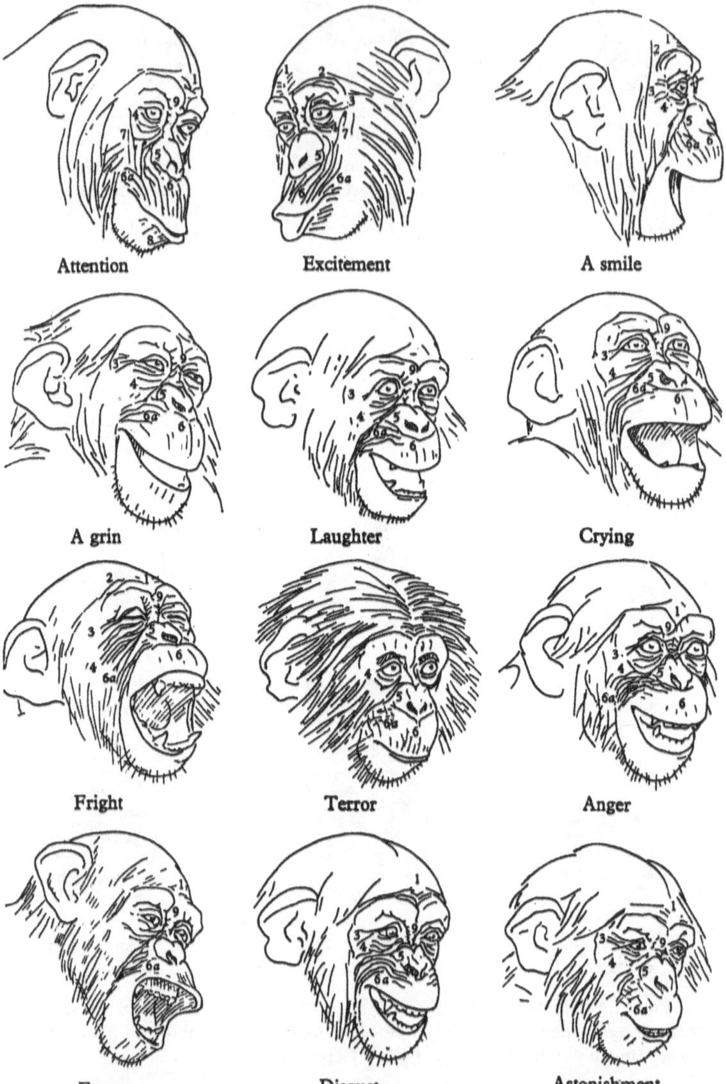

Fig. 14. Facial expressions of a young chimpanzee in various moods. Some of the creases are marked with numbers to emphasize that each is by no means confined to one expression. (After Kohts[45].)

a remarkable variety of displays, aided by the patterns of marks on the fur (fig. 13). Just as in the black-headed gull, these signals give a precise means of communicating subtle changes in mood.

When complex social habits are associated with a high intelligence, as in chimpanzees and ourselves, this trend of development may reach even further. Facial expressions become supremely important, and creases and hair tracts emphasize the disposition of the muscles. Here again expressions associated with different moods are not

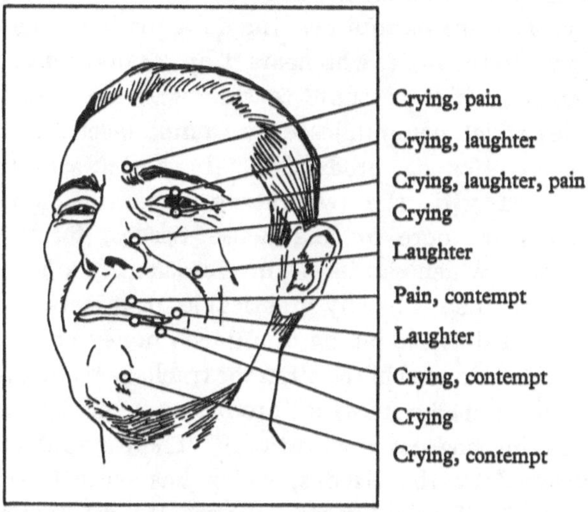

Fig. 15. A diagram to illustrate that, just as in the chimpanzee, the same fold or wrinkle in the human face may be involved in several expressions. (After Frois-Wittmann[46].)

sharply demarcated, but grade into each other. Furthermore, in her remarkable studies on chimpanzees reared by hand, Kohts[45] has shown that there are many elements shared by several expressions (fig. 14), a conclusion also suggested by studies of human expression (fig. 15).[46] This in turn helps to explain a point which Darwin noticed, that when people are asked to judge what emotion is portrayed in a photograph of the human face, they make a surprising number of errors. There is a great deal of ambiguity in expressions, so that some of them must be seen in their full context before they can be understood. The recognition of human expressions is thus an elaborate process, and it seems likely that visual signals are as important and as complex in man as in any animal, a fact which is sometimes forgotten in discussions of human communication.

Auditory Communication

Communication by sounds presents a new set of problems, particularly with regard to production. Certain animals can make use of sounds which arise incidentally from some other function of the body. In some mosquitoes, the noise made by the female serves as a signal to the male, who hears it by means of his specially adapted antennae, which resonate to this particular note.[47] But in most species which communicate by sound, special apparatus has been developed for its production. In vertebrates this is generally associated with the respiratory apparatus, since air columns, enclosed in more or less dense tissues, form ideal resonating chambers, which can be set in vibration by a flow of air.

This method is hardly practicable for insects, though it has been suggested that the piping of a queen honey-bee may arise through vibrations induced in the air in the trachea.[48] Instead many of them have developed devices for producing sound by friction, which sets resonating surfaces in vibration.[49] There is another unique method, developed by the cicadas, which has recently been explored by Pringle.[50] A pair of drums or tymbals is set in vibration by special musculature. Under their influence the tymbals buckle and then click back at a high rate to produce a loud buzzing sound, a large tracheal air cavity beneath each acting as a resonator.

Fish have also solved the problem of sound production in a variety of ways. Some again revert to a kind of stridulation, grinding their teeth or vibrating some part of the skeleton against another.[51] The sculpin, for example, is able to produce a low-pitched hum by vibration of bones in the neck.[52] The resonating chamber is provided by the air bladder. It may be set in vibration by muscles in the wall, or by squirting air from one part of the bladder to another through a constriction. The trigger fish goes even further by beating an external membrane with the pectoral fins. As a result of these various devices, the sea is full of grunting, booming, croaking, crackling, whistling and buzzing sounds, a fact only realized since the last world war, through the investigation of marine sounds for military purposes.[53]

Our main concern is not with methods of sound production, but with the nature of the signals produced. Before we can consider this, it is necessary to review some of the fundamentals of sound and of hearing. The rate of passage of a sound wave varies with the nature of the medium; water is ideal for this purpose, transmitting sound four times more quickly than air. A given sound signal will thus also travel further in water, and it is hardly surprising that fish should have exploited the many possibilities which this situation presents.

Any sound wave has two basic properties, one the displacement of the particles or molecules of the medium, the other the change of pressure with which this is associated. While the ears of vertebrates are adapted to perceive the pressure change, those of insects register particle displacement, and this difference has far reaching effects on the ways in which sounds are heard. Consider the process involved in the most elementary function of sound signals, enabling two animals to find each other when other senses are not available as, for example, in an opaque environment. Sound location is achieved in quite different ways with the two types of ears.

Particle displacement necessarily involves direction, and this fact can be exploited in order to find the sound source. The hearing organ of a grasshopper will respond much more strongly when it is placed at right angles to the direction of a sound than when in line with it. So all that the grasshopper needs to do is to turn to and fro, listening to the sound intensity in various directions. Although it can do this with only one hearing organ, it normally uses two, and since these are directed towards the sides of the animal it only needs to steer along the line of weaker stimulation to find the source of the sound.[54] It will be obvious that, since any sound involves particle displacement, grasshoppers can locate sounds of all types equally efficiently, as long as they are within the audible range.

For vertebrates the situation is quite different. The source of a sound striking a hearing organ which responds only to pressure cannot be so easily located. An ear separated from the head would respond equally strongly however it was placed in the field of sound. The nearest approach to the method used by insects is when the head prevents the sound reaching one ear and the direction from which the sound comes is judged by the changes in intensity as the head is moved. However, this is at best an inaccurate method and normally location of sound in vertebrates relies on comparison between the stimulation of the ears on each side of the head.

A difference in the stimulation may come about in three ways. First, the head may hinder the sound from reaching one ear, resulting in a sharp difference in intensity at the two ears, from which the direction of the sound can be inferred. Secondly, if the sound comes from one side of the head, there will be a time difference between arrival at the two ears. This is very brief, but nevertheless can be perceived. Finally, there may also be a difference of phase at the two ears—that is to say the rapid changes of pressure making up the sound will occur at different moments in the two ears, and this again can give information about the direction of the sound. We can show that in contrast with insects, the detailed characteristics of a sound have a considerable effect on the efficiency of these different methods of location.[55]

Consider sound intensity at the two ears. The head acts as an obstruction in the sound field, placing the ear farthest from the source in a kind of sound shadow. However, Lord Rayleigh showed long ago that this shadowing only becomes appreciable when the obstruction is of the same order of size as the sound wavelength, or larger. In ourselves this effect only becomes appreciable with quite high-pitched notes, about an octave above the highest note of a violin. So for low-pitched sounds, this method is of little use in location.

With phase difference the reverse is true, this method being confined to low-pitched sounds. There are two reasons for this. Ambiguity arises when the wavelength is less than the distance between the ears, differences of phase becoming difficult to interpret because several oscillations may succeed before a given wave reaches the further ear. The other reason concerns the characteristics of the nerve leading from the ear, which carries only a one-to-one representation of the sound oscillations when the sound is low-pitched. After each oscillation, the nerve is unresponsive for about a thousandth of a second. So if there are more than a thousand vibrations per second in the sound, some of them are not represented by impulses in the nerve. So once again comparisons of phase will break down with sounds pitched above this critical value. In man, the limit tends to be between 1 and 2 kilocycles per second, and may be somewhat higher in birds.

The last method, relying on time difference, is unaffected by sound pitch, as long as it is within the audible range. However, it is affected by the way in which the sound starts and ends. To compare accurately the time of arrival, it is essential that there should be sharp discontinuities which can be perceived at the two ears. So a short repetitive sound will be most readily located. If we combine this with the requirements of the other two methods of location, we can derive the ideal sound needed for a vertebrate to find its source. It will include a high pitch for location by intensity difference, a low pitch for location by phase difference, and it will be sharply broken and repetitive for location by time difference.

While sounds which insects produce for locating each other need no special characteristics, in vertebrates we may expect them to find direct adaptations for this purpose. Obvious examples are the songs of birds. Many of these serve to help females locate a mate and to enable males to keep rivals from their territory.[56] In both cases, ease of localization is an advantage, and we can show that the great majority of the songs of small birds tend to conform to this ideal. The same is true of many of the calls which birds produce.

When small birds discover an owl roosting during the day, they start mobbing it conspicuously, attracting the residents from quite a wide area. It appears that the function of this behaviour is to alert others to the possibility of danger at this particular place, and once again any calls used need to be readily locateable. Many ornithologists will have noticed that these mobbing calls often have a ticking or clucking quality, involving just those characteristics which are needed for easy location (fig. 16). One could list many other calls with adaptations in this direction.

It is also the experience of ornithologists that some bird calls are exceedingly difficult to locate, and it was the writer's encounter with this, in the call given by a male chaffinch when a hawk flies over, which first suggested the importance of questions of sound location. If we return again to this subject for a moment, it will be evident that it is possible to design a sound which presents the maximum difficulty for localization. First, the clues for time difference can be eliminated by using a longer sound which starts and ends gradually without any sudden discontinuities or breaks. Secondly, the pitch may be narrowed down to a particular level, somewhere above the value at which location by phase difference breaks down, but below the point where intensity differences become appreciable.

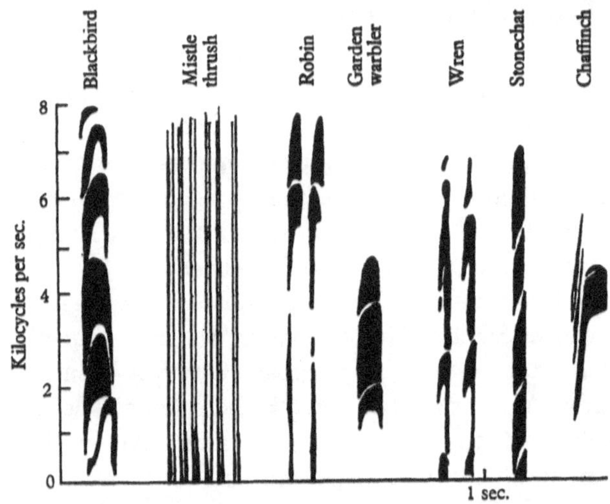

Fig. 16. Sonagrams of the calls of birds from several families given while mobbing an owl. The pictures should be read in the same way as a musical score, from left to right. The frequency in kilocycles per second is proportional to pitch, and the time scale on the base-line shows the duration of each note. Thus a thin vertical line corresponds to a short note with a very wide range of pitch, like a click or tap, which is the easiest of all to locate. On the kilocycle scale, middle C occurs at about 0·25 kcyc., two octaves above at about 1 kcyc. and four octaves above at about 4 kcyc., corresponding with the highest notes of the piccolo.

It has been shown that in ourselves this intermediate zone occurs around four octaves above middle C,[57] and notes pitched in this region are quite difficult to locate. If there is a similar weakness in sound location by hawks and owls, it could be of considerable value for small birds to exploit this. There seems every reason to think that they are affected in the same way. Their heads are smaller than ours, with the ears closer together, both increasing the difficulties of sound location. Probably the refractory period of the auditory nerve prevents the one advantage they might otherwise have, of being able to use phase difference location for higher-pitched sounds.

PLATE IV

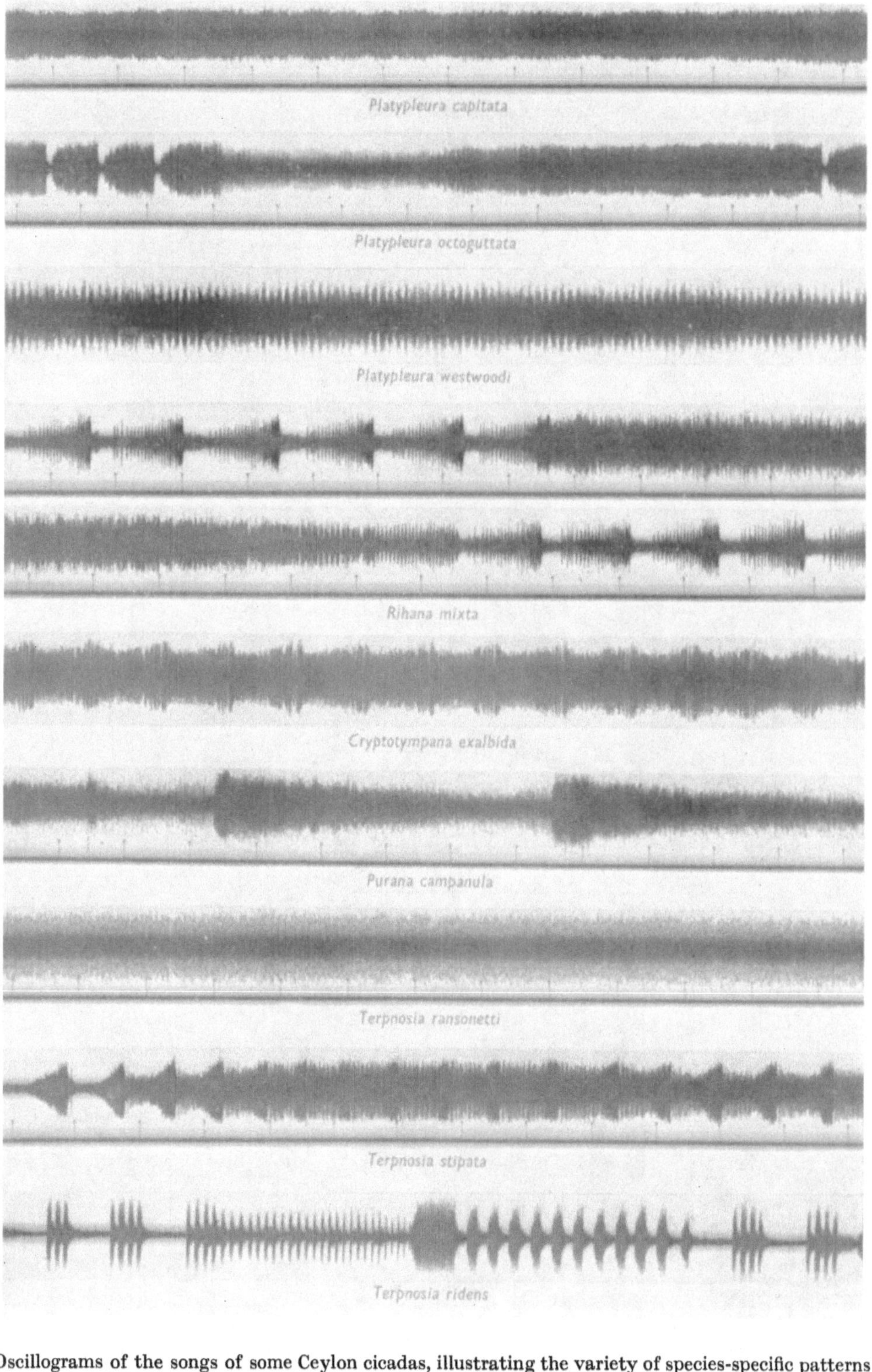

Oscillograms of the songs of some Ceylon cicadas, illustrating the variety of species-specific patterns of amplitude modulation. The time markers are 0·5 seconds apart. From Pringle.[64]

And even if they are more sensitive to intensity differences, which Schwartzkopff[58] has shown to be the case in song birds, there must still be a zone of minimum efficiency.

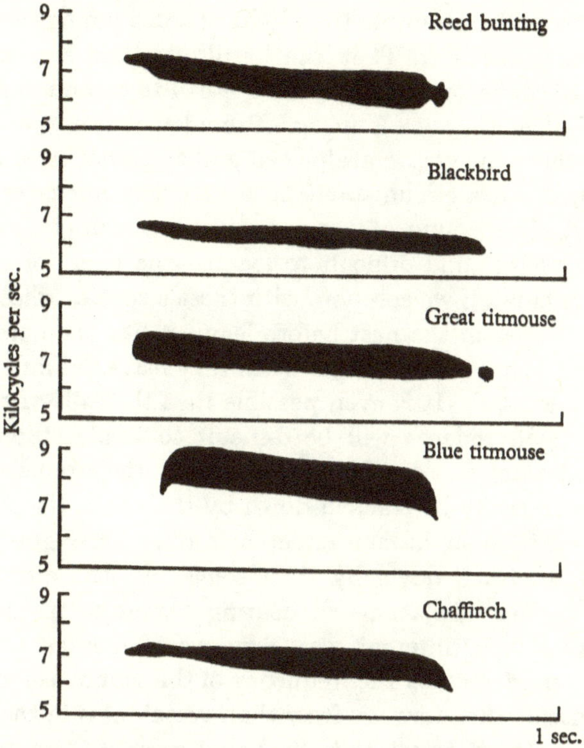

Fig. 17. The calls of five different species given when a hawk flies over. Each has a relatively narrow frequency range, somewhat over-emphasized in this and all other illustrations of sonagrams, as a result of the use of wide-band pass filters. They all sound like a high, thin whistle, and are difficult to locate.

Study of calls given by various song birds, including the chaffinch, when a hawk flies over, have precisely the structure we have predicted, namely a high-pitched pure tone, beginning and ending gradually, sounding rather like the squeak of a finger on glass (fig. 17). We must conclude that these small birds have evolved a call for this particular situation that is capable of warning others of their peril while at the same time exposing themselves to a minimum of danger. Even though small birds must also have difficulty in locating this call, this is no disadvantage, for their immediate response is to fly to the nearest cover, irrespective of the direction of the call or of the hawk.[59]

We can imagine other contexts in which this type of call would be useful. Consider for example the plight of some young birds when they first leave the nest. They can hardly fly, and are extremely vulnerable to predators, yet they need a call to communicate with the parent when they are hungry. Provided that the adults can remember where the young are placed and so do not need a call to guide them, this is a circumstance in which the 'unlocateable' call can be used. The young of various species certainly have notes which we ourselves find difficult to locate—the wren for example. On the other hand, if we compare with these a species whose young can remain longer in the nest before leaving, like the hole-nesting great tit, and so are in less danger when they leave, we find that the call is easily located. It is even possible that the calls of the more vulnerable small rodents will be difficult to locate, though it is striking that shrews, somewhat protected by their repellent taste and smell, can easily be tracked down by their calls (fig. 18).

It is also a familiar human experience that many grasshoppers are difficult to track down by their songs. As has already been pointed out, the 'displacement' hearing organs of insects locate sound sources in a different way from vertebrate ears, and the efficiency is unaffected by the frequency of the sound, as long as it is audible. Insects therefore are freer than vertebrates in their choice of frequency. It is interesting to note that most of them, nevertheless, lie between about 6 and 14 kilocycles per second.[60] In the absence of breaks to serve as time clues, this is probably a quite difficult frequency range for those birds which prey on grasshoppers to locate. One may speculate on the possibility that some insect songs are adapted so that they are both easy for insects to track down, and difficult for vertebrates. It is noteworthy, too, that crickets, with songs which are often broken and lower-pitched, around 2 to 4 kilocycles per second and therefore easier to locate, are largely crepuscular and nocturnal. Because of this, and their cryptic and often subterranean habits, they may be less exposed to predators than some other species.

Another curious characteristic of insect hearing, discovered by

Pumphrey,[61] is that, except with low-pitched notes, there is no response to the frequency of the sound. Instead they are adjusted to be sensitive to variations in the volume or amplitude. This has important implications when we consider the function of insect songs, which, like that of bird songs, is important in helping the sexes to find each other for mating. It is a prerequisite for such signals that they must differ distinctively from species to species, to avoid the possibility of confusion. Since sound frequency is not

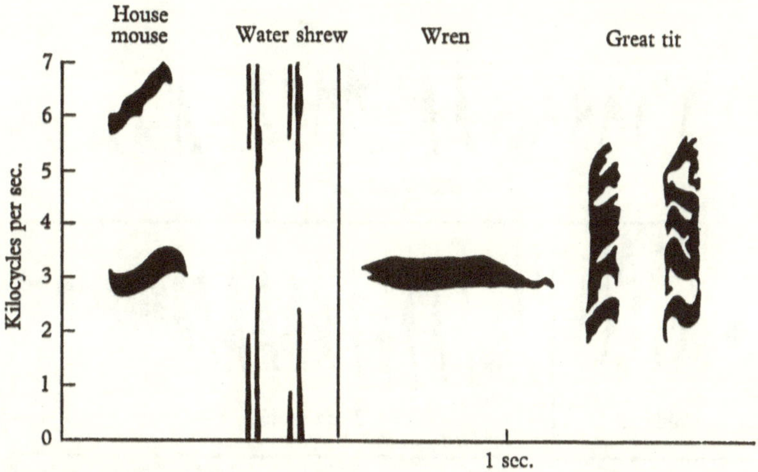

Fig. 18. A comparison of a house-mouse squeak and the chirp of a water-shrew, the first difficult to locate accurately, the second quite easy. On the right is a similar comparison of the calls of a newly fledged young wren and a great tit. Again the latter is easy to locate and the former is difficult, perhaps connected with the wren's greater vulnerability when it leaves the nest.

directly perceived, the specificity must lie in some other characteristic, and the pattern of amplitude variation is the obvious choice. There are two lines of evidence to suggest that this is the case. Comparative study of the songs of insects shows that there are more conspicuous differences between species' songs in the pulse pattern than in any other characteristic.[62] Haskell, for example, finds that in two very closely related bugs, *Kleidocerys resedae* and *K. ericae*, the pulse rate of one is eight per second, the other sixteen per second. The same is true of grasshopper songs[63] and also of cicada songs, as Pringle has recently demonstrated.[64] Here the various species exploit a wide range of variations, as they need to if each is to be distinguished without confusion (Plate IV).

Another line of evidence comes from experiments on the recognition of artificial sounds by insects. Various workers have shown that responses can be evoked by a variety of sounds produced mechanically, electrically or by mouth, often sounding to our ears quite different from the song they are supposed to imitate. In each case, it does not seem to matter which frequency is used, as long as the pattern of amplitude variation is true to type.[65]

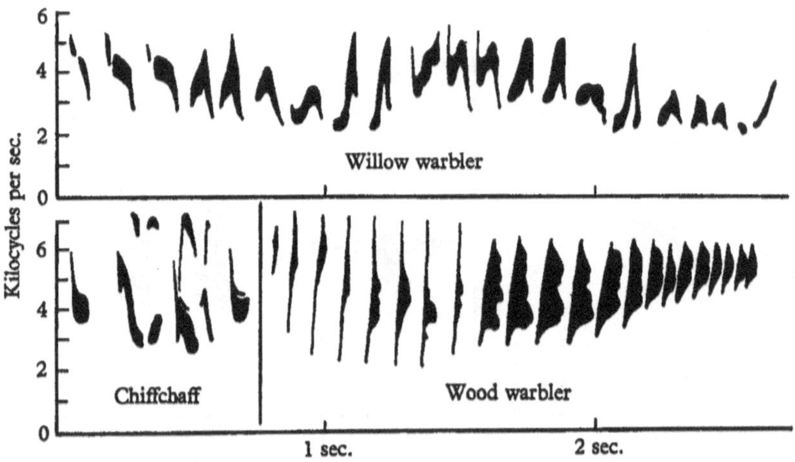

Fig. 19. Songs of three very similar warblers, first properly distinguished by Gilbert White on the basis of their song differences. Four chiffchaff phrases are shown, 'chiff-chaff-chaff-chiff'. The 'plaintive' song of the wood warbler is not given.

These observations imply that there must be a strong selection pressure favouring specific distinctiveness in these communication signals concerned with reproduction. There is also good evidence for this in birds. Ornithologists are familiar with the tendency for birds living together to have distinctive songs. In at least one case this has given the first clue to taxonomic separation, for Gilbert White[66] first distinguished three species of warbler on the basis of their song (fig. 19).

The Evolution of Bird Song

The possible effects of natural selection on bird calls would no doubt have received Darwin's attention, had he had the technical assistance which is available to us. We can still look to him for the kind of approach to adopt. In most cases our evidence can only be indirect, but we can learn a great deal from comparative study on the one hand, and from the study of variation on the other. There is obviously no direct evidence from the fossil record, but Darwin pointed out that domestication can sometimes give us information equally valuable, if we regard natural selection as represented by the animal breeder's conscious or unconscious selection of his breeding stock.

By a lucky chance we have one example of the domestication of bird song. Canaries have been bred in captivity in Germany at least from the sixteenth century. Since that time they have been transformed into a great variety of shapes and colours hardly identifiable with the greenish brown ancestor from the Canary Islands. There was also a great interest in bird song at this time, and, as an offshoot from the main stock, there arose a special breed, the roller canary, selected largely or entirely for its singing capabilities. To this day they are still matched in singing contests, championships being awarded to those which best satisfy the current fashion.

We can compare the song of the roller canary with both the wild canary and the border canary, the latter one of the breeds which has been selected for other characters (fig. 20). The first thing we notice is the resemblance between the wild and border canary songs. The only major difference between these two is the absence from the latter of the jumbled dissonant phrases which occur in the middle of the song of the wild bird. These are sufficiently harsh to the ear that we might expect even the fancier indifferent to song to eliminate them from his stock. Apart from this there are few consistent differences between the two, though of course both vary a great deal in detail.

Roller canary song, however, is very different from the other two. The pitch is lowered, the time pattern is transformed into long repetitive trills, the quality changed from the fluid notes of the wild and border canaries into harder sounds, sometimes with an almost metallic quality. This transformation is as great as any which Darwin describes in the form and plumage of doves or poultry, and it seems equally justifiable to conclude that bird song is also susceptible to change under the influence of selection.

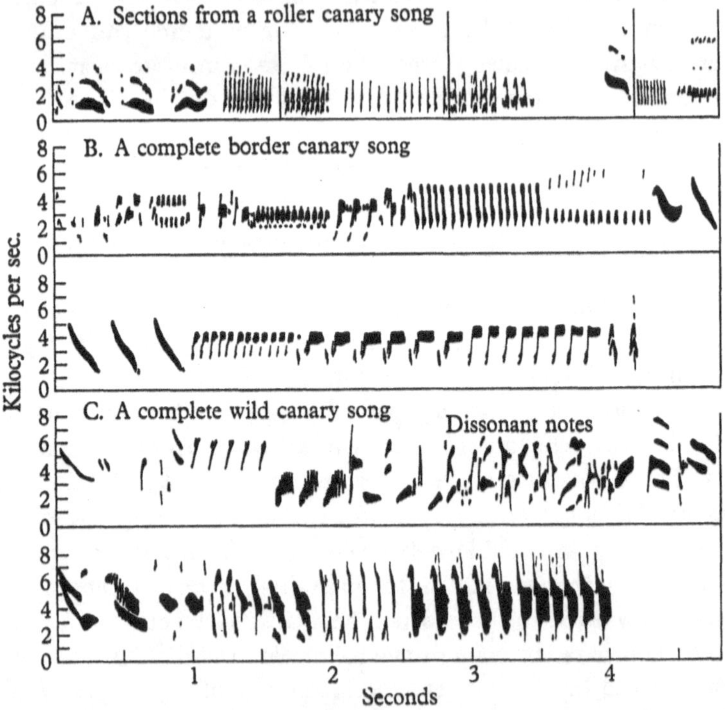

Fig. 20. Samples from the song of the roller canary, bred for many generations for song characters, the border canary, bred for appearance, and the wild canary, recorded in its home in the Canary Islands. Comparison reveals relatively little difference between the wild and border canary songs, except for the disappearance of the curious dissonant notes in the first half of the wild bird's song. Roller canary song, however, has changed radically, towards a lower pitch, a simpler and even more repetitive structure, with shorter intervals between notes, and as a result has a quite different sound.

The other prerequisite for evolution to occur is a certain degree of variation upon which selection can operate. With the signals of visual and chemical communication, variation is difficult to detect, since they cannot easily be described in quantitative terms. For sound signals this difficulty can be overcome with the aid of modern analytical methods. Though little has yet been done, the results are sufficiently promising to be reviewed in detail.

When we examine the repertoires of individual birds, it is remarkable how few sing exactly the same theme all the time. In Europe, perhaps the wood pigeon does, but this cannot be said of any of the song birds which the author has yet examined. The most monotonous and sterotyped singers among the warblers are perhaps the willow and wood warblers and the lesser whitethroat, but study reveals several themes for each bird. The corn bunting seemed at first to

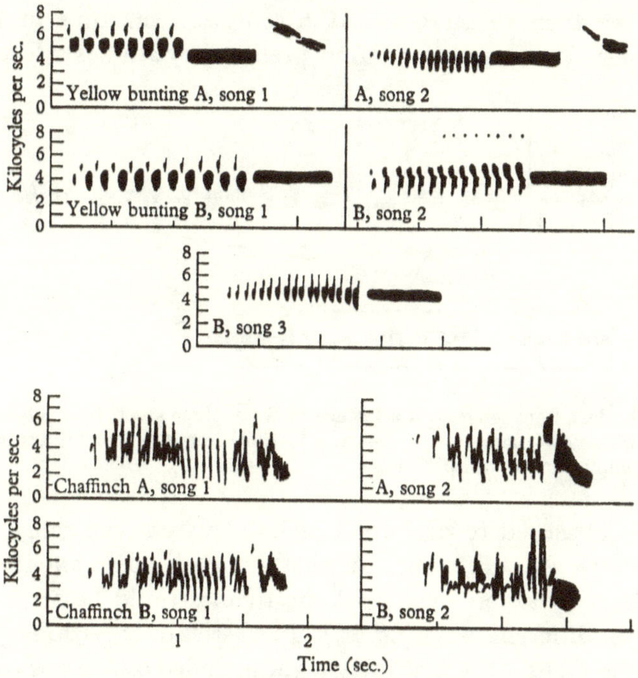

Fig. 21. Examples of yellow buntings with repertoires of two and three songs, and two chaffinches, each with two songs. The songs are given in bursts, several of one type, then several of another.

be an extreme example, but even here variations were discovered, though they could not easily be detected by ear.

In fact it almost seems to be a rule that these simple singers have several themes given in bursts, first of one theme, then of another. Typical European examples are the yellow bunting and the chaffinch, each individual having a repertoire of up to six songs (fig. 21). Among a total of seventy-one chaffinches, the mean number of songs per bird was 2·3. Many North American birds sing in the same way,[67] one of the best known being the song sparrow. In her intensive studies of this species, Nice[68] found from six to twenty-four songs in the repertoires of individual birds. Repertoires in the Carolina wren may be even larger; one bird is reported to have sung twenty-two different songs in twenty-four bursts of singing.[69] These are surprisingly large totals when we consider that these species give a first impression of being relatively uniform singers.

When we turn to more versatile songsters the repertoires are much larger. For example, in some preliminary studies a European

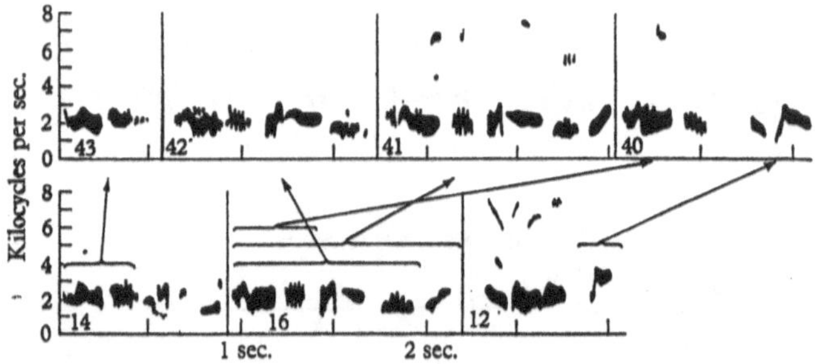

Fig. 22. The last four songs in a sequence of forty-three given by a mistle thrush compared with earlier songs in the sequence to show how new themes are produced by the recombination of old phrases.

robin was observed to perform a series of fifty-seven songs, every one of which was different. A mistle thrush gave twenty basic themes in forty-seven songs, but the themes could be arranged in a variety of different ways (fig. 22). In a series of fifty-three phrases from a song thrush, the first forty-eight included only a few repeats; then number forty-nine was a repeat of number two, number fifty was like number four, fifty-one like five, fifty-two like six and fifty-three like seven: so this bird probably had rather less than fifty themes (fig. 23). However, another bird gave only two repeats in eighty-five songs. And in the longest series studied so far, of 203 phrases, a song thrush gave sixty repeats; nevertheless, there were still signs that the repertoire was not exhausted, as can be seen from the distribution of the repeats:

Songs	0–50	50–100	101–150	151–200
Repeats	4	13	27	16

There was evidence of the same process which takes place in the mistle thrush, namely the recombining of old themes in new arrangements. The total of themes used at least once by this bird was 173, and allowing for up to three themes in each song there is a certain economy in building up 143 different songs in this way.

It is obvious, then, that there is ample variation of song in individuals, on which natural selection might operate. In fact the variation is sometimes unexpectedly wide. Is it simply that the limits imposed by selection are not very precise, and that some birds

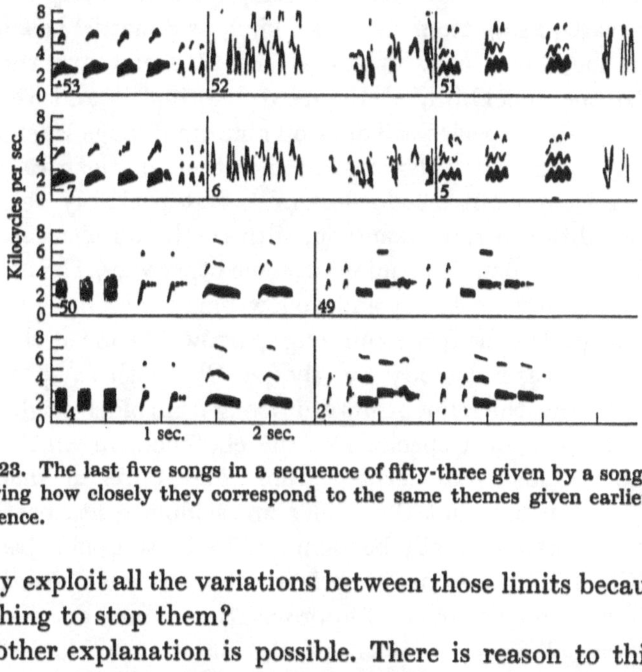

Fig. 23. The last five songs in a sequence of fifty-three given by a song thrush, showing how closely they correspond to the same themes given earlier in the sequence.

merely exploit all the variations between those limits because there is nothing to stop them?

Another explanation is possible. There is reason to think that birds recognize the songs of individuals they know. Male chaffinches distinguish between the songs of established rivals and newcomers, and it is possible that females recognize the song of a mate from the previous year. This could confer a definite advantage, for as well as the evidence that birds experienced in breeding produce more young,[70] considerable adjustment is needed when a pair forms for the first time. For individual recognition to occur, there must be a good chance that songs of any two birds will differ in some characteristic. The larger the repertoire, the greater will be the probability of achieving this. At least this is a plausible hypothesis, helping to explain the large number of themes which many birds sing, with perhaps some bearing on individual variation in the forms and markings of animals.[71]

For communication alone, it does not matter whether these variations are genotypic, as long as they remain stable for a reasonable period of time, so that companions can learn their characteristics. Evidence accumulating in recent years reveals wide variation in the relative contribution of learning and inheritance in the development of bird songs. Some develop quite normally when the birds are reared away from their own kind, as Sauer[72] has demonstrated in the European whitethroat and blackcap, and the same applies to the blackbird,[73] the reed bunting[74] and the corn bunting.[75] Other species will develop abnormal songs when reared in this way, as Scott[76] demonstrated in the American Baltimore oriole, and Lanyon in meadowlarks.[77] But probably the intermediate condition is most common, with the broad characteristics inherited, and the details acquired in some other way. The domestic canary, the European linnet and yellow bunting[78] have songs of this type, as well as the American song sparrow[79] and the chaffinch, the subject of careful study by Thorpe.[80] In these species the young bird learns the refinements of its song from older neighbours.

However, even in a species like the chaffinch, in which young males can be observed learning from older birds in their first spring,[81] we notice that the copies are seldom quite precise, so that each bird can generally be identified by its song in subsequent years. Either it makes an imperfect copy, or what seems more likely, there is some element of improvisation. Many species which develop normal songs in isolation also show individual characteristics. In the blackbird, Messmer[82] has shown that they arise both by learning from various neighbours, and by spontaneous recombinations of old phrases. Probably something similar occurs in the song of mistle thrushes (page 184). Even the simple and stereotyped inherited songs of the corn and reed buntings vary from bird to bird, and again an element of improvisation is probably responsible, unless, perhaps, one should regard the precise details as a matter of chance. If individual song characters are one of the ultimate objectives of this development, any of these methods will suffice.

The tendency to learn songs from neighbours results in the development of local dialects in many species,[83] including, among those mentioned above, the blackbird, yellow bunting, song sparrow and the chaffinch. Here the change from one dialect to another is most clearly marked across boundaries forming barriers to movement. In some of the isolated glens of Scotland, for example, chaffinch songs form quite distinct dialects. Reasonable stability is assured, since not only is the song learned after the male first sets up territory, but he also returns to the same place in subsequent years. Furthermore, there is a tendency to reply to rivals with a song of the same type, so that songs which are uncommon in the area may fall into disuse.

While variation in song may assist individual identification, it conflicts with the need for specific identification. Here a stereotyped song is the ideal requirement, differing as distinctively as possible from those of other birds in the area. The studies of Dilger[84] and Stein[85] have demonstrated how this specific identification is achieved in North American thrushes, both from comparative study of the songs and by observing the birds' responses to sound recordings. The most distinctive characters lie in the general timing of the song, the duration of notes, and the way in which the relative pitch changes during the song. Absolute pitch seems to be less important. The same is true of the songs of European thrushes (fig. 24) and probably of other birds such as the chaffinch.[86] Obviously the ease with which experienced ornithologists identify birds by their song also implies fairly rigid specific characteristics, even in those species with great individual variation.

It appears that the conflict between individual and specific recognition has been resolved by relegating the variation to characteristics of the song other than those which mediate specific identification. In the thrushes, the time pattern is of specific importance, while individual characters are represented rather in the details of pitch within the general pattern. In other birds the roles might be different, and this may even be true for another member of the thrush family, the small European robin. Possibly the main specific character here is the high pitch and unusual quality of the notes, for the time pattern shows much more individual variation than in some other members of the family.

If it is correct to infer that bird songs are subject to selection pressure encouraging specific distinctness, a strong mutual influence will be exerted by species which sing together. This may explain the considerable geographical variation of song which ornithologists have described,[87] since the neighbours with which a species lives may vary in different parts of its range. Unfortunately, variations of this kind are difficult to interpret, because changes in the avifauna are so complex. However, there is a comparison which can usefully be made, namely that between the songs of a species in a dense

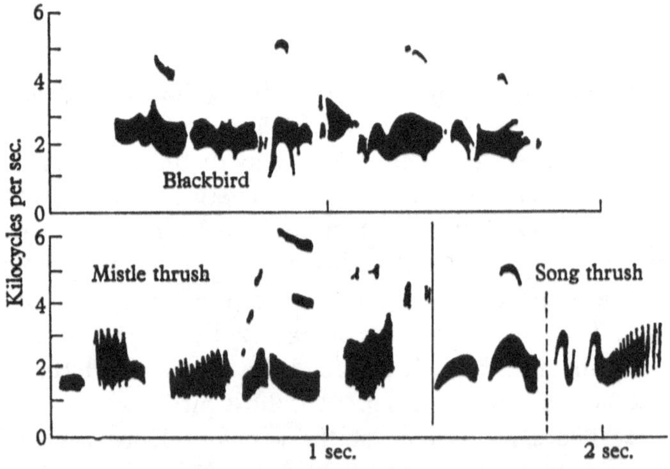

Fig. 24. Typical songs from a British blackbird, mistle thrush and song thrush, showing the essential similarity in the detailed structure of the notes. The species songs differ mainly in the pattern in which the notes are arranged.

continental avifauna and those of the same species in the sparse avifauna of a small island.

Lack and Southern[88] first drew attention to the Canary Islands as a subject for such a comparison, since the fauna consists of a sparse sample of predominantly European species or genera. They suggested that differences in bird songs between Europe and Tenerife might be explained by the reduced pressure for specific distinctness in the smaller avifauna, and visits to the Azores suggested a similar situation.[89] In 1956 the author visited Tenerife with a tape recorder, and was able to compare the bird songs with their European counterparts in detail.

The songs of some species hardly vary at all. Those of the turtle dove, great spotted woodpecker, corn bunting and blackbird do not seem to be affected significantly by island conditions (fig. 25). Others vary considerably, and there seem to be two main types of

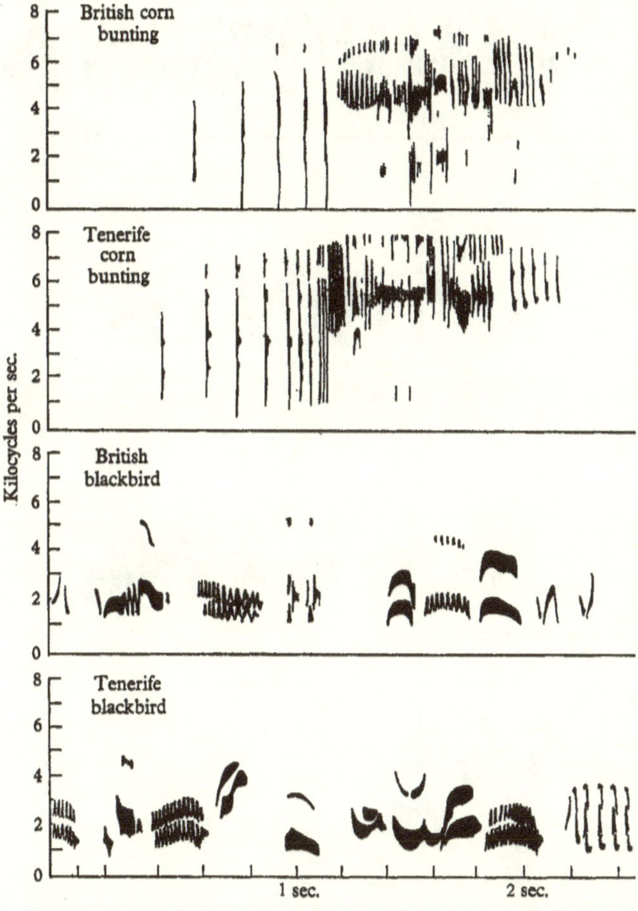

Fig. 25. A comparison of the songs of corn buntings and blackbirds in Britain and in the Canary Islands on Tenerife. There seem to be no consistent differences.

change, some becoming simpler and more uniform, others becoming much more variable. An example of the first group is the chaffinch, which occurs in two forms on Tenerife: the blue chaffinch, restricted to the pine forests, and the Canary Island chaffinch which occurs in the laurel woodlands. As already mentioned, the European

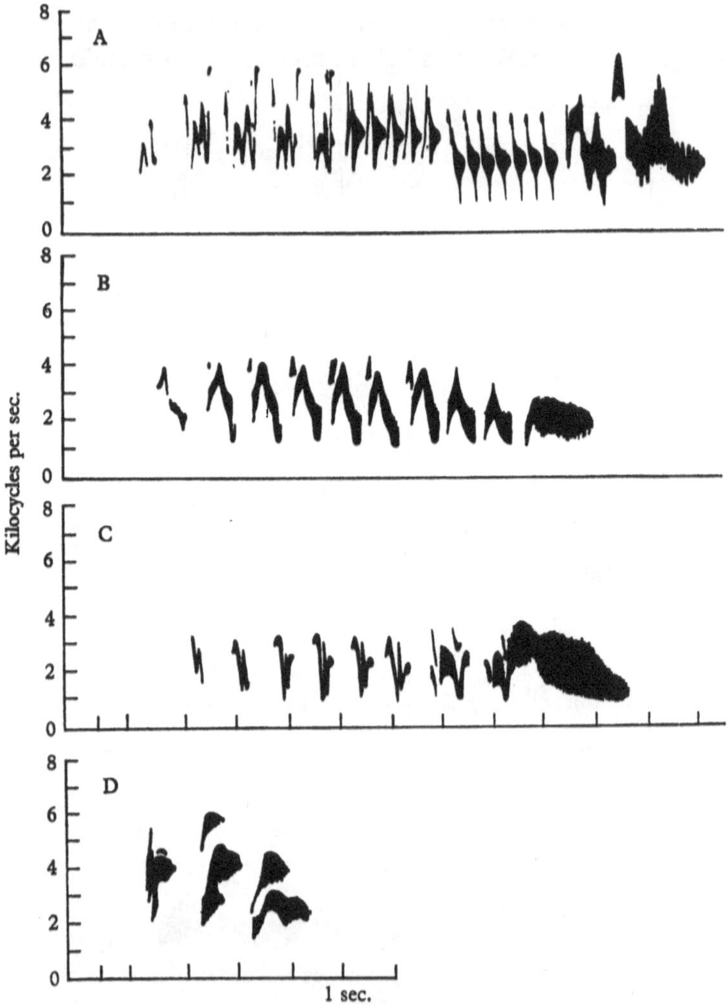

Fig. 26. Examples of chaffinch song. A is a typical song of a wild British bird. B is the song of such a bird reared out of hearing of any other chaffinch song (from experiments of W. H. Thorpe), which lacks many of the normal details. C is a song of the blue chaffinch (*Fringilla teydea*) on Tenerife. The resemblance with B is striking. D shows the 'chink' calls of a normal British chaffinch, a chaffinch reared in isolation and a Canary Island chaffinch (*Fringilla coelebs canariensis*). Again we see a resemblance between the calls of the island bird and the isolated British form.

chaffinch elaborates its song considerably by learning, and an isolated bird develops only a simple one.[90] Apparently the blue chaffinch has reverted to this simpler innate song, with a 'churr' or 'buzz' at the end, instead of the more musical flourish (fig. 26). The same has occurred in the Canary Island chaffinch, and also in the other subspecies found in the Azores.[91] The social call of the chaffinch has met a similar fate. In Britain this call is often abnormal in isolated birds, and the Canary Island, Azores and Madeiran chaffinches all have only this innate form. It appears plausible that the smallness of the island avifauna removes the need for elaboration by learning, which takes place in Europe, and consequently the simpler and less distinctive innate forms alone occur. Yet the capacity to learn may still exist, for blue chaffinches retain slight individual characteristics in their songs.

The song of the blue titmouse shows another kind of change. In Britain it is relatively uniform, and can easily be distinguished from the five other members of the same genus with which it lives. On Tenerife it is the only titmouse present, and has a bewildering variety of songs. Some sound more like the coal tit than any other, perhaps suggesting a closer relationship with this species than has been assumed. But the variation is very great, and the same bird may give several different songs in succession, many of them reminiscent of the songs of other species of titmouse in Britain (fig. 27). In the Azores, the voice of the goldcrest changes in a similar way.[91]

The chiffchaff is another case, the island bird's songs becoming more variable, sometimes sounding rather like its close relative, the willow warbler (fig. 28). The relationship between these two species has been of the greatest interest since Gilbert White first distinguished them by their songs. So similar are they in appearance, that the only sure way of identification apart from the song is to examine the leg colour or the wing formula. The birds themselves must rely heavily on voices for identification, which in Britain are clearly distinguishable. But where the willow warbler is absent, as on Tenerife and in southern Spain,[92] the chiffchaff's song becomes more varied, presumably because the need for a rigidly defined song pattern is reduced.

Unlike the chaffinch, the chiffchaff and the blue tit seem to be restrained from expressing their full vocal potentialities in the dense

European avifauna. On Tenerife these strictures are removed, and the songs become more varied. Yet we can relate this to what happens to the chaffinch, for both types of change would involve a loss of efficiency in communication, if they took place in Europe, by hindering specific identification.

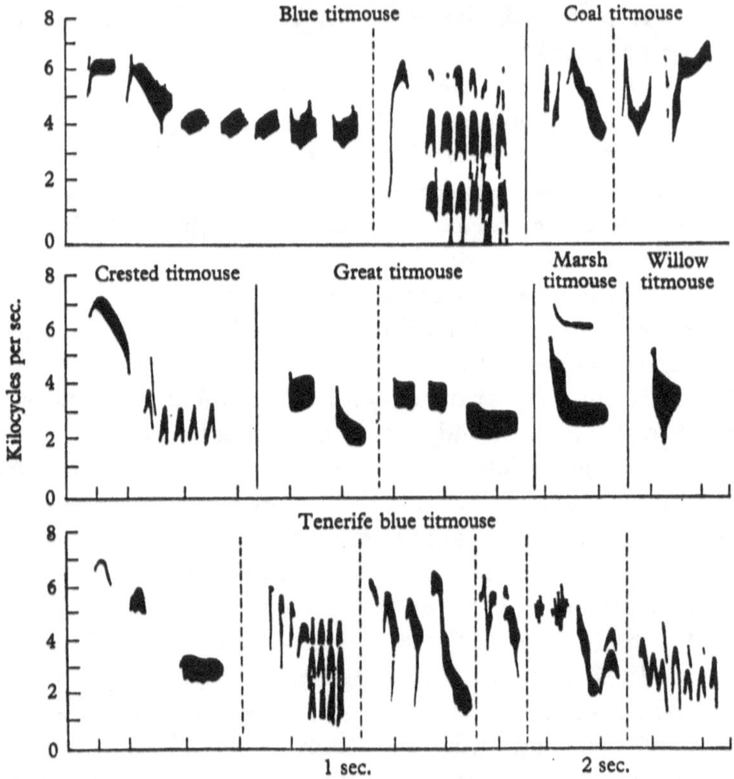

Fig. 27. Examples of songs of six British titmice, compared with a sample of the varied sounds of the Tenerife blue titmouse. Note the resemblance to several British species, especially the coal titmouse.

There seems to be no relationship between the tendency for song to change on islands, and the relative contributions of inheritance and learning to song development on the mainland. The corn bunting and the chiffchaff both inherit normal song in Europe, yet while the song of the former is unchanged on Tenerife, that of the latter is different. Evidently there must be plasticity in the genetic mechanism controlling chiffchaff song. While learning plays more part in the normal song of the blackbird and the blackcap than in the chiffchaff, their songs on Tenerife are not significantly changed. However, we do notice marked change in the two species in which learning plays a dominant part: the chaffinch and probably the blue

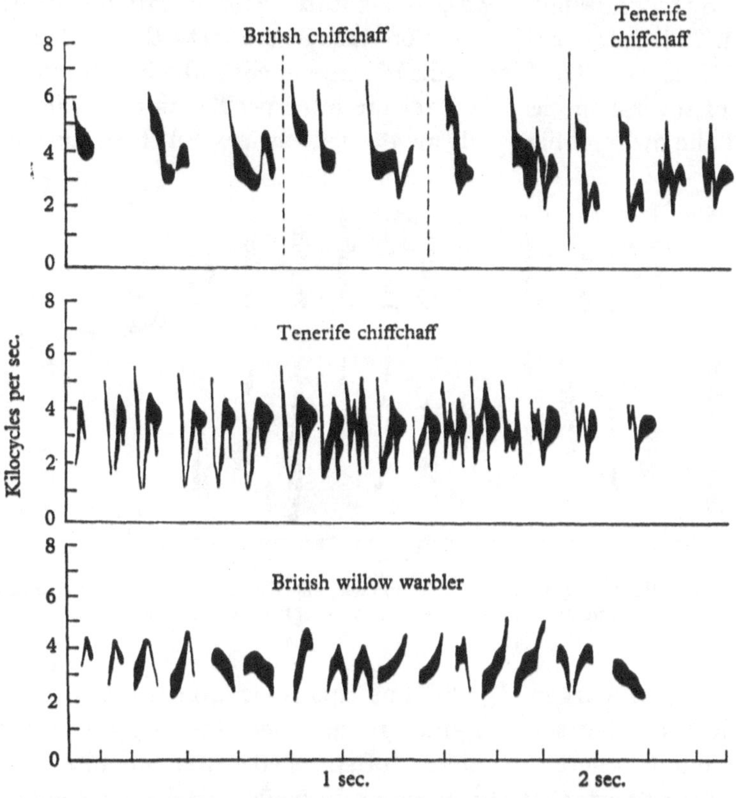

Fig. 28. Three samples of chiffchaff song, and one of willow warbler song from Britain, to be compared with two types of song from the Tenerife chiffchaff. The tempo of the second type agrees more closely with the British willow warbler than with the chiffchaff.

tit.[93] However, our knowledge of the roles of inheritance and learning in song development is still very limited.

While discussing bird voices, it is natural that songs should receive most attention, for their function requires that they be conspicuous and distinctive. It should not be forgotten that there may be different selection pressures operating on other sound signals.[94] Consider the alarm calls of small birds discussed earlier. Here specific distinctness has been discarded altogether, and we can see that this is consistent with their function. Not only is there no advantage in having a specific call, but, if these species live together and are endangered by the same predator, there is much to be gained by having similar calls, to facilitate communication between species. This interspecific function may also affect other calls in a more subtle way. The flight calls of some finches which form mixed flocks of several species in winter are interspecific, and the same is true of the owl-mobbing calls mentioned earlier. But here the con-

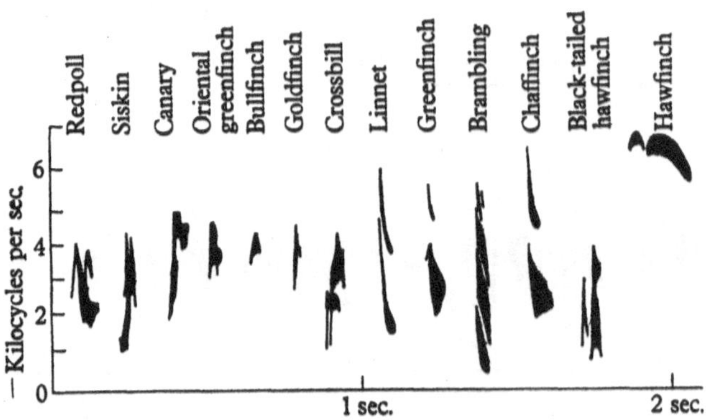

Fig. 29. The 'flight' calls of a number of finches, showing how they tend to conform to the same basic pattern, with the exception of the hawfinch.

formity is less marked (fig. 29), and this is also consistent with the fact that in other seasons and circumstances they play a part in reproduction, where some degree of specificity is an advantage.

Thus it appears that the direction in which vocal signals evolve is affected by a number of factors which can exert a selection pressure. As well as the obvious requirements of audibility, the need to avoid masking by other natural sounds, and the adaptations to facilitate or hinder localization, the relative advantages of contrast with or resemblance to the signals of other species will vary with the function of the signal. The kind of 'sound environment' in which the signal has to operate will exert a strong influence here. The more complex this is, the more elaborate the signal has to become if it is to remain distinctive. And finally, the advantages of individual identification by voice may help to encourage the high degree of variation in some aspects, which is such a prominent characteristic of many bird songs.

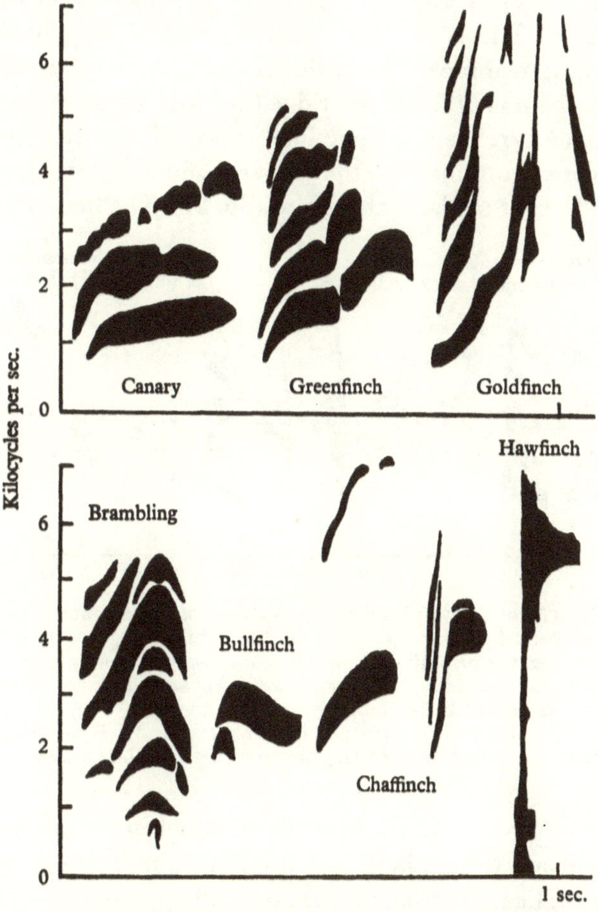

Fig. 30. The 'social' calls of various finches, all showing affinities with the same basic pattern. Those of the chaffinch and hawfinch, also used as mobbing calls, are the most divergent, but in the chaffinch at least this can be traced back to the same type (cf. fig. 31).

There is one other method by which we can sometimes uncover clues about the evolution of sound signals, namely by study of the way in which they develop in the young bird. The chaffinch can serve as an illustration. In some respects this species is a rather divergent finch, and there has been discussion about whether it is really a close relative of the Carduelinae at all. Evidence from morphology and behaviour suggests that it is more closely related to the finches than to any other group[95] and studies of its voice support this.[96] The social and mobbing call of the chaffinch is rather unlike that of most Cardueline finches, and at first sight the gap between them is difficult to bridge (fig. 30). However, study of the development reveals a complete series of transitions, some obviously corresponding to the general finch type, from the juvenile call to that of the mature bird (fig. 31). The first call in this

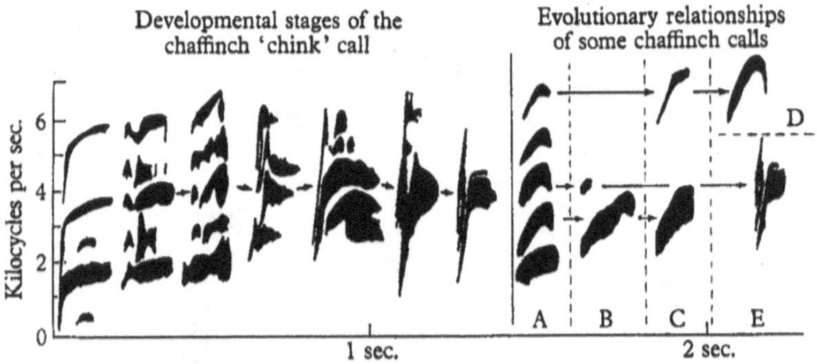

Fig. 31. On the left are shown developmental stages of the chaffinch 'chink' call, relating it to the basic finch type (fig. 30). On the right are suggested evolutionary relationships between the primitive alarm call A, used only by young chaffinches but widespread among other finches, and certain adult calls. It appears that different harmonics have been selected for the 'huit' call (C), the 'seee' hawk alarm (D; an unusual form of the call seen in fig. 17), and the 'chink' call (E). B shows an unusual form of the 'huit' call, which may correspond to an intermediate stage.

series, which exists only in the juvenile chaffinch, gives another clue, for it corresponds to an alarm call used in a similar form by most adult finches. The male chaffinch has instead adopted the call which is difficult to locate. Once again we can trace a developmental relationship between the original and final forms (fig. 32).

There are still other factors affecting the evolution of sound signals which have yet to be explored. The recent studies of Cullen[97] on the visual signals of the kittiwake have shown the extent to which communication methods may be modified by changes in ecology. With vocal signals, the differences between the songs of birds in open and wooded country (the former usually long and sustained, the latter shorter and repetitive) are so consistent that they must have significance. And it may be that body size and feeding habits affect the voice. If a bird is adjusted to a rapid metabolic turnover, and must be constantly collecting food, it cannot afford to interrupt this with long bursts of singing. Gibb[98] has pointed out that among the British titmice, the little coal tit feeds and sings at the same time, the larger blue tit sometimes stops

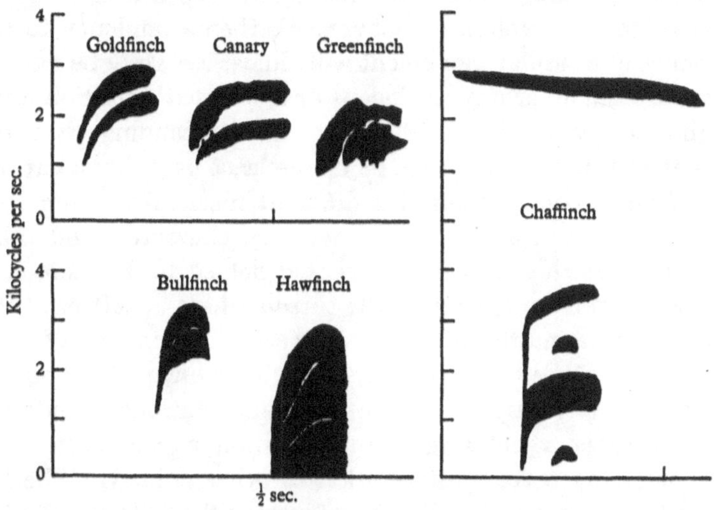

Fig. 32. Hawk alarm calls of various finches, showing the essential similarity from species to species. The chaffinch only uses this call when young (below), and in the adult male it is replaced by the type of call difficult to locate (above), which none of the other species has developed.

other activities to sing, and in the still larger great tit this is the normal practice. It may be no accident that the great tit has the most elaborate vocabulary, and the coal tit perhaps the simplest. There is no doubt that many relationships will be discovered in the future between the ecology of animals and the detailed form of their communication signals.

The Origins of Communication

Darwin suggested three principles governing the origin of communication signals, each of which still plays an important part in our approach to this problem to-day. The first he called the principle of associated habits. The second, the principle of antithesis, and the third, the principle of direct action, reflecting the constitution of the nervous system. Each of these coincides with ideas advanced by Lorenz,[99] Tinbergen,[100] and others, anticipating them in a remarkable way, though we can see now that Darwin was led astray in two directions. Instinctive or habitual movements, he thought, become fixed or inherited through continued practice. As he put it, 'when any sensation, desire, dislike, etc., has led during a long series of generations to some voluntary movement, then a tendency to the performance of a similar movement will almost certainly be excited whenever the same, or any analogous or associated sensation, etc., although very weak, is experienced; notwithstanding that the movement in this case may not be of the least use'. It is curious that he should have visualized a different mechanism from that proposed for the evolution of morphological characters and there is no doubt that the confusion so caused helped to dissuade subsequent workers from taking up the thread where he left off.[101]

There is also a subtle difference between the principle of antithesis and the other two which seems to have escaped his notice. It is clear that any response which has manifestations outside the animal's body is a potential communication signal, since it is capable of giving certain information to other animals. Whether natural selection encourages it in this function depends on whether the information carried has any significance for others. Thus there are two distinct problems, one concerning first origins, the other the effects of natural selection on these origins, and it is essential to keep them separate. While Darwin's first and last principles are concerned primarily with origins, the principle of antithesis, suggesting that some movements occur because they are in complete opposition or antithesis to some other attitude, is concerned with the effects of natural selection. The significance of this difference will become clearer later, when some examples have been considered.

What is known of the origins from which communication signals may have sprung? With sound signals we know very little. No doubt the stridulation of insects first arose by way of vibrating movements of the wings and other parts of the body during courtship.[102] Similarly, the sudden intake of breath of many vertebrates in alarm or pain to which Darwin drew attention may have been a precursor of some vocal signals. More is known of the origin of chemical signals, and Haldane[103] has recently drawn attention to the light this may throw on some of the mechanisms of internal communication within the body by the endocrine system. The resemblance in chemical structure between substances used as signals and the steroid sex hormones suggests that the latter were originally concerned with communication between individuals. Furthermore, the chemical receptor in tunicates which responds to these inter-individual signals is the ciliated pit, believed to be homologous with part of the vertebrate pituitary, the centre controlling the endocrine system.

Some chemical signals have undoubtedly arisen from habits of excretion, defecation and perspiration.[104] In many mammals there are special methods of distributing urine and faeces, and it has already been suggested that excretory products can provide specific and even individual odours, as a result of variations in diet or metabolism (page 153). The scent glands which mammals carry on various parts of their body[105] may well have originated as modified sweat glands.

Most attention has been given to the origin of visual signals, both by Darwin and subsequent workers, and several recent papers have reviewed this field.[106] At least four different categories have been suggested. Most widespread are the preparatory, introductory or incomplete movements which characteristically prelude certain activities. Many elements in the visual displays of birds must have originated in this way. These are the actions now known as *intention movements*, which Darwin described as 'serviceable associated habits'.

There is a second group, explained by Darwin as results of the direct action of the excited nervous system, somewhat neglected by recent workers (though not by psychologists[107]) until Morris[108] drew attention to it once more. During emotional excitement the autonomic nervous system produces many effects which are manifest outside the body. Defecation, urination and perspiration have already been mentioned above, as well as the changes in respiration. Blushing and weeping occupy almost whole chapters in *The Expression of the Emotions*, and Darwin gives special attention to the erection of hairs and feathers, which plays an important role in many visual signals.[109]

There is another similar category of movements, perhaps not strictly under autonomic control, which Darwin included under the same heading, and which Huxley[110] and Lack[111] refer to as 'general excited movements'. Though more difficult to identify, they may play an important role in some species. Eibl-Eibesfeldt[112] has shown how the tail shivering of many excited rodents, apparently without any signal function, can come to serve in communication, as in the house-mouse.

Lastly, visual signals may originate, as described by Selous and Huxley,[113] from movements, called by Tinbergen *displacement activities*,[114] which are quite irrelevant to the situation. Good examples are the incomplete preening movements which pigeons and ducks make during courtship,[115] and there are many others.[116]

If these four categories provide the origins of visual signals, what general statements can be made about the ways in which they evolve? We may regard it as a basic function of many signals to enable others to anticipate or predict what another animal is going to do in the near future. The development of intention movements in particular can be understood in this way, as, for example, when the animal signals readiness to attack by giving movements preparatory to fighting. But movements of any origin may serve in the same way, provided that they are regular preludes of a particular activity.

Some signals, however, seem to function by making possible, not a positive prediction, but a negative one, by signifying that a response which might be expected in the circumstances will not take place in the near future. It is here that the principle of antithesis operates. Darwin pointed out that the submissive actions of dogs and cats can only be explained as being the opposite of movements during fighting. Since then, numerous other examples have been described, all in the same kind of context. During aggressive encounters between paired black-headed gulls, the 'head flagging', in which they repeatedly turn their heads away from each other, negates certain aggressive movements (fig. 33). It thus reassures the partner and acts 'as a friendly gesture, meaning something like "no offence meant"'.[117] Here the movement was apparently initially an intention to escape.[118] Jackdaws and wolves have an analogous movement.[119] Another example of the principle of antithesis is the relaxed or 'fluffed' posture used by many birds to inhibit aggression (fig. 33),[120] the effects of which have been well described in Estrildine finches.[121] Most of the details of this posture can be regarded as the opposites of aggressive movements.

Fig. 33. Two examples of Darwin's principle of antithesis. Above are two black-headed gulls 'head flagging'. They turn away their faces from each other as a kind of appeasement gesture. (After Tinbergen.[39]) Below is a female chaffinch adopting the submissive posture in the face of an attack. The details of the posture are the reverse of those used in fighting and she faces away from the opponent instead of towards him. (From a photograph in Marler.[59])

It was probably originally a displacement activity[122] or even a direct result of a change of body temperature during conflict.[123] It is possible that the displacement activities of some finches (including feeding, preening and beak-wiping) may, by putting companions off their guard, have a similar social function.

There are elements in many visual signals which cannot be explained as assisting either positive or negative predictions. These are such characteristics as the colour of the skin or plumage displayed, the peculiar structures which often develop in some species, and also many of the details of the movements of hair, feathers and the limbs of the body. These are probably concerned with the effectiveness of the signal, as something to be easily perceived and recognized by others, and it is here that selection favouring specific and individual identification is likely to operate. Although these elements have received less attention than the others, they may ultimately prove of greater interest because of the more complex information they may carry. They may be imposed on any of the other categories, and once again may originate in a variety of ways.

Although the evolution in response to natural selection of any communication signal may be independent of its mode of origin, this need not necessarily be so. It is clear that many intention movements are selected because they enable others to anticipate the action which they normally prelude, as Darwin himself realized. Nevertheless, it can only hinder further progress if we are misled by this into confusing together the processes by which communication signals originate, and those by which natural selection controls their subsequent development.

Animal and Human Language

Throughout *The Expression of the Emotions*, it is assumed that the reader agrees with the original premise, namely that animals do communicate with each other by expressive sounds and movements. In *The Descent of Man* Darwin enlarges on this theme more fully, and his discussion of the relationship between animal and human language can hardly be bettered to-day. There are probably no more than two points on which we have significantly more information: the total size of the vocabulary of some animals; and the remarkable methods of social communication which von Frisch[124] has discovered in honey-bees.

Studies of visual communication have given some idea of the size of the repertoire available in certain species, but the only reliable estimates are for vocal signals, which are much easier to distinguish from the animal's general behaviour. There is a surprising consistency in the totals for the various species. Boutan,[125] in a penetrating study of the voice of gibbons, described thirteen cries, whereas howler monkeys have fifteen to twenty[126] and the black-tailed prairie-dog ten.[127] Among birds, the whitethroat has a vocabulary of twenty-eight,[128] the American song sparrow twenty-one,[129] the wren nineteen,[130] the chaffinch twenty-one,[131] and some other finches rather smaller repertoires. Although there is variation, these totals are all of the same order.

The most striking contrast between human and animal languages lies in the way in which they develop. While ours has an innate basis, the detail is very largely learned. That of animals is largely innate. Boutan[125] thought that in gibbons reared in isolation all of the calls developed normally, though his experience of the wild animals was limited. The complete vocabulary of the whitethroat is certainly innate[128] and the same probably applies to most bird calls. However, it was known to Darwin, through the work of Daines Barrington[132] that the songs of some birds are normally learned from other members of the species, as discussed earlier. And of course there are many examples of the capacity to mimic alien sounds both in birds, and to a limited extent in mammals such as the chimpanzee.[133]

Thus the difference is not that animals are incapable of learning sounds, but that this learning is restricted to certain narrow contexts. If we consider the function their language has to perform in the natural state, there is a certain consistency. The lives of most animals, and particularly birds, are generally dominated by a limited number of factors which overwhelm all others in importance. This is as true of their social life as of their interaction with the environment. For such activities as helping others to avoid danger (especially the mate and young), competing for food, keeping in contact with members of the same species and, above all, reproducing with them, it is probably most efficient to have a relatively limited vocabulary, which is instinctively used and understood the first time it is needed. Apart from certain exceptions such as bird song, where learning may have an important part to play, it appears that the more subtle advantages of elaborate communication by learned signals need a highly complex society for their value to be felt.

The dividing line between the nature of human and animal language becomes clearer when we consider the kind of information which is communicated. With animals this poses a problem, for the information can only be determined in an indirect way by observing the responses to a signal and inferring what information has been received. Consider, for example, two vocal signals of the chaffinch: the call given when a hawk flies over and the song of the male. On hearing the hawk signal, chaffinches respond just as though they had seen a hawk—by flying to a bush and hiding. We seem to be justified in inferring that the signal conveys the presence of a hawk, or at any rate of acute danger. Since the signal is difficult to locate (page 177) it does not convey information about position, and since other species use identical signals it does not convey that the signalling bird was a chaffinch. The essential simplicity of the information which we deduce to be carried by this signal becomes clear if we compare it with similar deductions about the song.

We know that the male song helps female chaffinches to find a mate in spring. Since a suitable mate needs to be both male and a chaffinch, we may infer that the song conveys both species and sex. She will ignore males not in full breeding condition, so we deduce that the song carries some information about physiological state. At times it even conveys the individual identity of the singer, as we have seen. And finally it conveys accurate information about position. It is clear that the song conveys more complex information than the hawk call, and it is significant that this difference in complexity is paralleled in the physical structure of the signals.

Of course one should not infer that chaffinches consciously communicate, or that listeners make a conscious analysis of the information which signals carry, nor that the different items are contained in different parts of the song. In fact the production and the response seem generally to be quite instinctive, except for the responses to individual characteristics. Yet the birds do respond as though they have received certain information. However, it appears that this information is available to them only from the complete set of signals. Given a limited number of signals in the vocabulary, it is most efficient to include the maximum of information in each one. Chaffinches are unable to separate the various items, and use them for new purposes.

This may be a crucial difference from our own language. Once our distant ancestors had achieved a more complex society, helping to relieve the individual of his extreme concern with personal survival in a difficult environment, a larger and more varied vocabulary would become increasingly useful. This in turn would permit separation of the complex of information in each instinctive call into a number of sounds which could then be used as separate items. Haldane[134] and Pumphrey[135] have both suggested that an increase of vocabulary, perhaps of this type, may have taken place during the technological revolution of upper Palaeolithic times, when a wide range of tools were used for the first time. Given a greatly increased number of sound signals for these objects and the way in which they are used, we might expect in time to see the first development of abstractions, grammar and so on.

However, another more subtle change from animal language is required. There are two aspects to our own language, the emotive or evocative, intimately involved with our emotional responses, and the descriptive or intelligible aspect, to which we respond rationally. It is clear that the greater part of animal communication is of the evocative type. If we visualize ourselves as chaffinches, the essential information in the hawk call is 'here is something to flee from'. It is a tremendous step for the signal to denote a hawk, as something with a particular appearance. However, we should notice that this is not essentially a change in the information carried in the signal, but rather in the way in which the information is responded to, a change in word usage rather than word structure.

There is, however, one example of the development of the descriptive aspect of language in animals. The researches of von Frisch into the social communication of the honey-bee are widely known, and have now been summarized in his two books.[136] It is for this reason that no review has been given of these studies, which are surely the most important advance in the study of animal communication since Darwin's time. During their hive dances honey-bees can transmit information to others, about the direction, distance and nature of a food source, which must be regarded as descriptive, in the same sense in which we apply this to our own language. It can be no accident that this is used by an insect with perhaps the most complex social organization of all animals.

We must conclude that there is no single characteristic which divides human language unequivocally from that of animals. Instead, as Koehler[137] has suggested, its distinctiveness results from a particular combination of attributes which, considered alone, are not unique to man. We are thus a step further towards confirming Darwin's conclusion that the faculty of articulate speech does not offer any insuperable objection to the belief that man has developed from some lower form.

The author is deeply indebted to many people, and particularly to R. A. Hinde, J. B. S. Haldane and W. H. Thorpe, for discussion of the subject matter of this paper, and to P. R. Bell and T. H. Bullock for their constructive criticisms of the manuscript.

5

H. L. K. WHITEHOUSE

CROSS- AND SELF-FERTILIZATION IN PLANTS

Introduction

Darwin's *The Effects of Cross- and Self-Fertilization in the Vegetable Kingdom*, published in 1876, was a landmark in the study of the biology of flowering plants. The sexual function of flowers had been established much earlier, and although the fact had been recognized for some time that the conspicuousness of flowers is an adaptation to attract insects, it remained for Darwin to show that the general law underlying flower structure is an adaptation for fertilization by pollen from flowers of different individuals of the same species. He realized that there are some exceptions to this law, notably the closed (*cleistogamous*) flowers of certain plants such as the violet, but even these showed traces of former adaptation for cross-fertilization and were to be regarded as secondarily adapted to self-fertilization for the sake of greater certainty of seed production.

Darwin carried out many experiments on the effects of self- and cross-fertilization, using numerous species of flowering plants, and extended these experiments for up to ten generations. He was surprised at the striking contrast both in vigour and fertility between the selfed and crossed lineages, because his knowledge of the close inbreeding practised on domestic animals without producing harmful consequences had not led him to expect anything very different in the plant kingdom. Yet his results, with few exceptions, showed a remarkable superiority, whether measured by height, weight or fertility, in the progeny from cross-fertilization compared with those from self-fertilization. Moreover, he showed that crossing different flowers on the same plant had no such beneficial effect, and similarly, there was no advantage over self-fertilization on crossing different individuals of the same variety of a habitually self-fertilized plant such as *Pisum sativum* L. (garden pea). From this evidence and from the diminished benefit of repeated intercrossing in successive generations, he deduced that the advantages of cross-fertilization lay not in the mere act of crossing, but followed from the crossed individuals differing slightly in constitution. These important differences were thought not to be in external characters, but in the internal constitution of the reproductive organs. Conversely, the disadvantages following upon self-fertilization he attributed to the similar constitution of the uniting sexual elements.

Having thus established by experiment the great advantages to the species which follow from cross-fertilization, Darwin was able to appreciate the significance of many features of flowers previously little understood. The bright colours, scent, nectar and innumerable mechanical contrivances of flowers were seen for the first time in their proper setting, as admirable adaptations to achieve cross-pollination by insects. *Dichogamy* (the maturing of the anthers and stigmas of individual flowers at different times) was seen not merely as a device to reduce self-fertilization of individual flowers, but also as a means of favouring the crossing of different individuals. Thus, the most frequent dichogamous condition is one of *protandry* (the maturing of the male region before the female), and Darwin related this to the habit of bees of alighting at the bottom of an inflorescence and working upwards. The insects would then meet functionally female flowers first, subsequently removing pollen from the younger flowers above and taking it to the older and hence female flowers of another individual. But the feature of flowers which in Darwin's opinion was the most important in favouring fertilization by pollen from a different plant, is the prepotency of pollen from another individual or variety, that is to say, the fact that the fertilizing power of pollen from another individual is greater than that of the plant's own pollen. Darwin said that scarcely any result from his experiments surprised him so much as the discovery of this remarkable prepotency, not merely of pollen from a different variety, but even of that from another individual of the same variety. He found that even with a 24-hour delay after self-pollination before applying pollen of a different individual, the progeny were largely of crossed origin, even though the species was self-fertile if other pollen was excluded. From experiments of this kind, he concluded that pollen from a different plant will generally prevent the action of pollen from the same flower. Recognition of the importance of this outwardly invisible phenomenon is due largely to the pioneer studies of Darwin.

Somewhat paradoxically, the more extreme condition of sterility of a plant with its own pollen he regarded as of incidental occurrence and not evolved as an adaptation to prevent self-fertilization. His reasons for reaching this conclusion were that in the plants he studied the degree of self-sterility appeared, first, to show no close correspondence to the extent to which the offspring were reduced in vigour by self-fertilization; secondly, to differ considerably in different individuals of the same parentage; and thirdly, to be greatly influenced by environmental factors such as temperature. Furthermore, Darwin found it impossible to believe that either self-sterility or *dioecism* (separation of the sexes in different individuals) had evolved as a means of favouring cross-fertilization, because in his opinion the acquisition of self-sterility or unisexuality would have been injurious had not a means already existed for the habitual transport of pollen (either by insects or wind) from flower to flower. Thus, Darwin's main discoveries reported in *The Effects of Cross- and Self-Fertilization* are that cross-fertilization is beneficial to the progeny and self-fertilization usually injurious, that the advantages of cross-fertilization lie in the differentiation of the sexual elements, and the injury from self-fertilization in the lack of such differentiation. Biologists to-day accept the validity of these conclusions, although they qualify the generalizations to some extent.

Two fundamental questions are raised by these conclusions. First, how do the differences between individuals arise? Secondly, why is it beneficial to the progeny to cross such individuals? With regard to the first question, Darwin considered that the differentiation of the sexual elements of different individuals came about, either as a consequence of 'spontaneous variation', or, more probably, through individuals having been subjected to different environmental conditions during development. His evidence for the latter view was slight, but he was accepting the theories of variation and heredity current at the time. Although Mendel's classic studies of inheritance in *Pisum sativum* had been published ten years before Darwin's book appeared, they had been overlooked by the world of science and were unknown to Darwin. Thus it was that Darwin held to the 'common-sense' views on heredity which had been current for so long. These views were based on two observations: first, the characters of an organism vary according to the environment in which it develops; and secondly, variations in characters are inherited from generation to generation. From these two observations, it would appear plausible to suppose that inherited variations arise as a consequence of differences in environment during the development of different individuals. Darwin considered the case of two foxgloves (*Digitalis purpurea* L.) growing a yard apart in his garden. When these plants were crossed their progeny were greatly superior to plants of self-fertilized origin, and Darwin asks: 'How can two plants in the same garden be differently acted on by the environment?' His answer is that one of the plants, or its ancestors, may have come from a distance, or the mere persistence of the seed in the soil in a dormant condition, perhaps derived from a plant which matured in a different season, would be adequate explanation of the difference in its sexual elements. The fallacy in this 'common-sense' view of the cause of inherited variation was pointed out by Weismann soon after Darwin's book appeared, but was not clearly demonstrated until 25 years later, when Johannsen, with the knowledge of Mendel's work, performed his classic experiments on *Phaseolus vulgaris* L. (French bean). These will be discussed more fully later.

With regard to the second fundamental question—why it is beneficial to the progeny that the sexual elements of the parents should be differentiated from one another—Darwin was unable even to hazard a guess. He refers to the work of Sprengel[1] published more than 80 years earlier, entitled (in translation) *The Newly-Revealed Mystery of Nature in the Structure and Fertilization of Flowers*, and points out that despite this work and that of many others, including himself, the mystery remained. Sprengel had discovered the part insects play in the fertilization of flowers, but had mistaken the significance of various structures because he was unaware that flowers are primarily adapted for cross-fertilization. Darwin had now demonstrated the beneficial effects of crossing, and how flower structure can be adapted in relation to it, but, as he says, 'the veil of secrecy is as yet far from lifted; nor will it be, until we can say why it is beneficial that the sexual elements should be differentiated...', and furthermore, 'why, if the differentiation be carried still further, injury follows'. He was puzzled as to why considerable differentiation between the sexual elements, as in a cross between distinct species, should lead commonly to sterility, whereas a cross between different individuals within a species was beneficial to the progeny. Darwin concludes '...we stand in awe before the mystery of life.' Some of this mystery still remains after a further 80 years.

In *The Different Forms of Flowers on Plants of the Same Species*, published in 1877, Darwin discussed heterostyled plants such as *Primula vulgaris* Huds. (primrose) and *Lythrum salicaria* L. (purple loosestrife), and then went on to consider various types of sex separation, such as dioecism and *gynodioecism* (separate female and hermaphrodite plants). Finally, he discussed cleistogamous species such as the violet (*Viola* spp.), where completely closed flowers are formed in addition to normal flowers.

The work described in this volume is of a less fundamental character than that which forms the main theme of *The Effects of Cross and Self-Fertilization*. The earlier publication is concerned with the basic problem of showing that cross-fertilization is in general advantageous and self-fertilization usually disadvantageous, whereas in *The Different Forms of Flowers* Darwin is concerned primarily with describing mechanisms for favouring either cross- or self-fertilization. Nevertheless, the contribution to knowledge in the latter book is a very real one. Heterostyled plants had previously been regarded as merely showing variability of style length and stamen height, although Sprengel[1] as long ago as 1793 had been impressed by the exceptionally conspicuous dimorphism of style and stamens in *Hottonia palustris* L. (water-violet) (Plate V), and did not believe the existence of the two forms to be accidental, though he could not explain their purpose. The true nature of heterostyly remained a mystery until Darwin made the remarkable discovery that in *Primula veris* L. (cowslip), pollen from the stamens of a 'pin' (long-styled) flower is nearly sterile when placed on the stigma of the same or any similar flower, whereas it is quite fertile on the stigma of a 'thrum' (short-styled) plant; conversely, pollen from a 'thrum' plant is almost sterile on a 'thrum' stigma, but fertile on a 'pin'. This was a finding without parallel in the plant or animal kingdoms. The two forms could not be called sexes because each is hermaphrodite, yet as Darwin says '...they are to a certain extent sexually distinct, for they require reciprocal union for perfect fertility'. He found a similar situation in all the distylic species he tested. Indeed in *Linum grandiflorum* Desf. (red flax) he found that pollen of a long-styled plant would grow only on the stigma of a short-styled plant; when placed on the stigma of a long-styled plant, pollen-tubes were either not emitted from the grains, or if they were produced they grew only for a very short distance, even in several days. In consequence, long-styled plants when self-pollinated or intercrossed, were completely sterile. Darwin had thus made the astonishing discovery that in distylic species pollinations between individuals of the same style-length may actually be as sterile as are, usually, pollinations between different species.

In *Lythrum salicaria* he found an even more remarkable situation. In his own words '...nature has ordained a most complex marriage-arrangement'. There are three different kinds of flowers which differ in style-length, and each has stamens with anthers at two positions, corresponding to the positions of the stigmas of the other two kinds of flowers (Plate V), all the flowers on each individual plant being of only one kind. Darwin found that of the 18 possible ways of inter-pollinating *Lythrum* flowers (the six kinds of anthers with the three kinds of styles), only the six pollinations between anthers and stigmas at the same height gave full fertility; the twelve pollinations between anthers and stigmas at different heights gave greatly reduced fertility and approximately one-third the number of seeds. Thus in *Lythrum* a curious situation exists in which the pollen from anthers at different levels in the same flower behaves differently on the stigmas of the other two kinds of flowers.

The conclusions which Darwin reached from his study of heterostyly are of considerable interest. He saw heterostyly as a device to favour cross-fertilization, with advantages in this respect over both dichogamy and dioecism, since with heterostyly all the flowers on the same plant are inter-sterile, and all the individuals set seed.

Nevertheless, as with the evolution of ordinary self-sterility and of unisexuality, he did not believe that the self-sterility and intra-class sterility associated with heterostyly could have evolved as a mechanism for favouring cross-fertilization since without efficient cross-pollinating mechanisms already in existence, self-sterility would be a disadvantage to the species. The intra-class sterility (for example, 'pin' primroses are more or less sterile with all other 'pin' primroses) appeared to him to be an incidental and purposeless outcome of evolution, for he could not believe that it was of any advantage to the species to prevent an individual from crossing with half the population (or with one-third in the case of a tristylic species). He compared this intra-class sterility with the inter-sterility of two related species. In each instance, he regarded the sterility as an incidental consequence of the mutual adaptation of the sexual elements: co-adaptation for reciprocal cross-pollination between 'thrum' and 'pin' in the case of heterostyly, and co-adaptation of male and female within each species in the case of inter-specific sterility. Thus he was led to the conclusion that the sterility of species crosses must depend 'not on any general difference in constitution or structure', but 'exclusively on the incompatible nature of their sexual elements', just as he believed the sterility within heterostyled species was due merely to such incompatibility. This conclusion about interspecific sterility, which is incidental to the main subject of the book, is nevertheless of considerable importance, and later work to be discussed presently has tended to support Darwin's view in many instances.

The second main subject of study in *The Different Forms of Flowers* is dioecism and gynodioecism. With regard to the origin of these conditions, Darwin tentatively suggested that the unisexual condition was advantageous because it put less strain on the powers of each individual. In a gynodioecious species, such as *Thymus vulgaris* L. (garden thyme) and *Satureja hortensis* L. (savory), he found that the female plants produced twice the weight of seed of the hermaphrodites. In this he saw the advantage of the condition, though he was unable to say whether it was initiated by a variation giving more seed, which then led to reduced pollen output; or whether the cause and effect were reversed, the initial variation being towards less pollen and this being followed by increased seed output. This 'compensation' as an explanation of how unisexuality evolved was a weak hypothesis since it left unanswered the question of why the great majority of flowering plants were successful as hermaphrodites, apparently not handicapped by the strain of producing both pollen and ovules. Unlike most biologists of to-day, Darwin did not believe, for the reasons already mentioned, that the unisexual condition could have evolved as an adaptation for cross-fertilization.

The final topic of this book is cleistogamy. Darwin showed how these flowers which never open can be regarded as showing arrested development: in many respects they resemble an early stage in the development of normal flowers, although they also have some special modifications such as the much-reduced number of pollen grains per anther. He pointed out how such plants are never exclusively cleistogamous, but always produce normal flowers as well (for example, the violet), though often at a different season of the year (Plate V). He also showed that the condition is widespread in flowering plants, though more frequent in insect-pollinated than in wind-pollinated species. It had been suggested that cleistogamy was an adaptation against seed failure, since in an unfavourable season the normal flowers, particularly where insect-pollinated, might not be fertilized. In cleistogamous flowers the pollen germinates within the anther and the tubes grow directly to the pistil. Although Darwin agreed with this suggestion, he thought the primary significance of cleistogamy lay in its efficiency in producing a large number of seeds with the minimum expenditure of energy on floral development and pollen production. The normal flowers borne on the same plant gave opportunity for occasional cross-fertilization, with its attendant advantages for the progeny. Darwin thus regarded cleistogamous flowers as secondarily adapted for self-fertilization, and hence no exception to the general rule established in *The Effects of Cross- and Self-Fertilization* that flowers are primarily adapted for fertilization by pollen from another individual.

The developments in the study of cross- and self-fertilization in plants since the time of Darwin may be considered under the following four headings: first, there is the fundamental question raised by *The Effects of Cross- and Self-Fertilization*—the causes of inherited variation between individuals, which gives cross-fertilization its advantage. This was the weakest topic in Darwin's book and subsequent developments have been profound. Secondly, there is the other fundamental question of why it is beneficial to the progeny to cross individuals which differ. Here also there have been important developments, though it is still not possible to give a complete answer. Thirdly, there are additions to knowledge of the mechanisms for achieving cross-fertilization—additions which were themselves dependent upon progress in the knowledge of heredity and of the mode of origin of differences between individuals. Finally, there is the question of how the various devices, such as heterostyly and self-sterility, for achieving cross-fertilization evolved. This presented many puzzles to Darwin, and is still to-day largely a matter for speculation.

The Development of Ideas on the Cause of Inherited Variation

In tracing the development of ideas on how the differences between individuals arise, it is necessary first to recall the views prevailing in 1876 when Darwin's book on cross-fertilization appeared. Eight years previously Darwin[2] had published his theory of the mechanism of inheritance—namely, that during the life of an organism, substances pass from the various organs of the plant or animal to the reproductive elements and so convey the characters of one generation to the next. There was nothing really new in this idea, which indeed, in all essentials, is to be found in the writings of Hippocrates and Aristotle. Moreover, it was not supported by experimental evidence. Numerous biologists during the nineteenth century had attempted to obtain experimental evidence to explain the mechanism of inheritance, but with the one exception of Mendel, they had all failed. Thus with the prevailing ignorance about heredity, it was natural for Darwin to accept the current classical theory.

Darwin was not entirely opposed to the view, associated with the name of Lamarck, that a character may be modified during the life of an individual and handed down to the next generation. Hence in considering the disappearance of a disused character in the course of descent, Darwin held that the very lack of use of the organ led directly to its gradual disappearance over successive generations, although he was unwilling to accept the inheritance of acquired characters in a positive sense as the cause of new inherited adaptations. So long as he held to the view that the reproductive elements were influenced, during the life of the individual, by substances moving to them from the various organs, then the idea remained tenable that the environment could directly mould the inheritance.

The first significant development in the theories of heredity and variation occurred about seven years after the publication of *The Effects of Cross- and Self-Fertilization* and therefore after Darwin's death. This was the challenging by Weismann[3] at Freiburg of the very occurrence of inheritance of acquired characters in any form. Previously the 'stream of life' had been thought to flow into each individual as it developed, and thence at maturity back to the reproductive cells and so on to the next generation. However, according to Weismann, the stream flowed direct from the reproductive cells of one generation to those of the next, and the individuals themselves (apart from their sexual elements) represented side-streams from which there was no return to the main stream. This obviously undermined the classical theory of inheritance in which something from the substance of each organ was thought to be conveyed to the reproductive elements. Weismann pictured just the converse situation: that the potentially immortal reproductive lineage carried in some mysterious way an exceedingly complex inheritance, which in each generation gave rise to mortal somatic offshoots—the individuals. This idea of the continuity of the 'germ-plasm' (as Weismann called that part of the reproductive substance which in his opinion remained unchanged during cell division) had an important influence on biological thought, even though at the time it was purely theoretical. It drew attention to the need for experimental study of variation, and to the value of studying the structure of the reproductive cells.

Two years later, Weismann[4] took his argument a step further. Hertwig,[5] van Beneden,[6] Strasburger,[7] and others, had shown that fertilization is associated with the fusion of nuclei derived from the male and female cells. Since the male gamete contains little else but nucleus and yet the paternal and maternal contributions to inheritance appeared to be equal, Weismann deduced that the nuclear substance, or more precisely 'the chromatin of the nuclear loops' (subsequently given the name 'chromosomes' by Waldeyer),[8] must be the bearer of hereditary tendencies. This view was supported by Strasburger and Hertwig,[9] though it was another 30 years before complete experimental proof of the significance of the chromosomes was obtained. Weismann pictured these 'nuclear loops' within the cell as being modified during the growth of the individual, in association with its development. He regarded the nucleus as controlling the development of the cell, and in order to explain the great diversity of cells and tissues which appear during the growth of a higher plant or animal, it seemed to him essential to suppose that the nuclei themselves undergo changes. It was known that the reproductive cells of an individual arise, not directly from the corresponding cells of its parents, but nevertheless from cells which have retained the power of cell-division, and which have never developed into any highly specialized form. This allowed Weismann to assume that these relatively undifferentiated cells, from which the reproductive cells arose, contained 'nuclear loops' some at least of which had not been modified and hence constituted 'germ-plasm'. In this way he was able to associate his hypothesis with definite physical structures. The idea that the nucleus was primarily responsible for heredity was soon supported by the experiments of Boveri[10] with echinoderms. He had managed to fertilize the enucleated egg of one species with the spermatozoid of another, and had found that the resulting individual resembled the male parent rather than the female.

In a later publication, Weismann[11] elaborated still further the hypothesis of 'the continuity of the germ-plasm', and introduced the term 'idant' for the nuclear loops or rods, and 'id' for components of these which he regarded as containing all the elements essential for the development of the individual. These terms were derived from the word 'idioplasm', which had been used by Nägeli[12] for the cell-nuclei. The term 'id' survives in 'haploid' and 'diploid', which refer to a single and a double set of chromosomes, respectively. Weismann regarded variation as due to different combinations of 'ids' arising at fertilization, and he predicted that there must be a special nuclear-division at some point in the life-history where the number of 'ids' was halved to compensate for the doubling of number at nuclear fusion. He suggested that this 'meiosis' (from the Greek, meaning reduction), as it is now called, occurred in the cell divisions preceding the formation of the gametes (spermatozoids and egg cells) in animals. This prediction was subsequently shown to have been correct.

Weismann's theory of the isolation and continuity of the germ plasm, with its implication that the environment during the life of an individual had no influence at all on its hereditary constitution, and hence on its progeny, had a profound impact on contemporary thought. It caused a sharp division of opinion among biologists: some agreed with Weismann, but others found it impossible to believe that all inherited variation arose through germ-plasm changes which were not related to the environment of the individual. Darwin had believed in a blending type of inheritance in which, at fertilization, the hereditary material of the two parents was supposed to amalgamate like mixing two fluids. This meant that there was a rapid loss of inherited variability, as one generation succeeded another. To counteract this supposed decay of variability, Darwin had to postulate that new variations were continually arising with high frequency, since inherited variations were essential to the theory of evolution by natural selection. The critics of Weismann could not believe that so many variations arose in the germ plasm alone, for its complexity would be impossibly great. Thus at this time, 15 years after *The Effects of Cross- and Self-Fertilization* appeared, some biologists still agreed with Darwin that the important differences in the sexual elements of different individuals were due to the different environments in which the individuals developed, whilst others supported Weismann and laid all responsibility on the germ-plasm and none on the environment. To resolve this difference of opinion, it was essential to know more about heredity, and it was not until some ten years later (1900) that a new era began with the discovery of Mendel's work.

Mendel's classic experiments with the garden pea (*Pisum sativum* L.) were performed in the monastery garden at Brünn (Brno), in what is now Czechoslovakia, between the years 1857 and 1865, and were published[13] in a local journal in 1866. The reason why Mendel's work was overlooked for 35 years is obscure. It may partly be attributed to his using an unsuitable genus (*Hieracium*) for further experimental work. These hawkweeds are among the few organisms in the whole of the plant and animal kingdoms where it is not easy to demonstrate what is now called Mendelian inheritance. This is because the pollen is frequently non-functional and the seeds arise directly from the parent plant without fertilization. Through his failure with *Hieracium*[14] to confirm his findings with *Pisum*, Mendel was perhaps discouraged from further experimental work and from bringing his *Pisum* work to the attention of other biologists. However, Mendel did correspond with the German botanist Nägeli, and copies of Mendel's *Pisum* paper reached libraries in other countries, including England. That the significance of his work was overlooked must therefore be attributed, in part, to the fact that he was 'ahead of his time' and that contemporary thought was directed along other channels.

Mendel's paper is concerned with the inheritance of distinct character-differences within one species, whereas after the publication of *The Origin of Species*[15] in 1859, most naturalists in the 1860's and 1870's were looking for evidence of how one species evolved into another, and hence were more interested in the groups of character-differences which distinguish different species. It is precisely because Mendel concentrated his attention on individual character-differences that he was successful in discovering the mechanism of inheritance. Moreover, he distinguished and recorded separately the results of breeding each generation of his pea plants, and took note of the numbers of individuals of each kind in each generation. Earlier experimenters had failed to make detailed records and so had 'all missed the clue without which the evidence so laboriously collected remained an inscrutable medley of contradictions' (Bateson).[16]

By this method of analysis, Mendel had obtained clear-cut results which enabled him to put forward the hypothesis that contrasting character-differences in the pea, such as tall and dwarf plants, round and wrinkled seeds, and yellow and green cotyledons, 'segregate' or separate from one another, at the time of formation of the reproductive cells, and come together again at the time of fertilization. Mendel used the term *Merkmal* (characteristic), or occasionally *Charakter*, for these hypothetical structures which separate and later come together again, but soon after the rediscovery of his work, the word *factor*, and later *gene*, came into use for these structures. Underlying this change of terminology was the recognition that Mendel's hypothesis demanded the existence of definite particles which determined the manifestation of the characters of the individual during its development and which were handed down from generation to generation.

It was perhaps partly because Mendel did not state in so many words that his hypothesis demanded the existence of definite particles that, during the first decade after the rediscovery of his work, a controversy took place over whether or not Mendelian inheritance could occur with quantitative characters such as height and weight, which show 'continuous variation' (that is, a continuous range of values). It was not at first realized that although the Mendelian method of analysis required discontinuity of characters, the Mendelian hypothesis does not. In other words, to demonstrate the existence of segregating particles, it is necessary that those particles should determine sharply contrasting characters, although it is quite possible for other particles to be associated with variable or overlapping characters so that no discontinuity is evident.

The mechanism of inheritance proposed by Mendel thus differed profoundly from that which had been current for so long and which had been accepted by Darwin. Instead of each organ of the individual contributing representative substances to the reproductive elements, so that characters were transmitted directly to the next generation, on Mendel's hypothesis what was transmitted was not the character itself (despite his terminology) but particles which determined the manifestation of that character during the growth of the individual. If an individual received from its two parents the particles responsible for the expression of both characters of a contrasting pair, such as tall and dwarf plant in peas, then only one of the pair of characters, called the *dominant* one, was expressed. In the present example, tall plant is dominant to dwarf. The latent character is spoken of as the *recessive* one. One of the basic features of the Mendelian hypothesis is that in such a heterozygote (an individual containing the pair of particles responsible for such a character-difference), the two particles do not modify one another in the slightest degree, so that when they separate from one another at the time of formation of the reproductive cells, each is quite untainted by the other. That this must be so was deduced from observations such as the true breeding nature of the dwarf plants which formed one-quarter of the progeny on self-pollinating a tall–dwarf heterozygote. These dwarf plants, although obtained from a heterozygote which was tall, showed no trace of the tall character in succeeding generations—up to six generations in the case of some of the characters studied by Mendel.

By 1900, when Mendel's work came to light, the outlook of many biologists was already prepared for the revolution of thought which its acceptance involved, since the theoretical views expressed by Weismann demanded that all new inherited variations should arise as changes in the germ plasm. That contemporary thought was moving towards the Mendelian viewpoint is further shown by the circumstances of the discovery of Mendel's work. During the last decade of the nineteenth century attention was directed to the study of discontinuous variation, particularly by the work of de Vries[17] and of Bateson.[18] By 1899, de Vries at Amsterdam, Correns at Tübingen and von Tschermak at Vienna had cross-fertilized different varieties of certain species of plants and found that the parental characters reappeared in the second generation in a constant proportion of the individuals. De Vries[19] had found approximately three-quarters of the second-generation plants showed the one character and the remaining one-quarter the other for at least 17 character-differences involving almost as many different species of plants. These species were representative of many different families of flowering plants. Correns[20] had likewise obtained 3:1 ratios for certain characters in *Zea mays* L. (maize) and *Pisum sativum*; and von Tschermak,[21] also with *Pisum*,* had unknowingly studied the inheritance of some of the same character-differences as Mendel had used, and had obtained similar results. Each thinking he was the first to discover the underlying law of heredity, the three authors had independently searched the literature for confirmatory data and had found Mendel's paper. Thus from the moment of its rediscovery, Mendel's work was not only known to be true, but the Mendelian theory of inheritance was known to be applicable to many other plants besides peas. Within the next

* It is of interest that von Tschermak had been led to work with peas from reading Darwin's *The Effects of Cross- and Self-Fertilization in the Vegetable Kingdom.*

decade it became evident that Mendelian inheritance was of extremely widespread occurrence, and Bateson[16] quotes over 100 examples involving a great diversity of characters in many different plants and animals.

The controversy over characters which show continuous variation and which in consequence are not susceptible to analysis by the Mendelian method was resolved during this decade. The idea that there was necessarily a fundamental difference between the mode of inheritance of quantitative characters like height and weight and qualitative characters such as those studied by Mendel, was disproved by Johannsen[22] in a now famous series of experiments with *Phaseolus vulgaris*. Johannsen showed that the weight of an individual bean seed was determined in the first place by its hereditary constitution—different strains or 'pure lines' maintained by self-fertilization differed in the mean weight of the individual beans which they produced; and, secondly, the weight was much influenced by environmental factors such as the position of the seed in the pod, so that the individual bean-weights of the different strains showed ranges of variation which greatly overlapped one another. Selection of the heaviest seed in each generation resulted in an increasing proportion of the beans being of the heaviest pure line, until eventually all were from this strain. Thereafter, selection was powerless to increase the mean weight any further. Johannsen thus showed that there was a clear distinction between hereditary and environmental variation—only the former responded to selection. Although bean weight showed continuous variation, there were well-marked discontinuities between the mean weights of the beans of the different lines. Thus the environmental modifications in individual bean-weight had obscured the hereditary differences in weight, and it was evident that such characters showing continuous variation did not differ in any fundamental way from characters showing discontinuous variation. The difference was merely that, with the latter, the effects of the environment during the development of the individual were insufficient to obscure the hereditary effects. That each of the nineteen strains of dwarf bean studied by Johannsen produced seed with a different mean weight, suggested that bean weight was controlled by a number of Mendelian factors or genes.

The Cause of Inherited Variation

It was therefore evident by about 1910 that Mendelian inheritance was probably the basis of inherited 'continuous variation' in nature, as well as of innumerable instances of 'discontinuous variation', and it began to appear as if almost all inheritance in living organisms was of this kind. Soon afterwards, it was established finally that the Mendelian factors or genes were carried in the chromosomes of the cell-nucleus. The chromosome theory, which had been favoured so strongly by Weismann 30 years earlier, was proved to be true primarily through the researches of Morgan[23] and his collaborators at Columbia University, New York, using the fruit-fly *Drosophila* as their experimental organism.

The differences of opinion that had existed 20 years previously concerning the source of the differences in the 'sexual elements' of different individuals had now been resolved. In the light of the widespread occurrence of Mendelian inheritance, and with the chromosome theory to explain it, one could no longer believe, as Darwin had done, that the differences between individuals, which were responsible for the advantages of cross-fertilization over self-fertilization, could arise through the influence of the environment during the development of each individual. If characters were directly transmitted from one generation to another, this would have provided a plausible hypothesis, but it was now known to be the chromosomes and not characters which were handed down. How could the environment modify these threads within the nucleus of each cell? It was known that the genes were remarkably stable, for they did not influence one another even in a heterozygote where a pair of corresponding or allelomorphic genes were present in each nucleus. It was thus impossible to believe that the environmental conditions during the growth of an individual could influence in any way its chromosomes in the reproductive cells (or the cells which were to give rise to the reproductive ones later in development). Johannsen's experiments had clearly demonstrated that environmental modifications were not inherited. What then was the source of these variations?

As already indicated, the critics of Weismann in the 1890's could not believe that the innumerable variations that must be continually arising could have their source in the germ plasm alone, since the complexity of the latter would have to be impossibly great. In accepting the chromosome theory of heredity one had to accept also the idea that in some mysterious way these threads of nucleoprotein contained a chemical code or message which during development was translated into the physiological and morphological characters of the individual. To this extent, Weismann's critics were justifiably sceptical, for the chromosome theory offered an explanation of heredity which required even the simplest living things to have a fantastically elaborate internal organization.

Weismann's critics, however, did not fully appreciate another aspect of the chromosome theory—one which was not fully analysed until 1930,[24] namely, that it was no longer necessary to assume a high frequency of origin of new variations. On the old blending theory variability was lost whenever fertilization of an egg cell occurred, but with particulate inheritance the genes separate again at gamete formation in the reproductive organs with their identity unimpaired, so that variability is not lost in the sexual life-cycle. Indeed, the converse is true, for new combinations of genes can arise at gamete formation and at fertilization, and so new variations can appear. There is, however, a slow process by which variations can disappear, when a particular gene is lost from every individual of an interbreeding population. Unless natural selection acts against a particular gene it can be eliminated only by chance fluctuations in its frequency in the population. It was shown mathematically in 1908, independently by Hardy[25] at Cambridge and Weinberg[26] at Stuttgart, that with Mendelian inheritance the frequencies of the genes in a population stay constant, except for the effects of chance fluctuations. This is true whatever the frequencies happen to be. The Hardy–Weinberg Law, as it is called, is one of the most fundamental propositions in population genetics. It is of course based on the assumption that the gene frequencies are not being modified either by natural selection or by mutation (one gene changing into another). It disposed of a popular fallacy that a dominant gene will automatically tend to spread at the expense of its recessive allelomorph. Hence if genes uninfluenced by selection can be lost from a population only by chance fluctuations in their frequency, such loss is likely to be of quite rare occurrence unless the interbreeding population is a small one.

The cause of the hereditary differences between different individuals which gives cross-fertilization its advantage over self-fertilization is therefore due to the genes carried in the population of the species. In any plant or animal which is usually cross-fertilized, many of the genes of each individual will differ slightly from those of other individuals of the species. Individuals from two different localities are likely to differ in more of their genes than are two from the same locality, since the latter individuals are probably closer cousins. Thus Darwin's 'differences in the sexual elements' are resolved into differences in gene content.

However, there still remains the question of how the genetic diversity within a population arose in the first instance, and how the slow loss of hereditary variability, due to chance fluctuations in gene frequency or to natural selection, is counteracted. With the realization that in Mendelian inheritance variability is not lost in each generation, it was evident that the process by which such variability arose must occur relatively infrequently. For this reason it was difficult to observe, and was described as due to 'rare spontaneous mutation'. The word 'mutation' came into use in biology, primarily as a result of the work of de Vries [27] on *Oenothera erythrosepala* Borbas (evening primrose). This plant appears to give rise to new species of *Oenothera* in a proportion of its progeny in each generation, and de Vries gave the name of mutation to this abrupt process of origin, and he used *Oenothera* as the basis of a theory of evolution. After a further 30 years' research in many countries, the reason for the anomalous behaviour of this and other species of *Oenothera* was eventually discovered—the plants are curious hybrids, and the new forms to which they give rise are mostly due to the reassortment of genes. Thus de Vries's theory of evolution of new species by mutation is no longer tenable as a general theory, but nevertheless, with the discovery of Mendelism, the idea of mutation, that is, of the abrupt change from one form to another, was readily transferred from the species to the gene.

The essential and remarkable feature of gene mutation is that the gene is perpetuated subsequently in the changed form, so that the associated change of character is inherited. Mutation is presumed to be a change in the chemical code in the nucleoprotein of the chromosome, but what causes 'spontaneous mutation' is still not known. It has, however, been shown that various physical and chemical agents will increase the frequency of gene mutation. The first discovery of this kind was made in 1927 at the University of Texas by Muller,[28] who showed that X-rays will cause mutation. It was subsequently found that ultraviolet light and all types of ionizing radiation have a similar effect. In 1941, Auerbach & Robson[29] at Edinburgh discovered that mustard-gas will bring about mutation, and since then a whole range of chemicals of diverse molecular structure have been found to be mutagenic. But discovery of the exact nature of gene mutation, whether spontaneous or induced by such mutagens, requires more information about the chemical structure of the gene than is at present available.

The main events which account for the advance in knowledge of the differences between individuals since the publication of Darwin's *The Effects of Cross- and Self-Fertilization* may be summarized as follows: first the discovery that much of inheritance is particulate, with the implication that the source of the differences between individuals lies, not in their environment during development, but in the particles or genes within the cell-nuclei which determine the characters of the organism; secondly, the demonstration that with Mendelian inheritance the frequencies of genes in a population remain virtually constant from generation to generation, apart from the effects of natural selection, so that there is normally only a very slow loss of hereditary variability, and indeed, new variability is generated by gene recombination in sexual reproduction; and finally, the realization that the ultimate source of most of this hereditary variation must lie in abrupt changes in the genes, or mutation. The precise nature and cause of gene mutation have yet to be discovered.

The Development of Ideas on Why it is Beneficial to the Progeny to Cross Individuals which Differ

Turning now to the second fundamental question raised by *The Effects of Cross- and Self-Fertilization,* it will be recalled that Darwin was unable to offer any explanation of why it is beneficial to the progeny that the sexual elements of the parent should differ in constitution. It was clearly essential to know more about the differences between individuals before one could say why such differences were important. With the discovery of Mendelian inheritance and the realization that differences lay in the Mendelian factors or genes on the chromosomes, an advantage of cross-fertilization over self-fertilization was at once apparent,[30] for, as already mentioned, cross-fertilization allowed new combinations of genes to arise and hence actually increased the amount of hereditary variability in the population. This recombining of genes could of course only occur if the individuals making up the population differed from one another in a number of their genes, but in species in which cross-fertilization occurred frequently there was reason to think that such differences did exist. Individuals of all species, whether of plants or animals, when examined with sufficient care, were seen to differ from one another in many characters, just as human individuals do, and although some of the differences could be attributed to environmental effects, many were inherited.

It was argued that although some of the new gene combinations were likely to lead to inferior adaptation, others might be superior to anything that had existed previously, and would lead to the individuals carrying them being favoured by natural selection. In this way cross-fertilization could result in better adaptation of the progeny to the environment. But it was also evident that this was a long-term advantage, for only a very small proportion of the progeny in any one generation could be expected to be better adapted than their parents, and it would be many generations before these favoured ones had multiplied and replaced the less well adapted members of the species. Hence, although gene recombination might be important for the ultimate prospects of survival of the species, it could not be the explanation of the immediate advantages to the progeny, following upon cross-fertilization, which Darwin had observed. The precise source of these advantages is still obscure. Before considering the possible explanations which have been proposed, it will be of value to recall some of the observations made by Darwin on the consequences of cross- and self-fertilization.

Darwin had studied the inheritance of characters in *Linaria vulgaris* Mill. (yellow toadflax) and had raised plants from seed, some obtained by self-fertilization and some by cross-fertilization. The latter were taller and more vigorous than the former. Such 'hybrid vigour' had been recognized for a long time—in the animal kingdom, the vigour of the mule had been known for thousands of years, and botanical writers at least since Kölreuter[31] had referred frequently to the increased vigour of cross-bred plants. Darwin knew that the yellow toadflax is nearly sterile when bees are excluded, and when it is artificially self-pollinated, and so he concluded that wild plants must have been intercrossed in the past for many generations. He found it 'quite incredible that the difference between the two beds of seedlings could have been due to a single act of self-fertilization', and consequently, the following year he repeated the experiment using another species, *Dianthus caryophyllus* L. (carnation), which is also almost sterile if insects are excluded. The result was similar, and in experiments with the self-fertile species *Mimulus luteus* L. (monkey-flower) and *Ipomoea purpurea* Roth (morning glory) he observed similar differences in vigour between the selfed and crossed progeny.

These results seemed so remarkable that Darwin began a series of investigations extending over a period of 11 years, which included detailed records of height, weight and fertility of the progeny and which were the first quantitative studies of the effects of cross- and self-fertilization ever to be made. He soon found that crossing by itself is not significant, for crossing different flowers on the same plant had no beneficial effect on the progeny, and thus he was led to the conclusion that the advantage depends on slight differences in constitution of the individuals which are crossed.

Although most of his data agreed with this hypothesis, Darwin recorded anomalous results in a few instances. In the sixth generation of self-fertilized *Ipomoea purpurea* he found one individual, which he called Hero, which was taller than the cross-fertilized plants. Moreover, the descendants of Hero by self-fertilization were similar to it, and a cross between these descendants showed no further beneficial effect in the progeny. This result was comparable to what he had found in *Pisum sativum*, which is regularly self-fertilized and where crosses between different individuals of the same variety of pea did not produce improvements in the offspring. Darwin concluded that Hero transmitted a peculiar constitution adapted for self-fertilization. A similar plant was found in the *Mimulus* breeding experiments which was more vigorous and more fertile than the cross-fertilized individuals. Darwin was thus led to conclude that self-fertilization was occasionally advantageous, though in general not comparable to cross-fertilization. However, he was unable wholly to resolve the paradox presented by plants such as *Pisum sativum*, *Lathyrus odoratus* L. (sweet pea) and *Ophrys apifera* Huds. (bee orchid) on the one hand, which are regularly self-fertilized and appear to suffer no ill-effects thereby, and such plants as *Viola tricolor* L. (heartsease), *Digitalis purpurea*, *Sarothamnus scoparius* (L.) Wimmer (broom) and *Cyclamen persicum* Mill. which are naturally cross-fertilized and 'suffer to an extreme degree from a single act of self-fertilization'. He was unable to decide whether or not the harmful effects of inbreeding such species increased in each generation, but he considered that if there was any increase it must be slight. He regarded the injurious effects of self-fertilization as due entirely to the lack of differentiation of the 'sexual elements'. He was unable to accept the view, commonly held at that time, that the harmful effects were due to an increase in some morbid tendency or weakness of constitution common to the parents, since the same individual plant was often used for self-fertilization and for cross-fertilization, and morbid tendencies were not shown by the crossed offspring. Thus, to him, the harmful effects of inbreeding, and the beneficial effects of crossing were two aspects of the same phenomenon.

No progress was made in understanding this phenomenon until after the widespread occurrence of Mendelian inheritance had been established. Important contributions were then made by East, Shull and others in America, primarily as a consequence of experiments on inbreeding and outbreeding in *Zea mays*. In this grass the sexes are widely separated on different parts of the plant, and it is normal for it to be frequently cross-fertilized. However, when inbred, the progeny are found to be reduced in vigour, and numerous abnormal strains appear which often lack chlorophyll or in other ways are less efficient than the normal maize. Their occurrence provided an explanation for at least part of the injurious effects of inbreeding. Plants which are normally cross-fertilized will usually be heterozygous for many genes, since every individual inherits a set of genes from each of its two parents. If the parents differ in a number of their genes, as they will do in a cross-fertilized species, then the two sets of genes in each individual will not be identical. However, on inbreeding, the progeny become homozygous for many of these genes; that is to say, a larger proportion of genes derived from one of the parents become identical with those derived from the other. It was thus established that the vigour of an individual varies directly with its heterozygosity.

As an outcome of his work with maize, and as an explanation of hybrid vigour, East[32] postulated in 1909 that when two strains differing in gametic structure are combined, development is stimulated. On this theory, hybrid vigour was a property of heterozygous individuals. It decreased on self-fertilizing because the progeny became more homozygous. This hypothesis was favoured also by Shull[33] and in order to emphasize the supposed connection between heterozygosis and hybrid vigour, Shull[34] coined the word 'heterosis' for the latter, as a shortened version of heterozygosis. He intended the word to be a substitute for such expressions as 'stimulus of heterozygosis', 'the stimulating effects of hybridity' and 'stimulation due to differences in uniting gametes'.

If this 'heterozygosity theory' of the cause of hybrid vigour was true, it would mean that plant and animal breeders would not readily be able to take advantage of hybrid vigour in attempting to obtain improved strains, since from its very nature the heterozygous condition will not breed true. An alternative theory, which may be called the 'dominance hypothesis', was also discussed at this time, but was not seriously put forward until 1919. It was known from the breeding work with maize that self-pollination led to the appearance of numerous aberrant forms. On further study, these were shown to be mostly due to single recessive genes, the dominant allelomorphs of which gave rise to the normal condition. The recessive character of such mutants explained why they became evident on inbreeding. If each individual of maize carried a number of recessive deleterious factors, very few of these would be evident in the progeny on cross-fertilization, since the two individuals that were crossed might be expected to carry different harmful genes and so each would bear the normal allelomorphs of the other's deficiencies. A possible explanation of hybrid vigour would thus be the absence (in homozygous form) of deleterious recessive factors. However, if hybrid vigour depends on the presence of a number of dominant genes, it should be possible to isolate pure-breeding strains from the initial heterozygote. These strains would be homozygous for all the dominants, and so would be as vigorous as the original heterozygote. However, it was well known that hybrid vigour could not normally be transferred to a pure-breeding strain, and it was for this reason that the 'heterozygosis theory' had previously been favoured, and the 'dominance theory' not seriously considered. East and Jones[35] in 1919 now suggested that the failure to obtain strains homozygous for all the dominants was because of 'linkage' between the genes concerned.

The phenomenon of 'linkage' had been discovered by Bateson, Saunders and Punnett[36] at Cambridge in 1906 when working with *Lathyrus odoratus*, and many instances of it were soon added when the study of the genetics of *Drosophila melanogaster* Meig. (fruit-fly) was begun by Morgan and co-workers at Columbia University. Although there was still opposition to the chromosome theory in some quarters, the theory may be said to have become established by the publication of *The Mechanism of Mendelian Heredity* by Morgan, Sturtevant, Muller and Bridges in 1915. It had been apparent ever since the rediscovery of Mendel's work that in every species of plant or animal the number of character-differences determined by genes must greatly exceed the number of chromosomes. It was to be expected, therefore, that some character-differences would be associated with others in inheritance because they were due to genes on the same chromosome. However, the puzzling phenomenon when such 'linkage' of characters was discovered was not that linkage occurred, but that it was not complete. In other words, it was found that linked character-differences stayed together in a majority of the progeny, but not in all. On the chromosome theory, this partial linkage implied that exchange of parts must occur between the chromosomes of one set and the corresponding ones of the other set (the two sets being derived one from each of the two parents of the individual). This idea naturally met with considerable opposition, for the chromosomes had not been seen exchanging parts, and it was fundamental to the concepts of Weismann, upon which the chromosome theory was based, that the germ-plasm showed continuity from generation to generation. If then the chromosomes were not handed down intact, could they really be the carriers of the Mendelian factors? However, it was realized that there was opportunity for exchanges to occur between corresponding chromosomes when they were closely associated at the beginning of meiosis, and by 1915 the *Drosophila* breeding work had provided overwhelming evidence in favour of the chromosome theory, including such an exchange of parts in meiosis.

On the 'dominance theory' of hybrid vigour, as presented by East and Jones, quantitative characters such as height and weight were controlled by a large number of genes distributed through all the chromosomes. Each individual carried a number of recessive genes causing reduced height or weight, but different individuals carried different deleterious genes of this kind. Hence, a cross-bred individual could be vigorous because it might possess all, or nearly all, the dominant genes for increased height and weight. But, it was impossible to obtain a pure-breeding individual carrying all these complementary dominant genes, because each chromosome contained harmful recessive factors as well as favourable dominants, and the process of exchange (crossing-over) between corresponding chromosomes at meiosis does not allow the free assortment of these genes. In consequence, it is not possible even by selecting the most vigorous progeny in each generation to avoid including some injurious recessive genes in the constitution when attempting to obtain a vigorous pure-breeding strain from the initial heterozygote. Nevertheless, the 'dominance hypothesis' offered much more hope to the plant and animal breeder than did the 'heterozygosis theory', for on the 'dominance theory' it might eventually be possible to incorporate hybrid vigour, to some extent at least, into a pure-breeding strain.

Darwin's data may be said on the whole to fit the dominance hypothesis better than the heterozygosity theory, for it is difficult on the latter view to account for the success of regularly self-fertilized species where there can normally be no heterozygosity. Darwin's anomalous findings, such as the *Ipomoea* which he called Hero, may be accounted for on the dominance theory as due to the chance selection of an individual containing most of the dominant genes for increased vigour. That it bred true, and that the progeny showed no heterosis on intercrossing, indicate that Hero was homozygous for the genes for vigour, a finding which is again in conflict with the heterozygosity theory.

For a full explanation of the effects of inbreeding and outbreeding in terms of the dominance theory, it was necessary to explain why disadvantageous genes should tend to be recessive. Fisher[24] proposed in 1930 a theory of the evolution of dominance which has now received wide acceptance. He suggested that disadvantageous genes are maintained in a population by the process of mutation, and that the frequency of such mutation is quite low, but sufficient to balance the slow loss of such mutations from the population by natural selection. Fisher further suggested that these mutations had occurred in the past, and that, probably over immense periods of time, natural selection had constantly favoured the existence of other genes which tended to suppress the effects of deleterious mutants when these are heterozygous, or, in other words, tended to make the mutant recessive.

It is nearly 50 years since the heterozygosity and the dominance theories of the cause of hybrid vigour were first suggested, but it is still not possible to decide which is nearer the truth. Indeed a sharp division of opinion on the matter still exists, and it may be that both theories are responsible in part for heterosis. In studies of the mechanism of gene action in Fungi, there are examples which fit each hypothesis. Numerous instances are known of biochemically deficient strains due to recessive genes. When two different strains of this kind are grown together, each has the dominant normal allelomorph of the other's recessive gene, and so each can supply the other's deficiency.[37] Thus the combined strains (equivalent to a heterozygote) show a vigour which may be compared with what would be expected in a hybrid on the dominance theory. On the other hand, a single instance is known where a pair of allelomorphic genes produce vigour, where either alone does not.[38, 39, 40] This provides support for the heterozygosity theory.

At the present time, the dominance theory of hybrid vigour is favoured by Mather[41] and Jinks,[42] who are impressed by the lack of heterozygosity in species which are normally self-fertilized, and by the inability to prove that any specific pair of alleles in a heterozygote is contributing more to hybrid vigour than is the dominant gene alone. Lerner,[43] Haldane,[44] and Lewis,[45] on the other hand, favour the heterozygosity theory. They have been impressed by the remarkable uniformity of the first-generation hybrids between two inbred lines in species which are normally cross-fertilized. It has been established in numerous species, including both plants and animals, that such hybrids are even more uniform than the individuals of an inbred line. It seems as if these heterozygotes are less influenced by the effects of environmental changes during development. This has led to the suggestion that in such heterozygotes both allelomorphs contribute to development in slightly differing ways and hence cause greater biochemical versatility. In support of this idea, a number of instances are known where two allelomorphic genes function together in a heterozygote. But do they in fact contribute to hybrid vigour? Jinks and Mather[46] suggest that stability in development leading to uniformity in the appearance of individuals of a population has evolved in response to selection. They find that stability of the homozygote in species which are normally self-fertilized may be as great as that of the heterozygote in species which are normally cross-fertilized. They therefore consider it unnecessary to postulate that recessive allelomorphs in the heterozygous state are contributing to such stability.

By making quantitative studies, Darwin took the first step in resolving the problem of the effects of inbreeding and outbreeding. To-day, 80 years later, despite the enormous advances in our knowledge of inheritance and variation, the problem still remains. It now appears as if hybrid vigour may be due primarily to the presence of complementary dominant genes for increased vigour, but perhaps also partly to the recessive allelomorphs contributing something to the development of the individual.

Developments in Knowledge of Mechanisms for Favouring Cross-fertilization

It will be recalled that Darwin was the first to appreciate fully the significance of flower structures as adaptations to favour cross-pollination between different individual plants, and not merely between different flowers, which might be on the same plant. He was thus led to regard the structural and colour adaptations of flowers and, in particular, dichogamy, in a new light. However, the phenomenon to which he attached the greatest significance in *The Effects of Cross- and Self-Fertilization* was prepotency, whereby pollen from one individual had an advantage on the stigmas of the flowers of another, compared with the latter's own pollen. He demonstrated this with *Mimulus luteus, Iberis umbellata* L. (candytuft), *Brassica oleracea* L. (cabbage, etc.), *Raphanus sativus* L. (radish) and *Allium cepa* L. (onion), by showing that when a flower was simultaneously self-pollinated and cross-pollinated with another variety of the same species, a majority of the progeny showed the character of the strain used for cross-pollination. This prepotency of cross-pollen, which he had discovered, and particularly the more extreme condition of self-sterility, which he established as occurring in many plants (e.g. *Passiflora caerulea* L. (passion-flower)), have since been the subject of much research discussed later. These studies have been made chiefly in the last half-century, following upon the recognition that Mendelian inheritance is widespread.

In *The Different Forms of Flowers*, Darwin discusses other devices for favouring cross-fertilization, and these have also been investigated further in the last 50 years. In particular, there have been considerable advances in our knowledge of heterostyly, dioecism and gynodioecism. Darwin had discovered that heterostyly is associated with partial or even complete self-sterility, and furthermore that such sterility extends to all individuals of the same flower type. Moreover, in long-styled *Linum grandiflorum* he showed that this sterility was due to a failure of the pollen-tubes to grow down the style. Similarly, Müller[47] had shown that in *Eschscholtzia californica* Cham., which is self-sterile, self-pollen tubes did not penetrate far into the stigma. Darwin thus reached the conclusion that self-sterility was due to a lack of differences between the pollen and the stigma or style, and the consequent inability 'to excite the mutual action of the sexual elements'. He found that there was no close correlation between the degree of self-sterility on the one hand and the magnitude of hybrid vigour on the other. This was understandable if self-sterility was due to a failure of pollen tubes to penetrate the stigma, and if the vigour of crossed progeny depended on the contents of pollen and ovules.

The first important advance in knowledge of self-sterility was made by Correns.[48] A few years after discovering that inheritance was particulate, he undertook a study of self-sterility in *Cardamine pratensis* L. (cuckoo-flower, family Cruciferae). He took two individuals, intercrossed them, and then back-crossed the progeny with both parents, possible in *Cardamine pratensis* because it is a perennial plant. Correns found that the progeny could be grouped, on the results of these crosses, into four classes according to whether they were fertile with both, with one, with the other or with neither parent. The numbers of individuals in each class were 16, 16, 14 and 14, respectively. This clear-cut segregation into four approximately equal classes led Correns to suggest that Mendelian inheritance with two pairs of factors was involved, such that sterility resulted whenever the individuals crossed had the same dominant gene in common. Further data of his did not fully confirm this hypothesis, and years later it was found that *C. pratensis* is unsuitable for detailed study of the inheritance of self-sterility because it is polyploid, that is, it has multiple sets of chromosomes, instead of only two, in the cell-nuclei. Nevertheless, Correns had made the important discovery that cross-sterility can occur in self-sterile plants when the individuals that are crossed have a particular sterility gene in common. Darwin did not encounter examples of cross-sterility between different self-sterile individuals, except in the special case of the self-sterility and intraclass sterility associated with heterostyly. Consequently, he regarded self-sterile plants (apart from heterostyled ones) as fertile with all other individuals of the species. Correns had now shown that some cross-sterility could be expected.

The next important advance in knowledge of self-sterility was the discovery of the genetic basis of this sterility in a *Nicotiana* hybrid (family Solanaceae) by East and Mangelsdorf[49] working at Harvard University. They used a hybrid between *N. alata* Link & Otto and *N. forgetiana* Hort. which had been inbred for several generations. The hypothesis which they put forward to account for their results had four essential features: First, failure of pollen-tube growth down the style occurred when the plants involved had a sterility gene in common. This was the principle first evoked by Correns. Secondly, there were many such genes, each causing self-sterility, and they formed an allelomorphic series. That the genes were allelomorphic meant that it was possible to have only one of them in each set of chromosomes, and so a normal individual would have two. But different individuals would probably have a different pair, because the whole interbreeding population apparently carried a large number of these genes for self-sterility.

The third postulate of East and Mangelsdorf was a remarkable and almost unprecedented one. It is usual in the formation of pollen-grains for their various characters to be predetermined by the influence of the anther tissue. However, East and Mangelsdorf postulated that in *Nicotiana* the self-sterility genes carried in the pollen act independently of the parent tissue. Hence at meiosis, when the chromosome number is halved, two kinds of pollen-grains will be formed, for half of them will receive one sterility gene and half another. This principle of the segregation of genes at gamete-formation was of course first postulated by Mendel.

The fourth feature of East and Mangelsdorf's hypothesis was that in the stigma and style, where of course there is no reduction in chromosome number, the two sterility genes in each cell act independently. The consequences of this hypothesis were that in a population containing about 100 different sterility genes, such as appear to be present in many instances, every individual would be self-sterile, but at least half of its pollen would be able to grow in the style of almost every neighbouring plant. Any particular individual plant would be expected to be cross-sterile with only about one in every 5000 individuals in the population. It was therefore no wonder that Darwin did not detect cross-sterility in the self-sterile plants with which he worked, if a similar mechanism applied.

East and Mangelsdorf's hypothesis received immediate confirmation, for Filzer,[50] working at Tübingen with *Veronica syriaca* Roem. & Schult. (family Scrophulariaceae), had independently put forward the same hypothesis. This genetic basis for incompatibility has since been found to occur in certain other members of these families of flowering plants, namely, in species of *Petunia* and *Solanum* (family Solanaceae) and of *Antirrhinum*, *Nemesia* and *Verbascum* (Scrophulariaceae). It has also been discovered in three other families: Papilionaceae, where it is known in species of *Trifolium* (clover); Rosaceae, in species of *Prunus* (cherry), *Malus* (apple) and *Pyrus* (pear); and Onagraceae, in species of *Oenothera*.[51] In many cultivated varieties of cherry, apple and pear, all the individuals of a particular variety are both self- and inter-sterile, so that in an orchard it is necessary to interplant different varieties in order to obtain good crops of fruit. The intravarietal sterility is a consequence of the vegetative method of propagation, by cuttings and grafting, so that every individual carries the same pair of sterility genes. A similar situation exists in *Veronica filiformis* Sm. (long-stalked speedwell), where individuals are self-sterile and largely inter-sterile owing to vegetative propagation.[52]

Another important development in knowledge of the genetics of self-sterility in flowering plants occurred in 1950, when it was shown that certain members of the family Compositae have a different genetic basis for self-sterility from that found in *Nicotiana*, *Veronica*, etc. Gerstel,[53] working in California with *Parthenium argentatum* A. Gray (guayule), and Hughes and Babcock,[54] also in America but using Yugoslav material of *Crepis rhoeadifolia* M. Bieb., had independently put forward the same hypothesis to account for their observations. Their hypothesis agreed with that of East and Mangelsdorf for *Nicotiana* with regard to the first two features mentioned above, that is, incompatibility of pollen and style was manifest when the plants concerned had a sterility gene in common, and many such genes, forming a series of allelomorphs, were involved. However, in *Parthenium* and *Crepis*, unlike *Nicotiana* and *Veronica*, the self-sterility genes acted in the more normal way in the development of the anther, with the result that all the pollen-grains produced in an anther behaved alike when placed on the stigma of a particular flower. The fact that the genes segregate at meiosis, prior to pollen-formation, has no effect on the behaviour of the pollen because the substances, whatever they may be, which cause the incompatibility, have evidently already formed in the pollen-mother-cells of the anther and are carried over to the pollen-grains in the cytoplasm.

A further difference from the *Nicotiana–Veronica* system is that the self-sterility genes may show dominance. This applies to both anther and style. It appears to be normal for some of the genes to be dominant, some to be recessive, and some to act independently of one another.[51] The type of gene action in any instance will depend on the tissue (anther or style), and on which particular allelomorphs are involved.

Mechanisms Favouring Cross-Fertilization

The *Parthenium–Crepis* type of incompatibility, with the anthers controlling the behaviour of the pollen, leads to greater cross-sterility than does the *Nicotiana–Veronica* system, with the pollen controlling its own behaviour. The cross-sterility is most frequent if the genes in the *Parthenium–Crepis* system show independent action, that is, without dominance. However, when, as normally, there are a large number of the incompatibility genes in the population, there is little more cross-sterility than with the *Nicotiana–Veronica* system, except between close cousins. *Cosmos bipinnatus* Cav., another member of the family Compositae, has been found by Crowe[55] to show the *Parthenium–Crepis* type of inheritance of self-incompatibility, and Bateman[56, 57] has shown that this system is of widespread occurrence in the family Cruciferae. He has demonstrated the occurrence of this type of self-sterility in *Iberis amara* L. (wild candytuft) and has shown how earlier work by various authors with *Cardamine*, *Brassica*, *Raphanus* and *Capsella* is in agreement. It will be recalled that Correns[48] postulated dominant genes for self-sterility in *Cardamine pratensis*. Such dominance is a frequent feature of the *Parthenium–Crepis* system, but is absent from the *Nicotiana–Veronica* system.

The phenomenon of 'prepotency' of cross-pollen over self-pollen in self-fertile species to which Darwin attached such importance has not been investigated as much as self-sterility. This is no doubt because the study of the inheritance of prepotency is more difficult than that of self-sterility owing to the ability of self-pollen to bring about fertilization when cross-pollen is excluded. The normal method of studying the inheritance of incompatibility is to cross two individuals and then if possible to back-cross the progeny with the parents, and also to intercross the progeny. The fertility of each cross is recorded, and the individuals classified accordingly. With 'prepotency' such a classification is impossible since all crosses are fully fertile. However, there is circumstantial evidence that 'prepotency' and complete self-sterility are two aspects of one phenomenon and merely differ in degree, for the two conditions tend to occur in related plants. Thus Darwin[58] found 'prepotency' in Scrophulariaceae (*Mimulus*), Rosaceae (*Fragaria*) and Cruciferae (*Iberis, Brassica, Raphanus*), which are families in which self-incompatibility also occurs. Bateman[59] has shown that another member of the Cruciferae, *Cheiranthus cheiri* L. (wallflower), is also 'prepotent', or, to use Bateman's terminology, shows 'cryptic self-incompatibility'. Darwin also found 'prepotency' in the family Liliaceae (*Allium, Tulipa, Hyacinthus*) and in Ranunculaceae (*Anemone*). Self-sterility is known to be of widespread occurrence in Liliaceae and also to occur in Ranunculaceae (East,[60] Fryxell[61]) although the genetic basis has not yet been established in these families.

As regards the mechanism of self-sterility and 'prepotency', Darwin rightly surmised that the poor growth of self-pollen-tubes was due to a lack of differences between pollen and stigma or style. No progress was made in understanding what took place when such a plant was self-pollinated until the genetic basis was known. Then East[62] suggested that when pollen and style carried the same sterility gene, pollen-tube growth was inhibited owing to a reaction between an antigen in the pollen and an antibody in the style. Lewis[63] has demonstrated that the incompatibility genes in the pollen of *Oenothera organensis* Munz. do produce antigens, following an earlier discovery that below 25° C. the growth rate of incompatible pollen-tubes diminishes with increase in temperature, indicating that growth is actively inhibited.[64] In *Petunia* a protein fraction, not present after compatible pollinations, appears in the style after incompatible pollinations and is presumably the antigen–antibody complex.[65]

With incompatibility of pollen and style of the *Parthenium–Crepis* type, pollen-tube growth is inhibited on the stigma, whereas

---

Plate V

Heterostyled plants. *Hottonia palustris* L. (i) Short-styled (thrum): the five stamens project above the mouth of the flower but the stigma does not (× 2); (ii) long-styled (pin): the greenish translucent stigma projects above the mouth of the flower (note shadow in left-hand flower), while the five anthers are half hidden in the corolla tube (× 2). *Lythrum salicaria* L. (iii) Long-styled, (iv) mid-styled, (v) short-styled. The long stamens have purplish anthers, while the mid and short have yellow (all × 1·5).

Dioecism. *Valeriana dioica* L. (vi) Male plant on the left, female on the right (× 1·5).

Gynodioecism. *Glechoma hederacea* L. (vii) Female plant on left, hermaphrodite on right. Two pairs of anthers can be seen in each hermaphrodite flower (× 1).

Cleistogamy. *Viola riviniana* Rchb. (viii) The capsule in the lower axil has developed directly from a cleistogamous flower such as that in the upper. This plant, photographed in July, had produced a profusion of normal flowers in the preceding April (× 1).

PLATE V

(i)

(ii)

(iii)

(iv)

(v)

(vi)

(vii)

(viii)

in the *Nicotiana–Veronica* type it is inhibited during growth down the style. This difference is probably related to the fact that in the former the control of pollen behaviour is initiated in the anther, allowing the antigenic substance or substances to be fully synthesized by the time the pollen is mature. In the *Nicotiana–Veronica* system the antigen must develop after pollen formation, since the pollen controls its own behaviour. Bateman[57] has suggested that in the anther-control type (e.g. as in the Cruciferae), the antigen may be present in the walls of the pollen-grain, for in some species self-pollen germination is inhibited on the stigma, and the pollen-tubes never develop.

As for heterostyly, it is clear that Darwin's breeding work,[66] particularly with species of *Primula* and with *Lythrum salicaria*, laid the foundation for later studies. Considering distylic species first, he found that in *Primula* (*P. veris*, *P. vulgaris*, *P. auricula* L. and *P. sinensis* Lindl.), *Pulmonaria officinalis* L. (lungwort) and *Fagopyrum esculentum* Moench (buckwheat), 'the long-styled form transmits its form much more faithfully than does the short-styled, when both are fertilized with their own-form pollen'. In these plants, unlike *Linum*, he found it possible to obtain seed by pollinating a long-styled plant with pollen from similar individuals, provided that the pollen from short-styled plants was excluded. Similarly, short-styled plants could be inter-crossed, provided the prepotent pollen from long-styled individuals was excluded.

Later work has confirmed Darwin's finding that in these plants the long-styled form breeds true. Only in *Primula veris* and *Fagopyrum esculentum* did Darwin find a few short-styled progeny from long-styled crossed with its own kind, and these he attributed to atavism. To-day they would be attributed to stray pollen from short-styled individuals. With the short-styled plants crossed with one another, Darwin's total data (*Primula veris*, *P. auricula*, *P. sinensis* and *Fagopyrum esculentum*) were 128 short-styled progeny and 44 long-styled. If he had known of Mendel's work, he would no doubt have recognized how well his own data fitted a 3:1 ratio. Indeed his data for *Primula auricula* fitted such a ratio perfectly, for there were 75 short-styled and 25 long-styled progeny. Later work has confirmed that in all these species the short-styled plants carry a dominant gene in heterozygous condition and long-styled plants the corresponding recessive gene in homozygous condition.[67] It is thought that the complex of characters associated with distyly, namely, style length, stigma-papilla size, anther height, pollen-grain size, and incompatibility reactions of each kind of pollen and style, may be due to several closely linked genes. The occurrence of homostyled individuals such as Darwin described in normally distylic species would then be due in the first place to crossing-over, that is, exchange of parts between the chromosome carrying the thrum gene-complex and that carrying the pin gene-complex at meiosis (pollen or embryo-sac formation) in a short-styled plant. Such homostyled individuals have the anther and pollen characters of one form and the stigma and style characters of the other, and consequently are highly self-fertile, unlike the normal long- and short-styled forms. Darwin was surprised at 'the remarkable fidelity with which the equal-styled variation is transmitted after it has once appeared'. This is now understandable, since once the homostyle gene-complex has arisen it will be inherited in that form, apart from the rare events of mutation or crossing-over within it. Naturally occurring populations of homostyled *Primula vulgaris* are now known in Somerset and Buckinghamshire.[68]

Darwin's breeding work with *Lythrum salicaria* (Plate V) established that the long-styled form bred true, as in the distylic species, and that the other forms gave a proportion of long-styled progeny when they were crossed with other individuals of their own form. On the basis of Darwin's data, together with the results of further breeding work, Barlow[69] and von Ubisch[70] independently put forward the hypothesis that there was a dominant gene for the short-styled character, that another dominant gene on a different chromosome controlled the mid-styled character, such that the latter gene had no effect in the presence of the former, and that long-styled plants carried in homozygous condition the recessive genes for 'non-short' and 'non-mid' style. However, further breeding work gave results which were not wholly in agreement with this hypothesis.[71] An alternative theory was put forward by East,[72] but eventually Fisher and Mather[73] showed that the Barlow–von Ubisch hypothesis was basically correct. The puzzling results, on this hypothesis, obtained from certain crosses were explicable in terms of 'tetrasomic inheritance', as the species is tetraploid (carrying four sets of chromosomes). The inheritance of the tristylic condition in *Oxalis* also appears to fit the Barlow–von Ubisch hypothesis, except that in *O. valdiviensis* Barn. the gene for short-style is linked to that for mid-style.[74] Furthermore, in *O. rosea* Jacq. (von Ubisch[75]) and *O. articulata* Sav. (Fyfe[76]), the gene for short-style is recessive, that for 'non-short' being dominant. This is the converse of the situation existing in *O. valdiviensis* and *Lythrum salicaria*. However, all instances of tristyly so far investigated appear to be controlled by two pairs of Mendelian factors.

Considering the mechanism of the incompatibility reaction between pollen and style in heterostyled plants, it seems probable that in *Primula* it is similar to that in self-sterile plants without heterostyly, since Lewis[64] has shown that the inhibition of incompatible pollen-tube growth in the styles of *P. sinensis* and *P. obconica* Hance becomes more rapid as the temperature is increased up to 25° C., as already noted in *Oenothera organensis*.

In *Linum grandiflorum*, Lewis[77] has shown that a different mechanism is involved. It appears to be normal in flax for the pollen to have a suction pressure about four times as great as that of the cells of the stigma. This occurs in both of the compatible pollinations, and it allows the pollen to absorb water from the stigma, and swell, prior to germinating. However, long styles have a suction pressure about double (value 2 in arbitrary units) that of short styles (value 1), and corresponding to this, the pollen of short-styled plants has a suction pressure twice (value 8) that of long-styled (value 4). In consequence, when a long-styled plant is self-pollinated, the pollen fails to germinate because the suction pressure of the pollen is insufficient (being only twice that of the style), and when a short-styled plant is self-pollinated, the pollen-grains burst because the suction pressure of the pollen is so great (about eight times that of the style). In both cases, the pollen fails to bring about fertilization. This mechanism is highly effective, more so apparently than that of *Primula*, where the mechanism tends to break down at low temperatures. In tristylic plants, there is a physiological differentiation of the pollen from different anthers in the same flower, such that the pollen is compatible only on a stigma at the same level.

Fertilization in Plants

In the light of what is now known of the cause of intraclass sterility in heterostyled plants, it is interesting to compare this sterility with that usually encountered when species are crossed. It will be recalled that Darwin[66] regarded these two phenomena as comparable, and even went so far as to deduce from their resemblance that the sterility of species crosses was due to the incompatibility of their 'sexual elements'. Recent work has certainly confirmed that the intraclass sterility of heterostyled plants is due to incompatibility of pollen and style, but how far can interspecific sterility be attributed to the same cause?

It would appear that there are two primary causes of interspecific sterility, first, incompatibility of pollen and style, and secondly, death of the young embryo after fertilization. Incompatibility of pollen and style may take any of the forms already mentioned with reference to intraspecific incompatibility, such as failure of pollen-grains to germinate, reduced growth-rate of pollen-tubes, or bursting of pollen-grains or tubes. It is probable that a variety of mechanisms operate, just as with intraspecific sterility, such as antibody formation or unfavourable suction-pressure differences. It is known that in some instances the genes for self-sterility also cause interspecific sterility. Thus, when a self-incompatible species is pollinated by a related self-compatible species, pollen-tube growth is commonly inhibited, while the reciprocal cross is often fertile.[78]

Death of the young embryo is a frequent cause of interspecific sterility, particularly with closely related species. Development of the embryo may break down for a number of reasons. Two of the commonest are a genetic constitution which does not allow normal development, and a lack of harmony between the development of the embryo and the endosperm. Such disharmony is particularly liable to occur when the two species crossed have different chromosome numbers. Valentine[79] has suggested that in interspecific crosses in *Primula*, sterility is due to disharmony between the diploid tissue of the ovule and the triploid tissue of the endosperm, during the period immediately after fertilization when both tissues are developing rapidly, since the endosperm, unlike the ovular tissue, contains a set of chromosomes from the male parent.

How far Darwin can be said to have been correct in suggesting that the intra-specific sterility of heterostyled plants on the one hand, and the sterility between species on the other, both 'depend exclusively on the incompatible nature of their sexual elements, and not on any general differences in constitution or structure' obviously depends on the meaning attached to the term 'sexual elements', but his idea was certainly correct at least in part.

Turning now to the condition known as 'dioecism', that is, the occurrence of unisexual individuals, the chief advance in knowledge since the time of Darwin has been the realization that like most other inherited character-differences, separation of the sexes in different individuals has a chromosomal basis. The first instance of sex-linked inheritance to be discovered concerned characters in *Melandrium album* (Mill.) Garcke (white campion).[80, 81] Subsequent research has shown that in most dioecious flowering plants an *XY* chromosomal mechanism is in operation. In one sex, usually the male, there is an *X* and a *Y* chromosome, and in the other sex, two *X* chromosomes. The *X* and *Y* chromosomes usually have a pairing segment, where exchange of genes between them can occur at meiosis, and a differential segment, where no exchange occurs. The differential segment of the *X* chromosome carries the genes for the characters of the one sex, and the differential segment of the *Y* those of the other sex. The latter are evidently dominant, so that the *XY* individuals have the sex corresponding to the genes in the *Y* chromosome. Such a mechanism exists in *Melandrium album*, where the male is *XY* and the female *XX*.[82]

In some other dioecious flowering plants, variants of this mechanism occur.[83] Thus, in *Rumex acetosa* L. (sorrel), the differential segment of the *Y* chromosome appears to be inert. Here the genes for the female characters are evidently mostly carried on the differential segment of the *X* chromosome as in *Melandrium*, but the genes for the male characters are carried on the other chromosomes and not on the *X* or *Y* chromosomes. Sex in *R. acetosa* is therefore determined by the proportion of *X* chromosomes to other chromosomes, the *Y* chromosome playing no part. Two *X* chromosomes give a female, one a male. In *Fragaria moschata* Duchesne (hautbois strawberry) and *Silene otites* (L.) Wibel (Spanish catchfly), it is the female which has the *X* and *Y* chromosomes and the male two *X* chromosomes. In *Spinacia oleracea* L. (spinach), no

X and *Y* chromosomes can be distinguished, and the sex difference appears to be determined by a single pair of genes, with the male dominant and heterozygous and the female recessive and homozygous. A third allelomorphic gene, of intermediate dominance, gives rise to a monoecious individual.[84] The evolutionary significance of these diverse genetic mechanisms for achieving dioecism will be discussed later.

The condition known as gynodioecism, in which there are hermaphrodites and females in the population, was studied by Darwin, who regarded the increased seed output of the females compared with the hermaphrodites as the key to the success of the condition. Nowadays, gynodioecism is looked upon as another device to favour cross-fertilization. The inheritance of this condition was studied by Correns[85] who found that in *Silene cucubalus* Wibel (bladder campion), *Cirsium dissectum* (L.) Hill (meadow thistle) and *Plantago lanceolata* L. (ribwort), the hermaphrodites on self-fertilization or on crossing with the females produced both types in the progeny. However, no simple Mendelian explanation could be found for these results. In *Satureja hortensis* and *Cirsium oleraceum* (L.) Scop. (yellow thistle), hermaphrodites gave only hermaphrodites on self-pollination, and only females on crossing to females. Correns explained these results as due to cytoplasmic inheritance, the male-sterility being controlled by substances in the cytoplasm of the female plants. East[86] proposed an alternative hypothesis involving Mendelian inheritance, but this explanation was necessarily somewhat elaborate.

Lewis and Crowe[87] have pointed out the difficulties of any theory of the Mendelian inheritance of gynodioecism, since, seed for seed, the hermaphrodites contribute three times as much nuclear material to the next generation as do the females. This is because the hermaphrodites supply their own pollen and egg nuclei and also the pollen nuclei for fertilizing the egg nuclei of the females. In consequence, the females will be rapidly eliminated from the population unless they are much more fertile than the hermaphrodites. However, these authors have found evidence that in *Origanum vulgare* L. (marjoram) gynodioecism is determined by a dominant gene which suppresses the anthers to give a female, and by another dominant gene, on a different chromosome, which suppresses the action of the first gene and so restores the hermaphrodite condition. Females will therefore always have the first gene, either in homozygous or heterozygous condition, but never the second one. Hermaphrodites, if they have the first gene, must also have the second. This curious genetical situation is further complicated by the occurrence of self-sterility and lowered viability in the hermaphrodite plants when they are homozygous. Lewis and Crowe suggest that the reduced viability of these homozygous hermaphrodites is an adaptation to favour the maintenance of the females in the population. They further regard the greater seed output of the females as an adaptation in the same direction. Thus, the high seed production in the females, instead of being the basic reason for the success of the gynodioecious condition, as Darwin had supposed, is now considered to be an adaptation to maintain a considerable proportion of females in the population, and hence to favour cross-fertilization.

It is remarkable that in gynodioecious species the male-sterile individuals usually have smaller flowers than the hermaphrodites, e.g. *Glechoma hederacea* L. (ground ivy) (Plate V). Darwin explained this as due to the tendency to abortion spreading from the stamens to the petals. Since in entomophilous species the females cannot set seed unless visited by insects, one might have expected the females to display more conspicuous flowers than the hermaphrodites, instead of the converse. Darwin noted this tendency for the females to have a smaller corolla in entomophilous dioecious species such as *Euonymus europaeus* L. (spindle-tree), where some individuals are hermaphrodite, and *Rhamnus cathartica* L. (buckthorn), and it is well shown by *Valeriana dioica* (marsh valerian: Plate V).

In *The Effects of Cross- and Self-Fertilization* and *The Different Forms of Flowers*, Darwin confined his attention to flowering plants, and indeed at that time, little was known of cross-fertilizing mechanisms in other groups of plants. However, in the intervening 80 years numerous such devices have been discovered, and these will now be briefly reviewed.

In the Pteridophytes, Wilkie[88] discovered incompatibility in *Pteridium aquilinum* (L.) Kuhn (bracken). Individual prothalli, if isolated, usually fail to produce sporophytes, but when pairs of prothalli, derived from spores from the same individual fern plant, are grown alongside one another, both individuals of about half the pairs produce fern plants (sporophytes), but both individuals of the other half usually fail to do so. This evidence suggests that individual prothalli are largely self-incompatible and that two mating types exist amongst the prothalli from a single sporophyte. Prothalli obtained from spores from different sporophytes were always fertile when paired, pointing to multiple mating-types in bracken populations, probably controlled by allelomorphic genes. The physiology of the incompatibility mechanism is not yet known, but is thought to be due to an influence of the archegonial mucilage on the movement of the spermatozoids towards the egg-cell, and perhaps involves an antigen–antibody reaction.

This mechanism prevents self-fertilization of prothalli, and it allows any particular prothallus to be fertilized by spermatozoids from only half the sister prothalli (derived from spores from the same parent fern plant). However, this is a less efficient device to prevent inbreeding than is incompatibility of pollen and style in flowering plants, since with the latter each individual plant (sporophyte) is self-sterile. The greater efficiency of flowering plant incompatibility is a direct consequence of the sexual differentiation of the diploid sporophyte generation. In particular, the diploid stylar tissue can prevent the growth of all self-pollen tubes, whereas in the fern the haploid archegonia on the prothalli can inhibit the approach of only half the self-spermatozoids, using this term for spermatozoids derived ultimately from the same parent sporophyte.

In the Bryophytes (mosses and liverworts), the most important mechanism for favouring cross-fertilization appears to be separation of the sexes in different individuals. About half the species of Bryophytes are reported to be dioecious, and it has been established in many instances that the differentiation of the sexes has a chromosomal basis. In liverworts of the genus *Sphaerocarpus*, the female has a large X and the male a small Y chromosome.[89, 90] These chromosomes separate from one another at meiosis in spore-formation in the capsule. Since the sexual liverwort plants are haploid, the genes in the X chromosome for female characters act quite independently of those in the Y chromosome for male characters. The diploid sporophyte generation contains X and Y chromosomes, but shows no sexual differentiation. Dioecious mosses have no visibly differentiated X and Y chromosomes, but the sexes appear to be determined in essentially the same way by genes which segregate at meiosis.

Within the Algae and Fungi at least six different mechanisms for favouring cross-fertilization are known, of which two may be classified under the general heading of dioecism. First, in two groups of sea-weeds, the plants are diploid and dioecism is found in some of the species, e.g. *Fucus serratus* L. (serrated wrack, a brown alga) and species of *Codium* (green Algae). It is probable that this condition has a similar genetic basis to the dioecism of flowering plants. Secondly, dioecism in the haploid Algae is widespread, occurring in *Volvox*, *Oedogonium*, etc. in the green Algae, *Polysiphonia*, etc. in the red Algae, and *Dictyota*, *Laminaria*, etc. in the brown Algae. It also occurs in certain Phycomycetes among Fungi, e.g. *Achlya*. In these forms the condition probably has a genetic basis similar to the dioecism of the Bryophytes.

The other four outbreeding mechanisms of Algae and Fungi may be classified under the heading of 'heterothallism' or incompatibility. First, many haploid Algae and certain simple Phycomycetes have gametes that are alike morphologically, but are differentiated sexually into two mating-types or sexes, such that all the gametes from one individual are of one mating-type. Examples of this condition of 'primary incompatibility'[91] are provided by species of *Chlamydomonas*, *Ulva* and *Cladophora* in the green Algae, and of *Ectocarpus* in the brown Algae. Secondly, many Fungi may have two mating-types, but there are no gametes in the ordinary sense, and nuclear fusion is preceded either by fusion of non-motile sex organs or by fusion of undifferentiated hyphae. This 'secondary incompatibility' is found in many Mucorales, Ascomycetes, Uredinales (rust fungi) and Ustilaginales (smut fungi). Nuclei of opposite mating-types frequently become associated early in development after hyphal fusion, but nuclear fusion is delayed.

Thirdly, multiple mating-types, comparable to those in *Pteridium*, occur in many species of Hymenomycetes (toadstools) and Gasteromycetes (puff-balls). Each mating-type is controlled by a different gene, and all the genes are allelomorphic. This mechanism provides greater opportunities for cross-fertilization than does the two-mating-type system, since with a large number of mating-types, any two sporelings from different toadstools are likely to be compatible. Fourthly, in many other species of Hymenomycetes and Gasteromycetes an even more elaborate system has evolved, in which each mating-type is determined by two genes on different chromosomes.[92] Again there are multiple mating-types controlled by allelomorphic genes, and for compatibility the two sporelings must differ in both their mating-type genes. This mechanism reduces the chances of mating between the spores of a single toadstool, since a majority of these spores will have one or other of the mating-type genes in common.[93]

Little is known of the physiology of incompatibility in Algae and Fungi, but the two-mating-type system may be of a 'lock and key' character,[51] such that each strain provides something which is necessary for sexual reproduction, while the multiple-mating-type system depends on the inhibition of nuclear association (and hence of nuclear fusion) when sporelings have a mating-type gene in common. The latter mechanism would probably be of the antigen–antibody type, as in pollen and style incompatibility based on multiple allelomorphs.

To sum up the developments in knowledge since the time of Darwin of mechanisms for achieving cross-fertilization in plants, it may be said that, apart from some additions to knowledge of the morphological adaptations such as flower structure in relation to insect pollination, the main advance has been in the knowledge of physiological adaptations, notably incompatibility. Prepotency of cross- over self-pollen is now seen as a part of the general phenomenon of incompatibility of pollen and style. This incompatibility appears to fall into two main categories, one showing two (or three) mating-types and diverse physiology, as in heterostyled plants, and the other showing multiple mating-types and diverse kinds of gene action, but probably dependent upon a relatively uniform physiological basis involving antigen-antibody reactions. A similar range of types of incompatibility is found in plants with sexual differentiation confined to the haploid phase of the life-cycle, namely, the Algae, Fungi and ferns, but in these organisms cross-fertilization between the haploids from one diploid is never wholly excluded. For the same reason, dioecism in haploid organisms such as Bryophytes is less efficient as a cross-fertilizing device than dioecism in diploid organisms such as flowering plants.

The Development of Ideas on How the Various Mechanisms for Achieving Cross-fertilization have Evolved

An important sequel to the study of mechanisms for achieving cross-fertilization is the question of how these various mechanisms have evolved. Darwin considered that the morphological adaptations of flowers and also dichogamy and prepotency of cross- over self-pollen had evolved as devices to favour cross-fertilization, but dioecism, gynodioecism and self-sterility, whether associated with heterostyly or not, he regarded as largely of incidental occurrence. He was unable to believe that unless the means for cross-pollination already existed these restrictions on self-fertilization could be anything but harmful to the species, although he admitted of course that once they had arisen, such conditions led to regular cross-fertilization.

Another reason, already mentioned, why he did not believe that self-sterility had evolved to prevent self-fertilization, was the apparent lack of any close relationship between the degree of self-sterility and the extent to which the progeny suffer on self-fertilization. Although the increased vigour and fertility of the progeny have no doubt helped to favour the development of cross-fertilizing mechanisms, it is now thought that the main advantage of cross-fertilization is the long-term advantage unknown to Darwin of increased opportunity for gene recombination and hence increased hereditary variability and therefore adaptability. Thus it is now accepted that under certain environmental conditions it can be advantageous to have less progeny, but showing new gene combinations, and it is generally agreed that dioecism, gynodioecism and self-sterility, in addition to morphological adaptations, dichogamy and prepotency, have evolved in response to selection favouring cross-fertilization. It was of course not possible for Darwin to give any detailed account of the evolution of these mechanisms, since their mode of inheritance was not then known.

Turning now to modern views on the evolution of cross-fertilizing mechanisms, Mather[94] has pointed out in connection with self-sterility that the *Nicotiana–Veronica* system, that is, multiple-gene incompatibility of pollen and style with the pollen controlling its own reaction, requires a minimum of three self-sterility genes before it could begin to function. For this reason he thought it would be unlikely to arise often in evolution. The distribution of its occurrence supports this idea, for it appears to be characteristic of certain families of flowering plants such as Solanaceae, Scrophulariaceae and Rosaceae, as if it had originated but once early in the evolutionary history of each family.[95] The self-fertile species which occur within these families may show incompatibility of pollen and style in a cryptic form causing prepotency of cross-pollen. Furthermore, species showing no incompatibility are usually thought to have been derived from ancestors which possessed it, and to have lost it in evolution. The argument here is that individuals which are regularly cross-fertilized are better equipped than self-fertilized individuals to adapt to new conditions and hence to be the progenitors of new species.[96] Consequently it is commonly thought that the regularly self-fertilized species represent side-branches of the tree of evolution, although some authorities hold the opposite view. Thus Williams[97] is impressed by the tendency for annual species to be self-fertile and perennial to be self-sterile. This was first established by Kirchner[98] from data given by Darwin,[58] and also from further data of his own. Muntzing[99] pointed out that perennial species often have double (or a higher multiple of) the chromosome number of related annual species, suggesting that the perennials have evolved from annuals. Hence Williams concludes that self-sterility has arisen many times by mutation in previously self-fertile forms.

A difficulty encountered with Williams's hypothesis, however, is that such mutation is apparently extremely rare. Indeed Lewis,[51] in studies of pollen compatibility after X-irradiation, has failed to find a single instance of mutation to a new incompatibility allele in 350 million pollen-grains belonging to four different species. A number of mutations to self-fertility have been recorded, and these have shown that the incompatibility gene consists of two parts, which usually mutate separately—one concerned with the stylar function and the other with the pollen reaction.

The arguments given above concerning the evolution of the *Nicotiana–Veronica* system of incompatibility apply also to the *Parthenium–Crepis* system where the anther controls the pollen reaction. The latter system appears to have certain evolutionary advantages over the former, namely, more immediate inhibition of incompatible pollen, which often fails even to germinate on the stigma, and more frequent intersterility between individuals with recent common ancestors. A further advantage is that the anther-control system seems to survive doubling of the chromosome number (tetraploidy) better than does the pollen-control system. Evidence for this has been brought forward by Lewis,[51, 78] who has shown that self-fertility nearly always follows artificial doubling of chromosome number in the pollen-control system. This is not surprising since with this system, the self-sterility genes normally act quite independently of one another, in separate pollen-grains. Tetraploidy then introduces a new situation and it is found that the two sterility genes in each diploid pollen-grain interfere with one another. However, with anther control, it is normal (since the anther is diploid) for two genes to be involved and hence tetraploidy has little effect. Since polyploidy appears to have played a big part in the evolution of many species of flowering plants,[100] it would appear that the anther-control system is advantageous in maintaining self-sterility in such species.

The two- (or three-) mating-type incompatibility associated with heterostyly is believed to have evolved independently in most or perhaps all of the families of flowering plants in which it is known to occur. The reasons for this belief are the diversity of its physiological basis and its scattered distribution in families which are not closely related.[66] From the wide distribution of heterostyly in the family Rubiaceae, Darwin[66] inferred that most of the heterostyled genera had acquired their heterostyly independently of one another and not from a common progenitor. However, more information about the genetics, physiology and distribution of heterostyly in this family is still required. In the family Plumbaginaceae, Baker[101, 102, 103] considers that self-fertility has arisen a number of times from self-sterility, and that dimorphism of pollen, dimorphism of stigmas, and heterostyly probably represent successive additions to the original system of incompatibility, in response to selection favouring reduced wastage of incompatible pollen. Thus the incompatibility of pollen and style which Darwin discovered in heterostyled plants is now seen as the primary feature of their breeding system, with the morphological peculiarities of flower di- or tri-morphism as secondary additions.

Baker[101] has also shown that within the species *Armeria maritima* (Mill.) Willd. (thrift, sea pink), where self-sterile (dimorphic*) and self-fertile (monomorphic) subspecies occur, individuals of the latter show great uniformity from any one locality but considerable variation between populations, while with the former the variation occurs between individuals irrespective of source. Baker correlates this difference in the pattern of variation with striking differences in habitat-range. Thus the self-sterile subspecies *maritima* is found in a great diversity of habitats (sea-cliffs, shingle, salt-marshes, inland mountain rocks), while the self-fertile subspecies *sibirica* and *californica* are highly restricted in habitat. This illustrates how cross-fertilization leads to greater adaptability.

Crosby[68] has shown how self-fertility can become established in the normally self-sterile *Primula vulgaris*, even though such a change may be harmful to the species. The self-fertile homostyled primroses (with high anthers and long styles) which occur in certain areas of England, appear to have an advantage over the normal pin and thrum forms, simply because homostyle pollen is better placed than pollen from short-styled (thrum) plants to pollinate homostyle stigmas, while equally well placed to pollinate long-styled (pin) stigmas. These primrose populations provide an excellent example of the opposing evolutionary forces associated with cross- and self-fertilization which were first revealed by Darwin,[58] and which are now known to be evident throughout the plant kingdom. It is remarkable that even in quite small groups of related species such as genera, whether of Algae, Fungi, Bryophytes or seed-plants, it is usual to find some species adapted for self-fertilization and some for cross-fertilization. Thus Bateman,[57] from a study of 182 species of the family Cruciferae, found a marked tendency for each genus to contain some self-sterile and some self-fertile species. He infers that the outbreeding species are providing the genetic variability essential for future evolution by natural selection, while new in-breeding species which are continually arising from the outbreeders are continually being extinguished when environmental conditions

* Either with stigmas weakly papillose and pollen-grains coarsely reticulate, or with stigmas strongly papillose and pollen-grains finely reticulate.

change. Crosby[104] and Bateman[57] suggest that these self-fertile offshoots may not merely be the passive consequence of breakdown of the outbreeding mechanism in response to the ever-present immediate advantage of easier pollination and hence greater fertility, but that the inbreeding descendants provide a means of revealing recessive genes which were rarely expressed in their outbred ancestors. Thus the inbred progeny through selection can become closely adapted to particular specialized environments. An example of how self-fertilization reveals such recessive genes is provided by a comparison of *Primula sinensis* and *P. obconica* in cultivation.[78] *P. sinensis*, as Darwin discovered, is partially self-fertile, and numerous varieties have appeared since its introduction to Europe, whereas *P. obconica* is strictly self-sterile and has given rise to only a few new forms.

Mather and de Winton[105] and Mather[106] have shown that both the two-mating-type incompatibility of *P. sinensis* and the multiple-mating-type system of *Petunia violacea* Lindl. are under the control of numerous genes in addition to those directly responsible for the mating-types. This is not surprising when one recalls that the incompatibility reaction involves two highly organized structures, the pollen and the style, each of which represents the end-product of a long series of developmental processes, in which many genes are no doubt involved. Mather[106] believes that incompatibility systems must arise gradually as selection favours the assembly of genes for pollen and style development allowing full efficiency of the incompatibility mechanism. He[107] has pointed out that systems promoting inbreeding or outbreeding can be regarded as mechanisms for controlling the degree of heterozygosis in the population, and that in this respect incompatibility mechanisms are adaptable, since by changing the assemblage of genes concerned with the development of the pollen and style, the mechanism may be made either more or less efficient as a cross-fertilizing device. The high frequency of 'prepotent' species suggests that partial incompatibility, with the possibility of self-fertilization if cross-pollination fails, is often advantageous.

Dioecism, that is to say the occurrence of unisexual individuals, is now generally considered to be another device which promotes outbreeding.[96, 108] Its comparative rarity in plants compared with animals is attributed to its relative inefficiency in sedentary organisms. Lewis[83] has presented evidence that this unisexual condition has evolved many times in flowering plants from the bisexual condition. This evidence consists of the scattered distribution of dioecism and the diversity of its genetic basis. Furthermore, Darlington[96] has shown how the genetic determination of dioecism apparently follows an evolutionary progression, starting simply with one or two pairs of genes and ending with the complex condition where the genes for one sex are in the X chromosome and those for the other sex in the remaining chromosomes. Lewis[83] has pointed out that most dioecious flowering plants represent an early stage in this evolutionary series, and from this he infers that in most instances the condition has arisen comparatively recently and in evolutionary terms is usually short-lived. It must not be overlooked that the Gymnosperms (e.g. conifers, yews, cycads, etc.) are frequently dioecious, and that the flowering plants must have evolved from gymnospermous ancestors. However, with the exception of a few families of flowering plants where dioecism is well established and might be primitive, for example Salicaceae (willows and poplars), it appears certain that this condition is of secondary origin.

The gynodioecious condition, with females and hermaphrodites in the population, has almost certainly arisen from the hermaphrodite condition a number of times. The sporadic distribution of gynodioecism in flowering plants supports this idea. Only in one family, the Labiatae, is the condition abundant. More information about its inheritance in various members of this family is required before its widespread occurrence here can be fully understood. Lewis and Crowe[87] suggest that the gynodioecism of *Origanum vulgare* may have evolved from incompatibility, since certain of the hermaphrodite individuals are self-sterile.

The arguments presented above concerning the evolution of multiple-mating-type incompatibility and of two-mating-type incompatibility in flowering plants apply equally to the corresponding conditions in lower plants. Thus, multiple-mating-type incompatibility (heterothallism) in Fungi may have arisen early in the evolution of the Hymenomycetes and Gasteromycetes, such that all the species which possess it have received it through a common ancestor in which it arose, and the self-compatible (homothallic) species found in many of the genera would then be recent offshoots from which the condition has been lost.[109] Alternatively, it might be that the incompatibility has arisen independently a number of times by mutation.[110] With two-mating-type heterothallism such repeated origin seems more likely, since the condition could probably arise more readily than multiple mating-types. Furthermore, the two-mating-type condition occurs in many diverse groups of Algae and Fungi. Homothallic members of these groups are again most probably recent derivatives of heterothallic species through loss of the condition. In certain instances, for example *Neurospora tetrasperma* Sh. & D., there is good evidence that the homothallism is secondary, for each spore from which the fungus develops contains nuclei of both mating-types.[111] This situation is comparable to cleistogamy in flowering plants, where inbreeding has been superimposed on a previously existing outbreeding mechanism.

Just as two-mating-type incompatibility in flowering plants is thought to accumulate morphological differences between the two forms leading to distyly, so the corresponding condition in Algae and Fungi is thought to evolve morphological or further physiological differences. Thus the 'primary incompatibility' of many green and brown Algae, in which the gametes are alike morphologically, is thought to give rise to dioecism, probably through the accumulation of other genes linked to the two mating-type genes. This implies the gradual differentiation of the gametes and of the structures in which the gametes are formed into distinct male and female organs. The 'secondary incompatibility' of the rust and smut fungi is associated with another physiological property, parasitism,[51] and complementary genes for parasitism probably occur linked to the two mating-type genes. In *Ustilago striiformis* (Westd.) Niessl. (stripe smut of grasses), parasitism appears to be controlled by more than two allelomorphic genes, any pair of which will allow infection of the host to occur,[91, 112] while in *U. maydis* (DC.) Corda (maize smut) a similar situation appears to exist, except that the factors for parasitism and for mating-type are not linked.[113, 114] Such multiple allelomorphic factors for parasitism are a possible evolutionary source of the multiple mating-types of the Hymenomycetes and Gasteromycetes.[110] However, it is evident that more information about the genetics, physiology and distribution of the various outbreeding mechanisms in plants is needed before ideas about the evolution of these mechanisms can be regarded as much more than speculation.

THE OUTLOOK

It is evident that there have been immense advances in knowledge of the significance of cross- and self-fertilization in plants, and of mechanisms for favouring cross-fertilization, since Darwin's books on these subjects were published 80 years ago. A great many of these advances have been built on the foundations laid by Darwin. Thus, although he did not understand the causes of inherited variation, he rightly inferred that the advantages of cross-fertilization could be related to such variation. Again, it was he who discovered the incompatibility of pollen and style associated with heterostyly, and the prepotency of cross- over self-pollen in many other plants.

It is now of interest to consider what further developments may be expected in these fields of knowledge. On the basic question of how the differences between individuals arise, there is still much to be learned about the chemistry of the gene and of gene mutation, and about how genes act in development.

A further aspect of inherited differences between individuals is the question of the part played by the cytoplasm. A few instances of non-Mendelian or cytoplasmic inheritance were discovered soon after Mendel's work was rediscovered. For some time the condition was regarded as of exceptional occurrence, applicable chiefly to certain character-differences in the chloroplasts of green plants. However, numerous instances of cytoplasmic inheritance are now known, involving a wide variety of character-differences. They have been studied particularly in micro-organisms such as *Saccharomyces* (yeast)[115] and *Paramecium* (a ciliated Protozoon).[116, 117] It has further been suggested that all the really fundamental characters of organisms are inherited in the cytoplasm, and that Mendelian (nuclear) inheritance is merely concerned with the features which distinguish related species from one another.[115] However, most biologists favour the view that the nucleoprotein of the nuclear genes carries some sort of chemical code which at the appropriate stage in the development of the individual is translated into specific protein molecules, and that these proteins, some in the form of enzymes, control the major activities of the cell. The analysis of the respective parts played by nucleus and cytoplasm is one of the most important aspects of inherited variation which future research may be expected to resolve.

The seemingly intractable problem of the cause or causes of hybrid vigour will no doubt also ultimately be solved, although the full understanding of this phenomenon will require detailed knowledge of how genes act and interact in development. It is now well established that such characters as height and weight, which are greater in cross-fertilized plants, are controlled by a large number of genes which individually have only a small effect. Mather[118] has shown that such genes display dominance and linkage like the genes for qualitative characters, but it is an even more intricate problem to discover the part played by individual dominant genes and whether or not their recessive allelomorphs are contributing something to the development, and in particular to the vigour, of the heterozygote.

It will be many years before our knowledge of cross-fertilizing mechanisms in plants is anything like complete. This is apparent when one recalls that experimental studies are usually required before self-incompatibility can be recognized, since apart from the special type associated with heterostyly, incompatibility is unaccompanied by morphological peculiarities. Moreover, the number of species of flowering plants alone runs into hundreds of thousands, of which a considerable proportion are probably self-sterile,[60, 61] but of which only a small fraction have yet been studied experimentally from this point of view. That such studies are likely to be rewarding is suggested by two recent discoveries.

First, Lundqvist[119] working in Sweden has studied self-sterility in *Secale cereale* L. (rye) and in order to explain his results has put forward the hypothesis that in this species the incompatibility is of the *Nicotiana–Veronica* type, except that there are two series of multiple allelomorphs on different chromosomes and that for incompatibility both the sterility genes in a pollen-grain must also be present in the style. Confirmation of the occurrence of this genetic basis for incompatibility in the Gramineae is required, preferably using a perennial grass where progeny can be back-crossed with the parents.

Secondly, Knight and Rogers,[120] working in Ghana with *Theobroma cacao* L. (cocoa), have investigated another kind of incompatibility. All types of self-sterility in flowering plants previously described (apart from a few exceptions that have not been fully investigated) have been attributed to the inhibition of pollen by carpel tissue, either stigma or style, but the condition in *Theobroma* appears to be one of inhibition of ovules by pollen. Self-pollination leads to abnormal developments in the ovules, followed by shedding of the ovaries. It would be appropriate to describe the *Nicotiana–Veronica* type of self-sterility as 'pollen-carpel' incompatibility, and the *Parthenium–Crepis* type, with that of heterostyled plants, as 'anther-carpel' incompatibility since it is the constitution of the anther rather than that of the pollen which determines the incompatibility. In this terminology the condition in *Theobroma* would be called 'anther-ovule' incompatibility, with the pollen merely acting as a means of transport of the self-ovule poison. Furthermore there appear to be multiple mating-types in *Theobroma* controlled by allelomorphic genes which may be dominant, recessive or independent in both male and female tissues, as in the *Parthenium–Crepis* system.

Considering the evolution of the cross-fertilizing mechanisms of plants, any general agreement amongst biologists on how and when these mechanisms have evolved must await fuller knowledge of their genetics and physiology. When such information becomes available, a more complete picture of the general evolution of plants will have been gained, for knowledge of the breeding system of a species or group of species provides a clue not only to its past history but also to its potentialities for future evolution. From his studies of cross- and self-fertilization, Darwin concluded that the most important means favouring the fertilization of flowers by pollen from a distinct plant is the prepotency of cross- over self-pollen. To-day this would be associated with self-sterility, which is regarded as essentially the same phenomenon as prepotency. Mather[94] has pointed out that the interposition of the diploid stylar tissue between pollen and egg is directly responsible for the greater efficiency of incompatibility in flowering plants than in Fungi. He suggests that the primary significance of the stylar tissue is as a means of regulating the mating system of the plant. He regards the ovule-inhibition of *Theobroma*, where the style plays no part, as so relatively inefficient that it is difficult to see how it could arise at all by direct selection, but must be the by-product of some other development. This argument is comparable to Darwin's views on the intra-class sterility of heterostyled plants, where self-sterility of individual plants has been bought at the price of inter-sterility of half (or one-third) of the population. The *Theobroma* system appears to be a degenerate form of the *Parthenium–Crepis* type of incompatibility, in which the reaction has been transferred from the stigma to the ovule, and reversed in direction so that the ovule instead of the pollen is inhibited. However, it is difficult to imagine what advantageous process could give rise to such a by-product.

It appears that 'pollen-carpel' and 'anther-carpel' incompatibility, with multiple mating-types, are the outstanding cross-fertilizing mechanisms of flowering plants, with advantages over any other devices known in the plant kingdom. Not only is self-pollination prevented, but each individual is fertile with almost every other in the population. It has been suggested that the advantages conferred by one or other of these two types of incompatibility were primarily responsible for the successful evolution of the carpel, the hermaphrodite flower and indeed the flowering plants themselves from their gymnospermous ancestors.[121] It is evident that the study of cross-fertilizing mechanisms in plants, based on the foundations laid by Darwin, may provide valuable information not only on how these mechanisms have evolved, but also on the evolution of the plants themselves.

6

J. S. WILKIE

BUFFON, LAMARCK AND DARWIN: THE ORIGINALITY OF DARWIN'S THEORY OF EVOLUTION

Empedocles, as is well known, produced an odd theory of the origin of living things. He suggested that, in the beginning, there were limbs and parts of limbs, heads and trunks of various kinds lying about loose; that these came together by chance, and that, of the combinations so produced, only the workable ones survived. It would be possible to attempt an account of the origin of the theory of evolution along the same lines: taking Maupertuis's brain, Buffon's hands and eyes, Lamarck's trunk and legs, and filling up with smaller parts of lesser-known authors, we might produce something having a certain resemblance to Darwin, and thus suggest that there was nothing original about him, and that the *Origin of Species* is not the original and epoch-making work it is commonly supposed to be.

That the publication of the *Origin of Species* did mark an epoch seems to be an indisputable matter of history, and it is not easy to see how this could be, if the book had not been in some important way original; but it is not at once clear wherein its originality lay, for certainly all its major theoretical positions had been advocated by one writer or another before it was printed. It would be possible to devote the rest of this essay to a consideration of the nature of originality in scientific work in general, and to ask whether Darwin's work on evolution had any claim to be considered original in any of the senses discovered in our discussion: I shall, indeed, attempt to state in a very summary manner what I suppose the nature of his originality to have been, but my principal task will be to examine the work done by his two great predecessors in the field of evolutionary theory, and I hope that the importance and originality of his own work will then appear in as clear a light as historical study can give. I say 'his two great predecessors' because I feel very strongly that the theory of evolution must be considered as a scientific theory, a theory, that is, proposed to explain or systematize a set of facts, and that no one has any claim to be considered as a serious rival to Darwin in the 'discovery' of this theory who did not conduct his evolutionary studies upon a reasonably wide basis of fact. To have ideas, *aperçus*, is not enough, and it is the overvaluation of such clever but uncontrolled guesses which is apt to produce the ludicrous situation to which I have referred in my opening paragraph: a fallacy of combination, in which fragments of the final theory are collected from widely scattered sources and are combined in such a way as to impugn the originality of him who was the first to see how such a synthesis was possible. I shall not, therefore, consider here all those authors, including Darwin's own grandfather, who may be considered to have anticipated Darwin on some particular point or points of theory.

Originality in scientific work may be of different kinds. Seeking for terms of comparison within the history of the biological sciences I am tempted to consider Darwin's work on evolution in comparison with the principal discoveries of Harvey, on the one hand, and of Mendel, on the other. What Harvey successfully attempted, was to establish an essentially synthetic theory, of which the two major terms had already been propounded by others, with, however, most inadequate empirical verification. These terms were: the notion of the circulation of the blood through the lungs, and the notion of its circulation through the body generally. In advocating the first of these, Harvey freely acknowledged his debt to Colombo,[1] and pointed out that even Galen[2,3,4] had suggested something of the kind: namely, that at least some of the blood might pass from one ventricle of the heart to the other through the lungs. Harvey either did not know, or had forgotten, that the existence of a general circulation had been suggested by Cesalpino;[5] but anyone who will read without prejudice the appropriate passage in Cesalpino's writings will, I think, readily admit that we have here the merest hint, and that the evidence given to substantiate the notion is grossly inadequate. Indeed, I cannot convince myself that Cesalpino had in mind a constant and rapid circulation; it seems to me that what he writes is consistent with the supposition that he imagined the blood to pass outwards from the heart during the day, and to return to it at night. But one may readily concede that both the major theoretical terms of Harvey's theory had in fact been suggested before Harvey's time. Most of the facts which he used in his verification were also common knowledge, and some had received partial theoretical explanation: Galen, for example, understood the valves of the heart as valves,[3,4] though his explanations of their functions were imperfect. All these anticipations, however, do not in any way invalidate the claim of originality made for Harvey. What he did was to assemble the existing evidence, to add to it, and to use it to establish a clearly formulated theory.[6] The simple historical fact is that the general acceptance of the theory was due to Harvey's work.

Mendel's discovery,[7] however, is of a different kind. His extremely bold theoretical simplification had not, I believe, been anticipated by anyone, and he certainly did not find a large body of well-established and generally accepted facts to challenge the application of his theory. Whereas it is possible to assert that all the important empirical evidence for Harvey's theory was either ready to his hand or was collected by him, the validity of Mendel's theory and its extremely wide powers could only be seen in comparatively recent times, many years after the theory was propounded. What is impressive in Mendel's papers is not the handling of great masses of varied evidence, but the extent to which he had developed the theoretical consequences of his hypothesis. It is on account of this development that we are so ready to consider Mendel as unquestionably deserving the title of originator; the originality of Harvey has to be judged by other criteria. One of these is the placing of Harvey's publications in the historical development of physiology. Where there is a high degree of internal development of a theory, itself essentially novel, we can assert originality even though, as in the case of Mendel, the work had no immediate effect upon contemporaries; but where the novelty of the theory is more questionable, or the development slight or absent, it is reasonable to deny the title of founder where no influence upon contemporaries can be shown. Thus it is justifiable to assert that Harvey's title rests partly upon the abundant evidence for the wide and immediate influence of his work, and to ask those who may wish to establish the rival claims of Cesalpino why it was that the Italian schools were so long accepting the theory,[8] if in fact Cesalpino, and not Harvey, was its discoverer?

Darwin's theory clearly resembles Harvey's rather than Mendel's, and by the test of rapid and widespread influence and, indeed, acceptance it unquestionably demonstrates its superiority to the theory proposed by Lamarck. Whatever may be the relative merits of the two theories, it is clear that Darwin contrived to make his presentation cogent, whereas Lamarck did not.

Consider the status of the general theory of evolution in 1859, when Darwin published the *Origin of Species*.* It seems certain that Darwin himself underestimated the popularity of the general theory among biologists,[9] but by no legitimate use of language could this theory be described as well-established at that time. Yet Lamarck's principal work upon the subject, the *Philosophie Zoologique*,[10] had been published just fifty years before, in 1809. If, now, we consider the state of affairs in 1909, fifty years after the appearance of Darwin's book, we see that it would be difficult to find any responsible biologist who would have questioned the validity of the general theory.

A test of the degree of cogency of Lamarck's arguments is supplied by the reception accorded to the general theory by the positivist philosopher Comte. In his *Cours de Philosophie Positive*,[11] published in 1838, Comte discusses the general theory of evolution and, surprisingly as it now seems, rejects it as without positive foundation. His rejection of the theory is based, it is true, partly upon his judgment that the causal mechanism suggested by Lamarck is totally inadequate, so that it is the Lamarckian theory which he rejects. But it would obviously have been open to him to reject the causal explanation as inadequate, while accepting the general theory; that he did not do so seems to me to be a measure

* By 'the general theory of evolution' I mean the theory which states that all living species have arisen by the modification of pre-existing species, and which generalizes this notion of the origin of species to include fossil forms, so as to postulate that all species, past or present, have arisen from one or a few primitive types. This theory says nothing about the causes of evolution.

of Lamarck's failure to provide the general theory with a convincing basis of fact. Comte was writing more than 25 years after the appearance of Lamarck's book; it is difficult to believe that he would have rejected the general theory of evolution had he been writing 25 years after the publication of Darwin's.

EVOLUTION AND THEOLOGY IN THE EIGHTEENTH CENTURY

Before commencing any study of the development of the theory of evolution in the eighteenth century, it is necessary to consider briefly the relationship of that theory to some of the theological ideas of the time.

Apart from certain traces of earlier theories, there is little in the *Origin of Species* which has any close relation to theology, and for the understanding of Darwin's ideas little or no knowledge of contemporary theology is required. In the *Origin of Species* itself, theology only intrudes in Darwin's unfortunate habit of contrasting evolution with 'special creation', as though 'special creation' were an alternative scientific theory. It is interesting to speculate on what might have been the reception of the *Origin of Species* had Darwin taken the supposition of spontaneous generation as the only scientific explanation of the origin of living forms which could be considered as an alternative to the theory of evolution. Open warfare with the theologians was by no means necessitated by the *Origin of Species*, though Darwin's later writings on the origin of man were bound to conflict with Christian theology.

It is clear, however, that in general, theology appears as something extraneous to Darwin's thought, not as a directing influence within it. With regard to Buffon, the position is somewhat similar, though he was necessarily more closely involved with theology than was Darwin, and this for two reasons. In the first place, orthodox theology wielded temporal power in France during the whole of Buffon's life, and the theologians of the Sorbonne, who exercised a form of censorship over books, caused him to issue, in 1753,[12] a recantation of some of his published opinions, which recantation he later characterized as 'sotte et absurde'.[13]

Secondly, although all the evidence inclines us to believe that he endeavoured to purge his mind of all influences from the side of theology, it was scarcely possible for a man of his time to do so completely. He, like all other intellectuals among his contemporaries, was to some extent infected with ideas originating in the heterodox theology of Deism. Though these influences were, in his case, neither powerful nor numerous, they can be more or less clearly observed, as will later be more fully explained, both in his early prejudice against schemes of classification, and in the traces of systematic optimism which prevented him from forming any clear idea of the struggle for existence.

In Lamarck, however, we encounter a theorist deeply and consciously imbued with the notions of Deism, and without some reference to the more influential of these notions it is really impossible to give an intelligent account of the structure and sources of his theory of evolution. That Lamarck seriously believed in the theological principles which he enunciated, there is no reason to doubt. All those of his works which we shall here examine were published before the final return of the Bourbons to power, so that we may suppose him free, when he was writing, from any external pressure: nor is there, in what we know of him, any indication of undue subservience to authority.

The central tenet of Deism is that the relation of God to the created world is essentially 'rational', that is, that the ways of God in dealing with his creatures are comprehensible to the unaided human reason. With this central notion were associated a number of subordinate propositions which could follow logically from the basic assumption only if it were also supposed that all God's purposes were known, and that to achieve them he would use only the means which recommended themselves to the common sense of the average enlightened man of the time.

One of the assumptions as to the purposes of God was that he must be supposed to create anything which is of its nature possible, or, in Leibniz's treatment of the subject, anything which is *compossible*[14, 15] with the existence of the other candidates for creation. Buffon, in an early essay, accepts this position,* which he expresses in the words, 'Il semble que tout ce qui peut être est'.[16] The idea

* Characteristically, however, he sees in it an argument against the deistic supposition that we can see the world as God sees it.

was commonly linked with the old axiom that 'nature makes no jumps', and when applied to taxonomy it suggested the existence of an infinity of forms and of infinite gradations between forms.

Why this principle should be supposed to result in the creation of a specifically linear series of forms, a Chain of Being, is not easy to understand. One would expect a Plenum of Being, rather than a Chain, and we do indeed find in the eighteenth century suggestions of other possibilities. These suggestions appear in connection with the search for a natural classification of plants and animals. Thus Linnaeus speaks of specific forms of plants being related taxonomically as the points on a two-dimensional surface, a map,[17] are related; and at least one writer[18] seems to have flirted with the idea that the relationships of species to one another could be expressed only by the use of all three dimensions. The linear scheme appears to have been suggested by rough empirical schemes of classification, for if the classes are sufficiently broad and sufficiently vague it is possible to arrange them in a descending scale, with man at the top, without doing too much violence to the facts. The easy and uncritical acceptance of this descending scale and of the general principle of linearity must be attributed to the extreme looseness of deistic thought. But, loose as it undoubtedly was, deistic thought was remarkably influential, and its influence was curiously tenacious. Among its tenets that of the Chain of Being[19, 20] was of cardinal importance in the development of biological thought in the eighteenth century. On the one hand, the transition from the notion of an infinitely graded series of separately created forms to that of a gradually developing temporal, or evolutionary series was extremely easy to those who, like the Deists, were not concerned with revelation, and consequently had not to fit their ideas to the Book of Genesis. On the other hand, the *a priori* assumption that the only natural classification of living forms must be a linear one was a considerable obstacle to the development of systems of classification, and it gave to Lamarck's evolutionary speculations an archaic framework which made inevitable a clean break between the Lamarckian and the Darwinian theories. The insistence upon a linear tendency in evolution, a kind of vector of development, was to the last a central feature of Lamarck's theory, long after he had realized that a linear scheme of classification was untenable.

Historically, this insistence upon linearity had the piquant consequence of constraining the development of the theory of evolution itself to become non-linear and discontinuous.

So tenacious was the influence of the Chain of Being that it was accepted as axiomatic by Comte,[21] and as late as 1851 Sir Charles Lyell thought it necessary to deny the existence of a linear scale of plant forms and to discuss that of a 'progressive chain' of animals.[9]

The idea of a Chain of Being, much as it was favoured by the Deists, did not originate with them: its roots can be traced far back into antiquity.[19] The same can be said of another axiom which they found well suited for incorporation into their system, the axiom that 'nature does nothing in vain'.[22] God, they supposed, could do nothing useless in creating the world, and since the main purpose of creation was commonly supposed to be the preparation of a convenient dwelling for man, the existence of anything not directly useful to man was apt to cause difficulties. Thus the possibility of any species becoming extinct, except through the destructive tendencies of ill-conditioned men, was frequently denied. Lamarck, who took his natural theology seriously, was much more hampered by this difficulty than was Buffon.

The relationship of Deism to the prevailing optimism of the first half of the eighteenth century is a subtle one. If God had made the world in accordance with human common sense and directed to human convenience, the expectation would certainly appear to be that the world would be a pleasant enough place to live in. If, however, we are to consider Leibniz's theology as a form of Deism, we have to admit that a facile optimism was not a necessary ingredient of deistic systems. There can, I think, be no doubt that Voltaire misunderstood the doctrine of Leibniz which is summarized in the expression 'the best of all possible worlds'. Voltaire supposed the stress to fall on the 'best,' whereas for Leibniz the stress fell heavily upon the 'possible', the implication being that the best of all possible worlds was still far from the most desirable of worlds.[23]

But though truth obliges us to be cautious in attributing optimism to this particular version of Deism, there is no doubt that the general climate of opinion tended to optimism in the early part of the eighteenth century, and that traces of optimism survived the shock of the Lisbon earthquake in 1755. To this optimism and to the closely related expectation of good housekeeping in the affairs of nature must be attributed, I believe, the almost total absence of any reference, in the works of Buffon and of Lamarck, to the struggle for existence.

The Theory of Evolution in the Works of Buffon

INTRODUCTION

Though it is, of course, impossible to give any precise date to the first stirrings of the embryo which was to develop into the theory of evolution as we now know it, yet it is possible to begin a description of its development with events occurring in 1749.

In this year Diderot published in London his *Lettre sur les Aveugles*,[24] in which, in a scandalously mendacious account of the death of Saunderson, the blind mathematician of Cambridge, he makes use of the ancient Empedoclean idea of the origin of living things. In the same year Scheidt printed, for the first time, the *Protogaea*[25, 26] of Leibniz, in which Leibniz at least hints at the possibility of the transformation of species.

Finally, it was in the same year that Buffon brought out the first three volumes of his great *Histoire Naturelle*,[27] the further publication of which was to occupy the whole of the rest of his life, and to be continued for some years after his death by a collaborator, La Cépède. Thirty-five quarto volumes were published during Buffon's life, and nine volumes were added to the series after his death in 1788.

It would be impossible, I believe, to exaggerate the influence of this work on European thought during the second half of the eighteenth century, during the whole of the nineteenth century, and even up to the present time. But to claim that Buffon was *un des fondateurs du transformisme*[28] seems to me to be completely wrong-headed and to do justice neither to Darwin nor to Buffon himself. His place in the history of the development of the theory of evolution is of the highest importance, for he was the first to subject the theory to extensive criticism based upon a wide range of empirical evidence, but he cannot be said to have elaborated any theory of his own. There is no Buffonian theory of evolution, in the sense that there is a Lamarckian theory. The reasons for this failure, if indeed it is a failure, are not difficult to discover.

In the first place, the alternative of either special creation or evolution did not appear to him as an inescapable dilemma, for he had, as has well been pointed out by Jean Rostand,[29] a third explanation of the origin of species: that they arose by spontaneous generation. It may seem to us an outrageous suggestion that a man, or even a mouse, could arise directly from inorganic matter, but when Buffon was young (he was born in 1707) it was commonly believed that 'worms' often arose by spontaneous generation, and at that time the class of 'worms' was most imperfectly delimited, and, at least for many naturalists, included animals as highly organized as insects. Moreover, those who to-day take an entirely 'mechanistic' view of all living things must believe that the synthesis of even the highest animal from inorganic matter is in principle possible, though in practice the difficulties of such a synthesis may be insuperable.

It is clear, as I shall show by citation from his works, that Buffon was prevented from giving whole-hearted support to the theory of evolution, not only by the possession of an alternative theory of the origin of species, but also by the inadequacy of the causal mechanisms which he could understand. It is simply not the case that he had 'a very clear appreciation of the struggle for existence';[30] such a notion was extremely difficult for a man of the eighteenth century to entertain, especially for one whose formative years were passed in the first half of the century, with its general tendency to optimism and its belief in a rational and economical ordering of the world.

I do not mean that Buffon was incapable of thinking that, on occasion, some one species might exterminate some other. But he did not think of the competition between species as an element in the causal processes underlying evolution. Passages in his writings which seem to suggest an appreciation of the struggle for existence commonly belong, in fact, to a system of thought which is far removed from that of Darwin: for example, one such passage terminates in nothing more than a reference to the ancient maxim, 'The corruption of one is the generation of another'.[31]

Buffon, it is true, calculated that, by the tendency to a geometrical ratio of reproduction, 'in 150 years the whole terrestrial globe might be entirely converted into one single kind of organic matter';[32] but to quote this in support of the thesis that he understood natural selection is to neglect not only the general tendency of his works, but also the particular context of the calculation; for he is arguing in this place that it is a mistake to suppose that there is any special difficulty in accounting for the origin of living beings. 'Reflecting upon this kind of calculation', he writes, 'we become familiar with the curious idea that the organic is the most ordinary work of nature, and apparently that which costs her the least.'[33]

Near the beginning of the chapter in which the calculation appears, Buffon argues that we cannot be sure what is simple and what is complex in the real world of nature; plants seem more complex than stones, and animals than plants.

> This notion is correct in relation to us [that is, from our point of view], but we do not know whether in reality the one kind may not be as simple or as complex as the other, and we cannot tell whether a sphere or a cube comes more difficultly to nature (coûte plus ou moins à la nature) than a germ or some particular organic part.[34]

In fact, the whole of the immediate context of the calculation, so far from being an argument in favour of evolution by natural selection, is an attempt to make plausible the existence of indestructible 'living molecules'. A belief in these made it particularly easy for Buffon to attribute the widest powers to spontaneous generation, which became for him merely the recombination of such molecules.

Finally, it seems that Buffon's reluctance to produce a general theory of evolution was due in large measure to the empirical tendency of his mind. He was fond of theorizing on the grand scale, but extensive acquaintance with his works suggests that he, unlike Lamarck, did not attach great importance to such exercises. Thus, having begun the biological section of his great work* with an all-embracing theory of spontaneous generation, he seems to have adhered to this theory in later life rather because he thought it no worse than another, than because he thought it better. His mind was always open to evolutionary speculations, but he simply did not think the facts sufficiently known to justify a firm adherence

* The first volume is devoted to physical cosmology.

to the evolutionary theory. We may, I think, take a sentence* from a late volume of the *Histoire Naturelle*, the third volume of the *Suppléments*, published in 1776, as his last word on the subject: 'In general, the relationship between species is one of those profound mysteries of nature which man cannot investigate except by experiments which must be as prolonged as they will be difficult.'[35]

THE ARTICLE ON THE ASS

The task which Buffon set himself in the *Histoire Naturelle* was in itself vast and diffuse, embracing not only the natural history of animals and plants, but also physical cosmology, geology, mineralogy and anthropology. From the first he displayed a marked disinclination for systematic presentation. In the *Premier Discours*, which opens the first volume, he examines the system of Linnaeus, as it then existed, and rejects it together with all other attempts at systematic classification of plants and animals. He proposes to treat of animals in the order of their interest for man, beginning with man himself and taking next the domestic animals. As the great work proceeded he became aware of the necessity of some scheme of classification, and convinced himself that such a scheme could be at least largely 'natural', that is, based upon fundamental resemblances between the organisms themselves and not merely formed for the convenience of the naturalist; but from first to last the plan of the work, though it gradually became more orderly, lent itself to digressions and remarks which might almost be described as asides. The reader must not, then, be surprised to find Buffon's earliest remarks upon the subject of evolution introduced, as it were casually, into a chapter upon the natural history of the ass, and even the most conscientious study which falls short of reading the whole of the *Histoire Naturelle* from the first volume to the thirty-sixth will probably pass over one or more casual references to the subject. However, I have done my best in another place[36] to present a full study of Buffon's treatment of the theory of evolution, and since many authors have scrutinized his works precisely to find evidence for the claim that he was a founder of

* In the same place Buffon gives what is, I believe, his only reference to the struggle for existence. He asks whether it may not be the case that 'the weaker species have been destroyed by the stronger'.

evolutionary studies, or to examine this claim critically,[28, 37, 38, 39, 40] it is improbable that anything of importance has been passed over.

I shall consider here only the two comparatively extensive passages in which Buffon considers the theory more or less systematically: the article on the ass, in which he appears to reject the theory altogether; and the essay 'On the Degeneration of Animals', in which he allows that evolution may have taken place to a very limited extent. The article on the ass has given rise to a great deal of controversy, several authors having supposed that Buffon did not intend his readers to take seriously the arguments it contains.[28, 40, 41] I shall later give my reasons for thinking this interpretation an entirely mistaken one.

At least one historian[42] of science has thought to find evidence of a revival of Buffon's interest in the subject in some of the latest of the volumes of his work, those concerned with the natural history of birds. This evidence, however, is extremely inconclusive. Buffon probably never entirely lost interest in the subject of evolution, but the passages in the articles on birds which seem to show a substantial revival or increase of interest were written at a time when he was becoming more and more dependent upon the help of collaborators, and was beginning to use whole passages written by them, contenting himself with, at the most, emendations and alterations which were merely stylistic.[43] All we can conclude from these passages is that Buffon did not violently object to the scraps of evolutionary lore which they contain.

The first three volumes of the *Histoire Naturelle*, which were published together in 1749, contain no references to evolution. The second volume contains the curious calculation already referred to, showing the effects which would result from the unchecked multiplication of even one species, but these calculations, as I have pointed out above, are shown by their context to have nothing to do with the theory of evolution. The first volume contains a passage of some length of which the general theme is that 'it is possible to descend by almost imperceptible degrees from the most perfect creature to the least informed matter, from the best organized animal to the merest mineral'.[44] To take this for a reference to an evolutionary series, however, would be most unwise, unless other passages could be found referring explicitly to genetic links between the members of the series. In fact there is no such reference in these first volumes, and the passage from which a few lines have been quoted here is merely an allusion to the Chain of Being, such as one would expect in any general work on natural history written at that time. Indeed, the whole passage might be, and perhaps is, a paraphrase of a similar one in the works of John Locke.[45]

If it is true, as I believe, that there is no reference to evolution in the first three volumes, what had happened between the publication of these volumes and that of the fourth volume, in 1753, to direct Buffon's attention to the idea? For it is in this fourth volume that we discover the article on the ass.

The question is not difficult to answer, for we know that Buffon was at this time in correspondence with Maupertuis,[46] and the article with which we are concerned bears a footnote[47] referring to a published letter of that author. This footnote gives no precise indication as to which letter is intended, nor of where it was published before 1753, but we may confidently identify it with one printed in the collected works of Maupertuis. It is clear that the letter contains information about a family of which several members had six digits on both hands and both feet. This is the Ruhe family about which Maupertuis was able to publish fairly complete information relating to four generations, and the letter describing this family appears in the collected works of Maupertuis published in Berlin in 1753.[48] It may have been published before that date, but it was probably also seen by Buffon before its publication.

In this letter Maupertuis makes only a very brief reference to evolution, but one which is entirely unambiguous. Having said that such characters as the polydactyly of the Ruhe family probably arose by chance in the first place, he adds, 'and it may be thus that all species have multiplied themselves', the context showing clearly that evolution and not merely reproduction is intended.

For the comprehension of the article on the ass, it is well to notice particularly two features of Maupertuis's letter. Although the genealogy of the Ruhe family shows a remarkable tenacity of the character in question, which appears to be a simple Mendelian dominant, yet, in the last of the recorded generations, the six digits appear only on one hand and one foot, at least in the case of one of the sibs. It was this, probably, which suggested to Maupertuis that the character 'becomes altered by alliance with those who have five digits, and by the repetition of such alliances would probably disappear, just as it would be perpetuated by the union of persons of the two sexes both showing it'.[49]

The second feature of importance in the letter is a most remarkable passage in which Maupertuis shows us a way of calculating the probability of parent and offspring both showing the character by chance, as opposed to causal inheritance.[50] When Buffon comes to consider the problem of evolution based upon chance variation, he seems to have this calculation in mind; for he objects that, if two persons having the character must be united in order to perpetuate it, and we must suppose that they come together by chance, then the probability of such chance characters being perpetuated so as to give rise to new species becomes vanishingly small.[51]

This letter, then, appears to be the one referred to by Buffon in his footnote. It must be admitted that the mention of evolution which it contains is of the briefest, but it is unlikely that Buffon had not also seen the curious tract, first published in 1751 in the form of a Latin *Dissertatio* and attributed to a fictitious Dr Baumann,[52] but already appearing (in a French translation) in the collected works of Maupertuis in 1756.[53] In this there is a longer and even more explicit reference to evolution by the accumulation of chance variations:

> There may be some arrangements so tenacious that, from the first generation they are prepotent over all previous arrangements, and efface their habits.* Could we not explain in this way how, from only two individuals, the multiplication of the most various species could have resulted? Their first origin would have been due simply to some chance productions, in which the elementary particles would not have kept the order which they had in the paternal and maternal animals: each degree of error would have made a new species; and by repeated deviations the infinite diversity of animals which we know to-day would have been produced; and this diversity, it may be, will increase still further with time, though possibly the succession of centuries only results in imperceptible changes.[54]

* Maupertuis is referring to arrangements of hypothetical reproductive particles, resembling the pangenes of Darwin's theory of heredity. To account for their regular arrangement in the egg or in the sperm, Maupertuis supposed them to have something analogous to memory or habits of association.

This passage, and the letter already discussed, together contain all that is of importance in Maupertuis's theory of evolution. The anticipation is remarkable, but since it is almost certain that Buffon derived his first ideas on the subject from this source, it is necessary to be quite clear as to what Maupertuis has contributed in these passages. There is in them absolutely no reference to any form of selection, whether artificial or natural; and in the absence of any suggestion of how the ratio of new to old forms of animal or plant might be changed, Buffon was totally justified in treating the theory with considerable scepticism. Authors who have supposed that Buffon's arguments against the theory of evolution, in the article on the ass, are to be considered ironical have commonly been unaware of the form in which the theory was first presented to him.

To make the deficiencies of Maupertuis's theory the more apparent, let us consider a full anticipation of the idea of evolution by the selection of chance variations, an anticipation acknowledged by Darwin himself.[55] This occurs in an essay by Dr W. C. Wells, a native of South Carolina who had come to live in England after the first rebellion of that restless colony, and who acquired a substantial reputation as a physician in the later years of the eighteenth century.

> Those who attend to the improvement of domestic animals, when they find individuals possessing, in a greater degree than common, the qualities they desire, couple a male and a female of these together, then take the best of their offspring as the new stock, and in this way proceed, till they approach as near the point in view, as the nature of the thing will permit.
>
> But what is there done by art, seems to be done, with equal efficacy, though more slowly, by nature, in the formation of varieties of mankind, fitted for the country which they inhabit. Of the accidental varieties of man, which would occur among the first few and scattered inhabitants of the middle regions of Africa, some would be better fitted than the others to bear the diseases of the country. This race would consequently multiply, while the others would decrease, not only from their inability to sustain the attacks of disease, but from their incapacity of contending with their more vigorous neighbours. The colour of this vigorous race I take for granted, from what has already been said, would be dark;* I do not however suppose, that their different susceptibility of disease depends,

* The passage between asterisks is inserted from another page of the same essay (p. 434), where Dr Wells also writes: 'Among men, as well as among other animals, varieties of greater or less magnitude are constantly occurring.'

properly, on their difference of colour. On the contrary, I think it probable, that this is only a sign of some [other] difference in them.* But the same disposition to form varieties still existing, a darker and a darker race would in the course of time occur, and as the darkest would be the best fitted for the climate, this would at length become the most prevalent, if not the only race, in the particular country in which it had originated.[56]

Returning now to Buffon's article, I believe it can be asserted with some confidence that the part of it which discusses the origin of species by the accumulation of chance variations consists of an entirely just criticism of Maupertuis's theory; for this theory, lacking any reference to natural selection, deserves to be rejected as fundamentally incomplete. We may note, however, that though Buffon found the theory unconvincing in the form offered by Maupertuis, he was quite unable to supply what we should now consider the missing element. This is one of the passages in which Buffon can be observed in the act of not discovering the principle of natural selection.

The part of the article which deals explicitly† with Maupertuis's ideas has not been much noticed by those who suppose that the whole article is nothing but an expression of Buffon's belief in the general theory of evolution, in the form of a sarcastic presentation of the arguments against the theory. It has been supposed, and the case has been persuasively argued by Samuel Butler,[57] that Buffon was obliged by his fear of the powerful theologians of the Sorbonne to express disbelief in the theory of evolution, but that he did so with irony and sarcasm so unmistakable as to leave no doubt of his wholehearted acceptance of the theory. Samuel Butler, however, as is often the way with rebels, was not as free from the all-pervading influences of his age as he supposed, and he argues as though there already existed in Buffon's time a fully elaborated theory of evolution which any enlightened person might be expected to accept; and this, of course, was not the case. However, it must be admitted that there are features of this article on the ass which at first sight are extremely puzzling, and which could be easily explained by the assumption that much or all of the article is to be taken as ironical.

† We may say 'explicitly', because of Buffon's footnote referring to Maupertuis.

Because the references to Maupertuis occur in the concluding section of the article, I have considered that part first. It will now be necessary to consider the opening section, which begins with the words:

> Considering this animal, even with attention and in detail, it appears to be no more than a degenerate horse...we might attribute the slight differences which exist between the two animals to a long-standing influence of climate, of food, and to the chance succession of many generations of small wild horses half-degenerate, which little by little had degenerated still more, had then degenerated as much as is possible, and had finally produced for our contemplation a new and constant species.[58]

It would be unwise to attach too much importance to the word 'degenerate', and to what appears to us the inversion of the evolutionary sequence which we should expect. It is clear that the word 'degenerate' is for Buffon a technical term which had lost some of its original meaning, and elsewhere he writes 'degenerated or perfected' in describing evolutionary change. Nevertheless, it has been correctly observed that, when choosing examples, he nearly always selects such as appear to common sense to be examples of degeneration.[59] This choice is partly explained by the historical circumstance that changes from one species to another, which were thought to have been observed during the previous century, clearly appeared to be degenerative; and partly by the importance which Buffon attached to the heat of the earth, which he believed to have cooled from a state of incandescence, in determining the degree of development of species.[60] He tended to think that the more highly developed organisms would be produced in the warmer conditions and climates,[61] and this, coupled with his conviction that the earth was and always had been growing colder, undoubtedly made it difficult to think of the temporal sequence of organisms as progressive. Though it would be unwise to attach too much importance to all this, yet it probably would be true to say that the use of the word 'degeneration' does express a certain tendency to think of changes of faunas with time as being on the whole changes for the worse, and that this tendency was a factor, if only a minor one, in the production of Buffon's scepticism regarding progressive evolution.

Having started his article with a statement of some of the arguments in favour of a genetic relationship between the horse and the ass, Buffon immediately begins to suggest doubts and difficulties; among others, 'the impossibility of uniting them [the horse and the ass] to form a common species, or even an intermediate species capable of reproducing itself'. This was for him an argument of very great power and one to which he attached the greatest importance during the whole of his life. I have quoted above a sentence which I have suggested might be Buffon's last word on the subject of evolution: 'In general, the relationship between species is one of those profound mysteries of nature which man can only investigate by experiments which must be as prolonged as they will be difficult.' Among the experiments which he regarded as necessary to establish or to refute the theory, he attached the highest importance to experiments in hybridization, a kind of experiment which he himself early began to make.[62] During the whole of his life he believed that different species of animals which had descended from some common ancestral species would always show an ability to produce interspecific and fertile hybrids. It is absolutely indispensable to an understanding of Buffon's evaluation of the general theory of evolution to appreciate the importance which he attributed to this test of common ancestry: to understand this will render clear what has appeared most obscure, and will show to be consistent views which might otherwise seem at variance with one another. Thus, in the article on the ass, he denied the possibility of the horse and the ass having a common ancestor, because, when he wrote the article, he believed mules to be sterile. Later, in the essay 'On the Degeneration of Animals', he allowed that these two species might have descended from a common stock, because, in the years intervening between writing the article and writing the essay, he had convinced himself not only that the mule is not entirely sterile, but that other forms which appear to be entirely distinct species, such as the sheep and the goat, can also produce together fertile hybrids.[63] Finally, the fact that he was never willing to accept as established any theory of evolution of wide application, however probable he might think such a theory to be, is readily explained by the difficulty or impossibility of obtaining evidence based upon the hybridization of widely dissimilar forms.

Up to this point, it might be thought that the article on the ass presents no very difficult problems of interpretation. But Buffon, having considered the problem of common ancestry in the particular case of the horse and the ass, passes immediately to a consideration of the theory of evolution in its most general form. He says that, if the horse and the ass are to be considered as belonging to the same family, because they resemble one another, we might, by the same argument, conclude that 'not only the ass and the horse, but even man, the ape, the quadrupeds and all animals might be regarded as making only one family'. He then asserts that to establish one case of common descent of diverse species would be enough to establish the possibility of common descent for all organisms:

> If it were true that the ass were merely a degenerate horse, there would be no limits to the power of nature, and we should be justified in supposing that, from a single being, she had been able to produce in the course of time all organized beings. But no! It is certain, from revelation, that all animals have participated equally in the grace* of creation.[64]

This certainly presents a major problem. Nowhere else does Buffon assert that one case of common descent would be enough to establish the general theory of evolution. The assertion is obviously nonsense, and Buffon was perfectly capable of appreciating this. Moreover, the appeal to revelation is certainly not sincere: even if nothing were known about Buffon's life and thought outside his works, it would still be certain, from the works themselves, that he did not care two pence for revelation.

It is true that when he wrote the article on the ass, or at least when he published it, he had been irritated by a condemnation launched by the theologians of the Sorbonne against certain propositions in the first three volumes of his work; and it is certain that he submitted to ecclesiastical authority only for the sake of peace. The supposition that all the arguments against the theory of evolution which Buffon has set forth in this particular place are to be regarded as ironical in intention is therefore extremely persuasive, particularly if no other explanation can be found for his sudden appeal to revelation. However, the thesis that he is saying in effect, 'Obviously all species have originated by evolution, and

* This odd expression is Buffon's own.

only those blinded by theological prejudice could fail to appreciate the fact', is somewhat anachronistic, because it could not have been obvious, in 1753, that the theory of evolution was the correct explanation of the origin of species. Then again, as I have tried to show, the article contains the expression of doubts and difficulties which are reasonable in themselves, and which, there is evidence to show, Buffon intended to be taken seriously.

We seem to be faced with an insoluble conundrum, but there is, I believe, a perfectly natural and satisfactory solution. If we turn to the very first essay in the *Histoire Naturelle*, the essay entitled 'Premier Discours de la Manière d'Étudier et de Traiter l'Histoire Naturelle', we find a long dissertation, written before Buffon had been irritated by the theologians and therefore, presumably, not to be suspected of irony due to this particular source at least. This essay or dissertation is largely an attack upon systematists in general, and upon Linnaeus in particular. Now, it can be shown[65] that Buffon's prejudice against systematists, or *nomenclateurs* as he calls them, was so violent as to result in a kind of feud, which constantly recurs, and so irrational as to persist even when he himself had adopted a scheme of classification which did not differ essentially from that used by Linnaeus.

If the *Premier Discours* is read immediately before the article on the ass, many points of similarity are discovered, even in detail, and the article appears as another incident in the recurrent feud. The mysterious passage ending in the appeal to revelation, for example, is introduced by the words, 'Do the ass and the horse come originally from the same stock? Are they, as the *nomenclateurs* say, of the same family?'

Then, after what seems a long digression on the subject of classification, Buffon concludes,

> Each species, each series of individuals capable of reproducing their kind and incapable of mixing with other species will be considered apart and treated separately, and we shall make no use of families, genera, orders and classes, any more than nature makes use of them.[66]

I have said that the central part of the article, which lies between these two references to systematists and to their devices, might seem to be a digression, because I believe it is in fact a mistake to regard the whole article as being principally concerned with the theory of evolution. It is easy for us, interested in the theory and in its origins, to misunderstand the intentions of any author who refers to the theory. Whereas for us the centre of interest in this article lies in the discussion of evolution, for Buffon, I suggest, this discussion is a digression and the centre of interest is in the problem of classification.

The explanation which I offer, then, for the paradoxical aspects of the article on the ass is that Buffon, concerned primarily with problems of systematics and with his feud against Linnaeus, is using what appeared to his contemporaries, especially to the more pious among them, to be an odd and even dangerous idea, as a means of discrediting his opponents. Immediately after the assertion that, if the ass and the horse are of the same stock all animals whatever might be considered to have arisen from a single species, *en se perfectionnant et en dégénérant*, he adds:

> Those naturalists who establish so lightly families among animals and plants do not appear to have been sufficiently sensible of these consequences, which would reduce the immediate productions of creation to any desired small number of individuals.... [And it is after this that we find the passage already quoted:] If it were true that the ass were merely a degenerate horse, there would be no limits to the power of nature, and we should be justified in supposing that, from a single being, she has been able to produce in the course of time all organized beings. But no! It is certain, from revelation, that all animals have participated equally in the grace of creation.[64]

The thunderbolt, intended to make the reader's flesh creep though the author thought it but stage-fire, is aimed at the systematists.

It is, of course, at this late date, impossible to prove exactly what was Buffon's full intention in writing the article we have been examining, but I think we can assert with confidence that it cannot be used as unimpeachable evidence that Buffon, when he wrote it, was convinced of and wished to advocate the validity of the general theory of evolution. This is important, because in other passages of his great work, where he examines the theory and rejects it as unproven, there is nothing like the same colorable suggestion of irony as there is in the article on the ass. Of the other passages which make any reference to evolution, by far the most important is the essay 'On the Degeneration of Animals', which I shall now discuss.

THE ESSAY ON THE DEGENERATION OF ANIMALS

The essay '*De la Dégénération des Animaux*', appears in the fourteenth volume of the *Histoire Naturelle*, in the volume, that is, published in 1766. By the time he wrote this essay, both the facts of geographical distribution of animals and his studies of hybridization had led Buffon to take a slightly more favourable view of the theory of evolution than that expressed, at least overtly, in the article published in 1753. Buffon tells us that he began to make experiments in hybridization with goats and sheep in 1751.[67] What the results of these were it is not easy to say. Buffon certainly believed that he had obtained hybrids of these two species, and that these hybrids were fertile; but authorities with whom I have discussed this question have expressed themselves as extremely sceptical of the possibility of even the first-generation hybrids. However, there seems no reason to doubt that Buffon honestly believed that he had obtained fertile hybrids from two animals which common sense would regard as belonging to different species. Whether he was deceived by dishonest assistants or really did obtain the hybrids must remain uncertain, but clearly some animals supposed to be hybrids were produced and were coupled and found to be fertile. All this would take at least two, and probably three or four years, so that the crucial result—the fertility of the hybrids—would not have been available to Buffon when he wrote for the volume of 1753.

His studies of geographical distribution concern the mammals of the Old and New Worlds, and he thus sums up his results:

> Thus, of ten genera and four isolated species,* to which we have tried to reduce all the animals belonging particularly to the New World, there are only two, the genus containing jaguars, ocelots and so on, and the species of the pecari, with its varieties, which one can refer with any certainty (avec quelque fondement) to the animals of the old continent. The jaguars and ocelots can be regarded as species of leopard or of panther, and the pecari as a kind of pig. Then there are five genera and one isolated species, the species llama, and the genera of monkeys with, and those without prehensile tails, the genera of skunks, of agoutis and of ant-eaters, which can be compared, but in a most uncertain and

* Isolated species, he explains elsewhere, are those which are the only species of their genus.

distant manner, with the camel, with those monkeys of the Old World which have tails, with the polecats, with the hare and with the scaly ant-eaters; and, finally, there remain four genera and two isolated species, the opossums, the coatis, the armadillos, the sloths, the tapir and the capybara, which cannot be referred to, nor even compared with genera or species of the Old World. This seems to prove sufficiently that the origin of these animals peculiar to the New World cannot be attributed simply to degeneration; however great, however powerful might be its effects, we could never persuade ourselves, with any appearance of reason, that these animals had originally been the same as those of the old continent; it is more reasonable to suppose that, at one time, the two continents were continuous or contiguous, and that the species which had taken up their abode in the New World, because they found the climate more suitable to their nature, were shut off and separated by the irruption of the sea.[68]

I have quoted this passage in full because it shows Buffon at his best as a conscientious naturalist, and seems to dispose completely of the supposition that any hesitation he had in accepting a whole-hearted theory of evolution arose merely from fear of further trouble with the theologians. But why, it is natural to ask, did he draw the line where he did? Why accept a common origin for the pig and the pecari, and not for the camel and the llama? The answer to these questions is, I think, that he did not believe that the causal mechanisms which his mind had assimilated were adequate to the production of more than the most trivial differences. He writes:

It would be very difficult to see how the tailed monkeys of the Old World could have taken on in America a differently shaped face, a muscular and prehensile tail, a long partition between the nostrils, and the other characters, both generic and specific, by which we have distinguished them from the monkeys of the New World....With regard to the agoutis and pacas...they can only be compared with the hare and the rabbit...but what makes it doubtful that there can have been anything common in their origins, is that the hare has spread in nearly all the climates of the old continent, without its nature having changed, and without any other alteration than that of the colour of its coat; one cannot, therefore, imagine with any justification that the climate of America was able to do what no other climate could, and that it could change the nature of our hares so far as to turn them into *tapetis* [*Lepus americanus*] and *aperea* [wild guinea-pig or restless cavy] which have no tail, or agoutis, with their pointed snouts, with short rounded ears, or pacas with thick heads, short ears, short rough hair and white bands.[69]

I have spoken of the causal mechanisms which Buffon's mind had been able to assimilate, because he does indeed mention artificial selection, and even a kind of 'Lamarckian' mechanism; but he nowhere even suggests natural selection, and he makes no use of the 'Lamarckian' mechanism when, as in the example just quoted, climate does not seem adequate to explain the production of the observed differences. The particular case he mentions of the inheritance of an acquired character is that of the callosities on the legs or chests of camel and llama,[70] but these he regards as merely the effects of 'servitude or domestication'; it does not seem to occur to him that anything of the kind could be an important cause of evolutionary change in wild animals.

Even artificial selection appears not to have received from him the attention it deserves. For example, in his article on the dog, which is in the fifth volume of the *Histoire Naturelle*, published in 1755, he mentions artificial selection as one of the causes of new breeds, but entirely forgets it (though the whole article is of only a few pages) in summarizing at the end of the article. In the course of his summary, he says that, of the thirty known breeds, seventeen are due to the influences of climate and the other thirteen are mongrels produced from these seventeen climatic races.[71] Having just stated that the twisted legs of the basset are due to the inherited effects of rickets, he concludes:

> All these races with their varieties have only been produced by the influence of the climate, joined to the comforts of shelter, the effects of food and the results of careful training (une éducation soignée); the other dogs are not of pure race, and come of mixtures of the other races.[72]

Again, in the essay 'On the Degeneration of Animals': 'Climatic temperature, the quality of food, and the evils of servitude, these are the three causes of change, alteration and degeneration of animals.'[73]

Here, then, we have three causes, of which only two are said to be operative in the wild state. Later in the same essay he makes an obscure reference to a cause of change linked with the number of individuals in the race, which should clearly constitute a fourth cause, but later still he speaks of 'the three causes of change'.[74] This looks like another lapse of the kind which occurs in the article on the dog. It may be, however, that, in the final reference to three causes Buffon means, though he does not say so, the three causes acting in the wild state. But, in any case, the mention of a cause of variation linked with the number of individuals is no more than a memory of Maupertuis's theory, which Buffon entirely fails to complete by any reference to natural selection. Had he possessed even an imperfect grasp of this concept, he surely would have spoken of it in this context.

I have tried in these pages, which are too few to do full justice to so voluminous an author, to give as clear an idea as possible of Buffon's treatment of the idea of evolution. I do not think he can be called an originator or founder of the idea, and on the whole there seems to be small reason to suppose that he even subscribed to it. He seems to have regarded it as an interesting speculation of which a lot might be made some day, but which certainly could not be established upon the basis of such facts as were known to him. This seems to me to be exactly the position which a competent naturalist should have adopted during the eighteenth century, and to say that Buffon did adopt it is, I think, in no way to detract from his stature, rather, indeed, to enhance it. But the relation of Buffon to the theory of evolution is by no means merely that of a prudent judge who regarded it as non-proven. His services to the cause of the theory were not unimportant. He attempted a natural explanation of the origin of the earth in general, and of sedimentary rocks in particular, and thus did for Lamarck much what Lyell did for Darwin: he fostered the idea of a gradual temporal development due to natural causes. He directed attention to the problems associated with the theory of evolution, and he was, as far as I am aware, the first to appreciate that the geographical distribution of animals contained an essential part of the evidence relating to the origin of species. Finally, he recognized that many species of animals had become extinct before the appearance of man on earth, and that there were, consequently, causes of extinction which owed nothing to human agencies. This is a not unimportant element in the case for evolution, and it is one which Lamarck failed to understand.

BUFFON AND LAMARCK

Buffon and Lamarck belonged to two successive generations which might, perhaps, be expected to differ profoundly. Buffon was born in 1707, and belonged to the time of Louis XV; Lamarck, who was born in 1744, was of the generation of Robespierre and Jefferson. It is, however, difficult to convince oneself that the two men would not have shown much the same differences of taste and temperament had they been born in the same year.

Buffon appears to have been expansive, avid, sensual and energetic. He was not of noble birth, and we feel no surprise on learning that he inherited his fortune from a *Fermier général*, and that he himself greatly added to what he had inherited.

The impression one has of Lamarck is that he was restrained, perhaps a trifle prim, and conscientious even to a fault: a poor aristocrat worried by the incomprehensible difficulties of practical life.

In praising the naturalists of antiquity, Buffon has left us more than a hint of what, we may suppose, he would like said of himself:

> The ancient authors who wrote on natural history were men of parts, who had not confined themselves to this study alone; their minds were elevated, their knowledge was various and profound; they had broad views, and if it appears to us that they lacked exactitude in certain details, it is not difficult to see that they did not suppose it necessary to give to trifles the attention which they have received in recent times.[75]

But though Buffon left the accumulation of detailed knowledge to others, he had a great respect for any facts which he had convinced himself were well established; and though he loved a 'broad view', as his two little essays called '*Vue de la Nature*'[76, 77] testify, he did not mistake such pieces of *bravura* for contributions to natural science. It is no anomaly, but entirely characteristic of him, that his contribution to evolutionary thinking is fragmentary; when he wrote as a scientist, if he saw only fragments he described only fragments.

Lamarck was certainly not less conscientious, but he distinguished less clearly between what he could prove and what he should have recognized as speculation with only the flimsiest basis of fact. He was thus able to elaborate a complete and, apart from one fundamental flaw, a consistent theory of evolution, unhampered by any too nice an attention to evidence. It is a great deal easier to expound his theory of evolution than to sift the fragments which make up Buffon's contribution to the subject. But an exposition of Lamarck's theory cannot be entirely simple, because, as I have hinted, the theory was in one important respect not simple.

To explain this complexity, we must return for a moment to consider yet another contrast between the two great naturalists. Buffon never seems to have had any sympathy for any monotheistic religion, though he might, perhaps, have had some intuitive understanding of some more ancient and more naturalistic religion; the relationships of his writings to theology are entirely external. Scepticism was Buffon's natural atmosphere, but it was one in which Lamarck's mind, essentially serious and orderly, could not breathe. In his works he shows himself a convinced Deist, and he even allowed that some truths might be known, and that man might approach the Supreme Being by means other than reason and observation.[78] In so far as he was a Deist, Lamarck belonged to an earlier age than did Buffon, and we find his theorizing hampered by elements of eighteenth-century, or even seventeenth-century thought which gave Buffon little trouble.

Buffon, it is true, began by accepting the Chain of Being, but the notion had no greater hold upon his mind than any other piece of metaphysics; the traditions which influenced his thought were rather those of the ancient world, especially Aristotelian and Epicurean traditions, than those of the seventeenth century. But the Chain of Being haunted Lamarck's thoughts like an inexorcizable ghost.

Lamarck's Theory of Evolution

THE GENERAL STRUCTURE OF THE THEORY

Lamarck gave three versions of his theory of evolution. In the first, published in 1801, which is a brief exposition originally given as a lecture, the whole of evolution seems to be attributed to the agency of the 'Lamarckian' mechanism, the inheritance of acquired characters. There is only the merest hint that Lamarck is concerned in any way with the Chain of Being, and the influence of the idea seems to be limited to the production of a scalar system of classification of animals, which one might easily suppose to be merely a convenient artificial device, having no relation to the animals' mode of origin.

In the second treatment of the subject, contained in the first eight chapters of the *Philosophie Zoologique*,[79] first published in 1809, the central and principal evolutionary process is one which cannot be analysed causally: a natural tendency to increased complexity in organisms. This unexplained tendency, which is described as 'a law of nature', is supposed to result in a linear development on which is based the linear arrangement of the main groups of animals in Lamarck's classification. This linear scheme running from the least to the most complex animals is often described by Lamarck as a 'chain', and it is in fact a part of the Chain of Being, differing from the whole Chain, as described for example by Locke, only in being confined to animals (including man). Lamarck, who was an excellent botanist, constructed a second linear series for plants, since he saw clearly that the old notion of 'zoophytes' was untenable. In the *Philosophie Zoologique* the inheritance of acquired characters is relegated to an entirely secondary role: this process is now used to explain the obvious deviations from a linear order. It is thus true to say that the 'Lamarckian' mechanism, so far from being used to account for evolution, is here only called in to explain deviations from an ideal evolutionary progression.

It must be admitted that when Lamarck discusses, in a chapter specially devoted to the subject, the results of the inheritance of acquired characters, he does to some extent give the impression that the process is central to his theory; but it is easy to show that this impression is a mistaken one, and even in this chapter, devoted particularly to the 'Lamarckian' mechanism, he is at pains to point out its subordinate character. He does so briefly, but then he had made the position abundantly clear in the earlier chapters of his book.

The first version of Lamarck's theory, to which I have referred above, was printed as an introduction to his *Système des Animaux sans Vertèbres*, which bears the date *An IX—1801*, and the introductory passage bears the superscription *Discours d'Ouverture*, *prononcé le 21 Floréal An 8.* In 1815 Lamarck published a much enlarged treatise on the invertebrates, *Histoire Naturelle des Animaux sans Vertèbres*,[80] which he describes as 'susceptible d'être considéré comme une seconde édition de mon *Système des animaux sans vertèbres*'. The first volume of this second edition is devoted to a restatement of his biological principles, and here we find the third exposition of his theory of evolution. This third treatment of the subject appears unaltered in the posthumous edition of the *Animaux sans Vertèbres*, of 1835,[81] so that, especially as the general doctrine is the same as that expounded in the *Philosophie Zoologique*, we may consider this third treatment as definitive. It is also, I think, the most stylish of the three, so that I may begin my account of Lamarck's theory by a brief consideration of this final exposition.

Lamarck has been vexed by the accusation that he has merely refurbished the old Chain of Being, and he rejects with some asperity the notion of a single *chaîne graduée* linking 'the different bodies which nature has produced'; he does not believe that there can be any gradation between living and non-living things, and he reaffirms that there are no such things as zoophytes.[82] He attempts to remain consistently mechanistic, and attributes the sharp break between the living and the non-living to a difference in level of organization: 'The individuality of the species [consists] in the union, the disposition and the state of the constituent molecules of various kinds which compose their bodies, and never in any of these molecules considered separately.'[83] Animals and plants do not form parts of a single chain of beings, but constitute two independent lines; the differences between these two being due to differences in the chemicals initially used by nature in the synthesis of the primordial plant and the primordial animal respectively.

> I shall prove [he writes] that there is no real chain, linking together the productions of nature in general, and that such a thing can only be found in certain branches of the series which they form; and even there it only appears in a general way, and not in details (encore ne s'y montre-t-elle que sous certains rapports généraux).[84]

In spite of these disclaimers, however, Lamarck was, as we shall see, very far from having freed himself from the influence of the Chain of Being. It is true that he did not attempt to force all living things into a single linear scheme, but he constantly supposed that such a scheme underlay the relationships of animals, and he constantly speaks of branching in the systematic arrangement as a deviation or even an 'anomaly'. Moreover, he often uses the word *échelle*[85] in describing the basic classificatory series, and even slips into the use of *chaîne*[86] here and there, though perhaps less often than in the *Philosophie Zoologique*.

In a *Supplément* at the end of the first volume of the *Animaux sans Vertèbres* of 1815, which appears unchanged in the edition of 1835, we find what must be almost if not quite literally Lamarck's last word on the subject of evolution. Here he allows two animal series, as though this were a new idea to him, though he had admitted in the *Philosophie Zoologique* the possibility of a short second chain or series,[87] and writes:

> After the spontaneous generations which began each series in particular, the later animals arose from one another. Now, though the laws which directed this production are always and everywhere the same, yet the diverse circumstances in which nature has worked, during the course of her labour, have necessarily produced anomalies in the simplicity of the scale (échelle) resulting from her operations. We should therefore try to form and to perfect two different tables:
>
> One giving the simple series, which we should use in our publications and lectures, to characterize, to distinguish and to describe the animals which have been observed; a series which we should in general model on the progression which occurs in the complexity of the various animal organisms, considering each in the totality of its parts, and making use of the directives which I have suggested.
>
> The other giving the particular series, with their simple branches, which nature seems to have formed in the production of the actually existing animals.[88]

The idea of two schemata, separable in thought but combined in nature, that is, in the actual process of evolution, both equally real because based upon two distinct 'causal' processes,* runs through Lamarck's writings with remarkable tenacity and consistency. Only in one place that I have been able to find does he seem to say explicitly that there is but one underlying process in evolution: the causal process usually called 'Lamarckian', the inheritance of

* One of the underlying processes is 'lawlike' rather than causal.

acquired characters formed in the first place by reaction to environmental changes or as a result of the spontaneous formation of new habits. But even this apparent deviation from his usual doctrine disappears on careful examination. The difficult passage is as follows:

> By the four laws which I have just mentioned, all the facts of organization appear to me to be easily explicable; the progression in the complexity of organization of animals, and in their faculties, seems to me easy to comprehend (concevoir); finally, the means used by nature to diversify animals, and to bring them to the state in which we see them, become easily determinable.[89]

Now, the four laws which Lamarck has listed immediately before this passage are these:*

1. Nature tends to increase the size of living individuals to a predetermined limit.
2. The production of a new organ results from a new need.
3. The development reached by organs is directly proportional to the extent to which they are used.
4. Everything acquired by the individual is transmitted to its offspring.[90]

Here, then, there seems to be no reference to the tendency to increasing complexity, and the whole of the evolutionary process appears to be attributed to the single causal factor of inheritance of acquired characters. This, however, turns out to be merely a piece of carelessness. In formulating the first of the four laws, Lamarck speaks of 'increase of size', but in commenting on the law he makes it clear that the phrasing of the law is defective. For he writes: 'This first law of nature, which gives to life the power of increasing the size of a body and of stretching out its parts... enables this power gradually to increase its forces in the complexity of the animal organization.'[91] And he describes a process of increase of complexity due entirely to the movement of internal fluids, in no way guided by reactions to environmental changes or by the imprint of new habits upon the nervous system.

This, as I have said, is the only passage I have found in which Lamarck even seems to depart from his characteristic two-factor theory.

* Paraphrased, for the sake of brevity, but nothing essential has been omitted.

The Originality of Darwin's Theory

THE ZOOLOGICAL PHILOSOPHY

Lamarck sets out his two-factor theory very clearly in the fourth chapter of his ***Philosophie Zoologique.*** The chapter begins:

Among the considerations of interest to Zoological Philosophy, one of the most important is that concerning the *degradation* and simplification observable in the organization of animals when we follow the animal series (la chaîne animale) from one end to the other, from the most perfect animals to those which are of the simplest organization.

Now, we have to find out whether this can really be established as a fact; for if so, it will throw a strong light on the plan which nature has followed, and will start us off on the way to the discovery of many of those of her laws which are the most important for us to know. I have here taken as my task to prove that what we are discussing is a positive fact, and that it is the consequence of a general law of nature, always acting in the same manner; to prove also, however, that a particular cause, which is easily recognizable, produces irregularities in one point or another within the whole extent of the animal series, disturbing the regularity which this law would otherwise have produced.

If the cause which is always tending to make organization more complex were the only one affecting the form and the organs of animals, the increasing complexity of organization would everywhere follow an extremely regular progression. But this is not the case. Nature is under the necessity of submitting her activities to the influences of circumstances which act upon them and everywhere (de toutes parts) these circumstances produce variety in her productions. This is the particular cause which gives rise here and there to the often bizarre deviations which the *degradation*, which we shall exhibit, shows us in its course. Let us try to make perfectly clear both the progressive *degradation* in the organization of animals, and the cause of the anomalies suffered by the path of this degradation within the animal series. Clearly, had nature produced only aquatic animals, and had all these animals always lived in the same conditions of temperature, the same kind of water, the same depth and so on, there can be no doubt that we should have found a regular and even a very fine (nuancée) *gradation* in the organization of these animals.

After having produced aquatic animals of all grades, and having varied them strikingly by the agency of the diverse conditions offered by the water,* those which she [that is, nature] has induced little by little to live in air, first on the shore and then on the dry land, found

* There is here an unconformity in the syntax of the French.

themselves in time in conditions so different from the original ones and so powerfully affecting their habits and organs, that the regular *gradation* which they should exhibit, in the complexity of their organization, was greatly disturbed thereby, so that it is hardly to be perceived in many points.

These suppositions which I have turned over in my mind, and which I shall establish by positive proofs, suggest to me the presentation of the following *zoological principles*, which seem to me incontestable.

The progression in the complexity of organization suffers, here and there, in the general series of animals, anomalies produced by the influence of the circumstances of the environment (circonstances d'habitation), *and by those of the habits contracted.*[92]

Now, although we no longer think in terms of a 'scale of perfection', the idea of advance in evolution is, of course, not totally to be rejected. An increase in complexity is indisputably present; but, more than that, one can also recognize certain well-defined advances which give to the animals possessing them an undoubted advantage over species less advanced. Some animals, for example, are at the mercy of comparatively small changes in the concentration of salts in their environment, while in others the concentration of salts in the body-fluids is independent, within wide limits, of changes in the environmental concentration. Again, the ability to preserve a constant body-temperature, while the external temperature varies widely, gives those animals which possess it a clear advantage over those which do not.

Where such clearly comprehensible advantages are concerned, it is reasonable to think of one group of animals as 'higher' than another. But Lamarck's desire to construct a linear scale of perfection carries him far beyond any such justifiable comparisons. Consider, for example, the following 'Observations on the Vertebrates', which occur later in the chapter from which I have just quoted:

Vertebrate animals, although they show among themselves great differences in their organs, appear to be all constructed upon a common plan. Passing from the fishes up to the mammals, we see that this plan becomes more perfect from class to class, and that it only reaches its final form in the most perfect mammals; but we also observe that, in the course of reaching its perfection, this plan has suffered numerous modifications, and even very striking ones, on account of the places in which the animals live and the habits which each race has been forced to acquire according to its circumstances.

Thus we see, on the one hand, that if vertebrate animals differ greatly from one another in the state of their organization, it is because nature only began to realize her plan for them in the fishes, that she improved it in the reptiles, that she carried it nearer to its perfection in birds, and that at last she only contrived to bring it to its final state in the most perfect of the mammals; on the other hand, we cannot fail to perceive that, if the process of perfecting the plan of vertebrate organization does not show everywhere, from the least perfect fishes to the most perfect mammals, a regular and fine *gradation*, it is because nature's work has been often modified, impeded, and even changed in its direction, by the influences which strikingly different, and even contrary conditions have exercised upon the animals exposed to them for very many generations.[93]

Again, in the same chapter, we find the following, under the heading 'Molluscs':

If the molluscs, in their general organization, which is inferior in perfection to that of the fishes, also prove, for their part, the progressive *degradation* which we are studying in the animal series (chaîne), the same degradation is not so easily perceived among the molluscs themselves; because, among the very numerous and diverse animals of this class, it is difficult to separate what belongs to the *degradation* in question from what is the result of the habitats and habits of these animals.[94]

At least three times during the next few pages we find references to 'anomalies due to environment'.[95]

The next chapter, the seventh of the *Philosophie Zoologique*, is entitled 'On the influence of conditions upon the activities and habits of animals, and of that of the actions and habits of these living bodies, considered as causes modifying their organization and their parts'. This chapter, therefore, is devoted to an exposition of the 'Lamarckian mechanism'; but even in this chapter which deals particularly with only one of the underlying processes postulated by Lamarck's theory, there is an explicit and unambiguous reference to the other:

In the preceding section we have seen that it is now an incontestable fact that, considering the animal series (échelle) in the direction opposed to the natural one, we find that there is, in the groups (masses) which make up this series, a sustained but irregular *degradation* in the organization of the animals of the groups; an increasing simplification in the organization of these living bodies; finally, a proportional diminution of the faculties of these beings.

This well-known fact can throw the clearest light upon the actual order followed by nature in the production of all the animals she has made; but it does not show us why it is that the organization of animals, in its growing complexity, from the least to the most perfect, presents only an *irregular gradation* of which the whole extent displays a large number of anomalies or deviations which have no apparent order in their diversity.

Now, looking for the reason for this peculiar irregularity of the increasing complexity of the organization of animals, if we bear in mind the cumulative sum of the influences which infinitely diversified conditions all over the earth exercise upon the general form, the parts and the organization itself of these animals, all will then be clearly explained.

In fact, it will become clear that the actual state of animals is, on the one hand, the result of the increasing *complexity* of organization, which tends towards the production of a *regular gradation*, and, on the other, that it is the result of the influences of a multiplicity of circumstances very different one from another, which tend continually to destroy the regularity in the gradation of the increasing complexity of organization.[96]

THE TWO-FACTOR THEORY AND THE IMPORTANCE OF DEISTIC ELEMENTS IN LAMARCK'S THOUGHT

To show how seriously Lamarck took his two-factor theory, it is necessary to return, even at the risk of some tedium, to his treatment of it in the 'Introduction' of the *Animaux sans Vertèbres*. Cuvier had objected that it would not be possible to establish 'une série unique', because 'if each particular organ is considered, we shall have as many different series as we have taken for our guidance different organs...to construct a general scale of perfection, we should have to calculate the resultant of each combination, which is hardly possible'.[97]

Answering this criticism, Lamarck writes:

This is the troublesome consequence of considering the data of observation as though there were but one cause responsible for the progression we are talking about....In fact all this has been looked at as though it were the product of one single cause...but it is easy to see that we are dealing here with the results of two very different causes, of which the one, although unable to destroy the predominance of the other, nevertheless frequently diversifies its consequences.

Nature's plan of campaign in the production of animals is clearly marked out by this primal and predominant cause, which endows animal life with the ability to complicate organization progressively, and to complicate and perfect gradually, not only the total organization, but also each system of organs in particular, as this cause has been able to establish each one. Now, this plan, that is, this progressive complication of organization, has really been effected by this primal cause among the various animals which exist.

But a quite separate cause, an accidental and consequently variable one, has here and there cut across the execution of this plan, without however destroying it, as I shall prove. This cause, in fact, has given rise to whatever real discontinuities there may be in the series, and to the terminated branches (rameaux finis) which depart from it, at various points, and diminish its simplicity, and finally to the anomalies to be seen in the various organ-systems of the different organizations.[98]

Lamarck then offers the 'proof', which he has promised, of the real existence of a scale of perfection. This 'proof' consists of an exhibition of animals arranged in a series so as to illustrate their gradual departure from the perfection of man. Lamarck also offers four 'facts on which are based the proofs of the existence of a progression of complexity of the organization of animals', these add nothing material to the series he has exhibited, but they are of interest as illustrating his way of thinking about animals and about their scalar arrangement. The 'four facts' are: the general resemblance of animals one to another; the resemblances between man and other animals; the perfection of human organization; the fact that some animals resemble man more and some less.

The third 'fact' is stated thus:

One can present as a positive fact, as a truth susceptible of demonstration, that, of all types of organization, that of man is the most complex and the most perfect, considered as a whole, and from the point of view of the abilities which it procures for him....[99] [This is further explained in a footnote:] Many animals present, in some of their organs, a perfection and a wealth of abilities of which some organs in man are devoid. Nevertheless, his organization is more perfect in its totality than that of any animal whatever; a thing which cannot be denied....[Having established this to his own satisfaction, he continues] The organization of man being the most complex and the most perfect of all those which nature has been able to produce, it can be asserted that the nearer an animal organization approaches that of man, the more complex it is, and the more advanced towards its perfection (son perfectionnement); and, similarly, the further it is away, the more it is simple and imperfect.[99]

This, again, has a long footnote in which Lamarck writes that 'many zoologists' have found the expression 'perfect animals' and 'imperfect animals' ridiculous; as though, he says, they could not see that he means these expressions to be taken as implying a comparison with human perfection.

> Who does not know that, in its actual state of organization, each living body of what kind so ever is a really perfect being, that is, a being which lacks nothing of that which is necessary for it! But nature having complicated animal organization more and more, and thereby having been enabled to endow the animals having the most complex organization with more and more advanced abilities, it is possible to see in the last term of her efforts a perfection from which those animals which have not attained it can be seen to recede by stages.

This passage seems to me to be important in showing how firmly Lamarck's thought was rooted in that of the eighteenth, and of preceding centuries; how seriously he took his scale of perfection; and, in his observations on the 'real perfection' of all animals, how generally his mind was imbued with the notion of a planned, and neatly planned, universe. This tendency to suppose that it must be relatively easy for any intelligent man to understand the intentions of nature appears again in Lamarck's slips into a teleology which really is vicious, and which can be only partly excused as a metaphor. Discussing the molluscs, he writes,

> If these soft and jointless animals make only slow and weak movements, it is because nature, preparing to form the skeleton,* has abandoned in them the use of horny teguments and the joints which she employed from the insects upwards,† so that their muscles have, under the skin, only very feeble points of attachment (points d'appui).[100]

He had used the same kind of language in the *Philosophie Zoologique*:

> Nature, on the point of beginning the plan of organization of vertebrates, was obliged, in the molluscs, to give up the device of an encrusted

* 'Skeleton' is used by Lamarck only for the skeleton of vertebrates.

† His scale is: Insects, Arachnids, Crustacea, Annelids, Cirripedes, Molluscs; the insects being the lowest.

or horny skin as supports for muscular action, and getting herself ready (se préparant) to carry these supports into the animal's interior,* molluscs happened to be, as it were, in the path of this change.[101]

The metaphorical element in these statements is limited to the personification of nature. Lamarck is elsewhere careful to assert that nature is not a conscious agent:

> The general power which holds in its domain all the things we can perceive...is truly a limited power, and in a manner blind; a power which has neither intention, nor end in view, nor choice; a power which, great as it may be, can do nothing but what in fact it does; in a word, a power which only exists by the will of a higher and limitless power, which, having founded it, is in truth the *author* of all that it produces, that is, of all that exists....
>
> And *nature*...is only an instrument, only the particular means which it has pleased the *supreme power* to employ in the production of the various bodies, in their diversification; to give them properties, or even abilities....She is, in a way, only an intermediary between God and the parts of the physical universe, for the execution of the divine will.[102]

From the point of view of natural science, however, it matters little whether we attribute conscious designs to nature, or consider her as an unconscious agent, if we too lightly assume that the designs themselves will be easily comprehensible to us. What is heterodox is not the supposition that the universe is planned, but that the plan was made by a mind not essentially different from our own. From the whole tone of Lamarck's writings it seems certain that his appeals to the notion of a divine plan are entirely sincere. It seems most improbable that, at the time he was writing, such expressions were required, as they were in the time of Buffon. There is internal evidence that they were no longer *de rigueur*, for Lamarck himself tells us that the spirit of his times was in general frankly naturalistic or atheistic: 'It has been thought that nature is God, indeed, it is the opinion of the majority....An odd thing! To confuse the watch with the watchmaker, the work with its author!'[103] The effects of his Deism are nowhere more obvious than in his treatment of the problem of extinction. There is one passage in which he makes use

* There is another syntactical unconformity here in the French. The reading is guaranteed by a parallel passage in the *Animaux sans Vertèbres*.

of the concept of the struggle for existence as a contributory factor in the causal explanation of evolution:

> Another cause of change of activity which has contributed to the diversification of the parts of animals and to the multiplication of races is the following: As animals, by partial emigrations (par des émigrations partielles) changed the place of their abode and spread to different points on the earth's surface, they were exposed, on arriving in new situations, to new dangers which required new activities, if they were to be avoided; because most of them devour one another to conserve their existence.[104]

But this is an isolated flash of insight. In general, Lamarck shows that the notion of wastage of species was profoundly foreign to his mind. Thus in the *Philosophie Zoologique*, which was published only four years before the passage just quoted, he discusses only one cause of extinction, and only one class of extinct animals: the extinction of large animals such as Megatherium and Mastodon by human agencies.[105] In the same place he also proposes his own peculiar explanation of remains which appear to belong to extinct forms. He says that few, if any, species have ceased to exist except by changing into other species. He considers it most probable that no fossil form is extinct even in this sense, but that all species which appear to have been lost may one day be found living in out-of-the-way places on the earth, or in the depths of the sea. Somewhat later in the same work he mentions the checks on the numbers of one species provided by the activities of other species, but on this his comment is:

> By these wise precautions, everything remains in the established order; the perpetual changes and returns (renouvellements) which are to be seen in this order are maintained within limits which they cannot transcend: the races of living bodies all continue to exist, in spite of their variations, the elements of progress in the perfecting of organization are not lost; all which appears to be disorder, anomaly, returns endlessly into the general order, and even enhances it: and everywhere and always the will of the sublime Author of nature and of all that exists is done without exception.[106]

There is, I believe, a defect in Lamarck's theory of which he seems to have been unaware. If there is an implanted tendency in some living form to evolve into some more complex form, one can see no reason why this tendency should not be universally present in the less complex form. Were this the case, no living representative of the less complex form is to be expected; and since Lamarck undoubtedly supposes this state of affairs to be the rule, we should expect that the only living representative of the animal series would be the most complex. Yet he supposes that most, if not all links in the chain, or members of the series, are in fact still to be found alive. He does indeed suggest a mechanism which would result in some descendants of a given form differing from others, the mechanism of the inheritance of acquired characters, but this is supposed by him to explain deviations from the direct line of advance, and consequently could not be used to explain the persistence of ancestral forms within the direct line. It may be that he believed living forms which are in the direct line, but not at the summit, to be the products of other lines which originated independently of the line which produced man, by separate spontaneous generations at different times;* but this clearly robs the theory of all its apparent simplicity and elegance.

Conclusion

Any satisfactory theory of evolution must be able to assimilate three great sets of data: the actual geographical distribution of plants and animals, the real similarities of structure which must form the basis of any system of classification of organisms, and, finally, the fossil record.

Buffon's most significant contribution in this field was his attempt to grapple with the problem of the distribution of mammals in the Old and in the New World. Not only does he approach the solution which Darwin gave for this problem, but even his methods are curiously like those of Darwin. No book of travels or memoirs is too remote; if it but contain some crumbs of information which appear significant and reliable, Buffon will find them. However, as we have seen, Buffon could not satisfy himself that the causal theory of evolution, as he understood it, was adequate to explain the differences he observed between the animals of the one World and those of the other.

It seems probable that Buffon suspected that it would one day

* I owe this suggestion to Mr H. G. Ll. Bevan.

be possible to elaborate a theory of evolution which would systematize, not only the facts of geographical distribution, but also those of comparative anatomy and palaeontology; but it is certain that he did not produce, or even attempt to produce, such a theory. What we have in his work is a fragment towards the construction of a theory of evolution; it is no more than a fragment, but it is one of considerable brilliance, and what gives it its peculiar merit is that it is no mere essay in uncontrolled theorizing, but a piece of inductive science which scrupulously respects the facts.

Lamarck's methods were, as we have seen, the obverse of those of Buffon. The theory which he produced had every appearance of completeness and consistency, though in fact, as I have endeavoured to show, it had not that internal simplicity which we look for in a good theory. But whether it was good or bad, considered in isolation as a theory, it certainly produced insuperable difficulties when applied even to the one set of facts upon which Lamarck attempted to base its justification. For the theory required a classification of animals which should be in principle linear, and such a system is at variance with the facts. Thus, although Lamarck had already produced a branching scheme in 1809, he still adhered to a linear one in 1815, and he found himself obliged to produce two conflicting arrangements, one to satisfy the basic assumption of an ascending scale of perfection, and another to accommodate the obvious deviations from this order.

Lamarck was not unaware of the important evidence to be derived from the geographical distribution of organisms, but he nowhere, I think, sets forth this evidence in detail. A brief reference to this subject occurs in the *Animaux sans Vertèbres*:

> Let anyone pass slowly over the surface of the earth, especially in the north–south direction, stopping from time to time to give himself leisure to observe; he will invariably see the *species* varying little by little, and more and more as he is farther from his starting point. He will see them follow, in some sense, the variations of the localities themselves, the conditions, exposed or sheltered, and so on. Sometimes he will even see varieties produced, not by habits required by the conditions, but by habits contracted accidentally, or in some other way. Thus man, who is subject by his organization to the laws of nature, himself shows remarkable varieties within his species, and among these some which seem to be due to the causes last mentioned.[107]

In general, Lamarck appears to have had little appreciation of the nature and amount of evidence required to establish any major theoretical position. Thus, in discussing the inheritance of acquired characters, he writes:

> Indeed, this law of nature, by which all that has been acquired by their progenitors during their life is transmitted to new individuals, is so true, so striking, so well attested by the facts, that there is no observer who has not been able to convince himself of its reality.[108]

Considering the three sets of facts given above which any theory of evolution must be able to systematize, we find that Lamarck gave full attention to the facts of comparative anatomy, that he considered those of geographical distribution only in a summary manner, and that he found nothing but difficulties in the fossil record. Indeed, the facts of palaeontology are particularly suited to throw into relief all the defects of his theory. These defects appear to me to be directly due to the fact that the theory was everywhere infected with bad theology. I think it can be asserted with confidence that Deism is bad theology, because it cannot sustain examination by its own canons. It affects to be greatly superior in rationality to orthodox Christian theology, but it suffers, itself, from a basic inconsistency. In common with all theologies of our era, Deism asserts that the mind of God is infinite, yet supposes that we can see things from God's point of view, and can easily discover what he would do in particular sets of circumstances. It is well enough to say that God would adopt the best means for achieving his purposes, and that he would avoid wanton waste: but even supposing us to know what his purposes are, how can we believe ourselves in a position to assign the best means for attaining them? And how are we to know what would appear wasteful to infinite power combined with infinite intelligence? It is, however, deistic assumptions of this kind which give Lamarck's theory its peculiar structure. He supposes that God, in directing the process of evolution, was aiming directly at the production of man, and that this must appear in an empirically discoverable linearity in the evolutionary series. Further, he cannot tolerate the idea that any species should be wasted on the way, so that he is unwilling to admit the possibility of extinction, in the simple sense of the word.

It was these presuppositions and their consequences which made the Lamarckian theory unable to adapt itself readily to the data of comparative anatomy; but they made it even less suitable to the assimilation of the fossil record, which shows no simple upward trend of organization. It is everywhere marked by evidences of the extinction of species, few of which can have been the direct ancestors of living forms.

These considerations help to explain a puzzling feature of Darwin's presentation of his theory. It must have struck many readers of the *Origin of Species* that Darwin seems to give unnecessary prominence to the causal part of his theory. Would it not have been possible, it might be asked, to assemble and sift the three main types of evidence, and to establish upon them the general theory, without necessarily propounding a causal theory at all? Yet it is clear that Darwin thought of his causal theory as an indispensable element in the establishment of the general theory, as, indeed, the full title of his book makes clear: *The Origin of Species by means of Natural Selection, or the Preservation of Favoured Races in the Struggle for Life*.

Though it is, I believe, correct from the point of view of analysis to separate the general theory of evolution from the causal theory, nevertheless, from the psychological and historical point of view the two are separated only with difficulty.

As I have tried to show, a major factor among those which prevented Buffon from elaborating a general theory was his inability to produce a causal theory satisfactory to himself. It is clear also that the common reaction to the suggestion that animals had been produced by evolution would be to ask, 'But how could that be?'

Plainly, Darwin thought it essential to answer this question before setting out the general theory, and this is reflected even in the arrangement of the *Origin of Species*, for the discussions of the Geological Succession, of Geographical Distribution and of the Mutual Affinities of Organic Beings are relegated to the last chapters of the book. But in one sense at least the general and the causal theories are really inseparably united within Darwin's theory, for it was necessary to present the general theory in a form capable of 'saving the appearances' of the geological record. This could only be done by stressing the element of chance in the actual course of the evolutionary process.

Thus at least a negative statement relating to the underlying processes was necessary. It had to be made clear that the theory offered by Darwin did not involve the postulation of any empirically verifiable directional tendency in evolution. Though Darwin did not view the matter in this way, we might say that only by stressing the element of chance could he finally exorcize the spectre of Deism, and present a theory consonant with the facts as they undoubtedly appear to the human observer.

The indirect and meandering course of evolution, as presented in the fossil record; the starting and stopping; the bizarre experimentation; the long periods in which forms of life predominated which proved to be far from the directions of advance; the preservation, here and there, of archaic species; above all, what must strike the tidy mind as the intolerable wastage; none of these things could be reconciled with the delusion that the human mind can directly and fully comprehend the plan of the divine mind in the production of living things.

I hope that the originality and peculiar merits of Darwin's presentation of the arguments in favour of the general theory of evolution will be clearly revealed by this consideration of the work of his two great predecessors.

I offer no summary of the *Origin of Species*, for no summary could do justice to a work of which the principal value lies in the ordering of great masses of varied data to build up a cogent argument. In such a work the argument and the facts upon which it is based are inseparable; and the reader acquainted with the accidental circumstances which caused Darwin to write the *Origin of Species* when he did will remember that he himself regarded the book as the shortest account which could usefully be given of his theory.

It is very far from my intention to minimize the achievements of Buffon and of Lamarck. If Darwin himself appears to have done less than justice to his predecessors, we must remember that the labours he undertook as a naturalist might well have exhausted a more powerful constitution than that which he possessed: it is unreasonable to blame him for not having added to these labours those of an historian of science.

Although Darwin was little conscious of the preliminary work done by Buffon and by Lamarck, he must have profited by it, if only indirectly. Viewing the historical development of the theory of evolution dispassionately we can now see that there really was a development, though a discontinuous one. The merit of being Buffon in the eighteenth century, or Lamarck between centuries, is not necessarily less than that of being Darwin in the nineteenth, and anyone who likes to do so can say that Darwin only completed the building which others, under greater difficulties, had begun.

The fact remains that it was he who completed it.

NOTES AND REFERENCES

1

THE MOVEMENT OF PLANTS IN RESPONSE TO LIGHT

[1] Darwin, C. assisted by Darwin, F. (1880). *The Power of Movement in Plants.* London.

[2] Anon. (1880). *The Times*, no. 30,044, p. 9. London.

[3] Hort, A. (1916). *Theophrastus. Enquiry into Plants*, vol. i, p. 10. Loeb Classical Library, London and New York.

[4] Storr-Best, Ll. (1912). *Varro on Farming*, p. 96. London.

[5] Rackham, H. (1949). *Pliny. Natural History*, Book xviii, p. 89. Loeb Classical Library, London and New York.

[6] Acosta, C. (1585). *Trattato della Historia, Natura, et Virtu delle Droghe Medicinali*, cap. 8. Venice.

[7] Alpinus, P. (1592). *De Plantis Aegypti*, cap. 10. Venice.

[8] Cornuti, J.-P. (1635). *Canadensium Plantarum Historia*, p. 113. Paris.

[9] Sharrock, R. (1660). *The History of the Propagation and Improvement of Vegetables by the Concurrence of Art and Nature.* Oxford.

[10] Ray, J. (1686–1704). *Historia Plantarum*, vol. i, pp. 2, 15. London.

[11] Hales, S. (1769). *Statistical Essays*, 4th ed., vol. i, p. 39. London.

[12] Linnaeus, C. (1751). *Philosophia Botanica.* Stockholm.

Pulteney, R. (1758). Some observations on the sleep of plants; and an account of that faculty, which Linnaeus calls *Vigiliae Florum*; with an enumeration of several plants, which are subject to that law. *Phil. Trans.* 50, 506–17.

[13] Linnaeus, C. (1755). *Somnus Plantarum.* Uppsala.

[14] —— (1737). *Flora Lapponica*, p. 222. Amsterdam.

[15] Bonnet, C. (1754). *Recherches sur l'Usage des Feuilles dans les Plantes.* Leiden.

[16] Dodart, D. (1700). Sur l'affectation de la perpendiculaire, remarquable dans toutes les tiges, dans plusieurs racines, & autant qu'il est possible dans toutes les branches des plantes. *Mém. Acad. R. Sci., Paris*, 47–63.

[17] Duhamel du Monceau, H. L. (1758). *La Physique des Arbres*, p. 154. Paris.

[18] Candolle, A. P. de (1832). *Physiologie végétale*, vol. iii, pp. 1082, 1083. Paris.

[19] Dutrochet, H. J. (1837). *Mémoires pour servir à l'histoire anatomique et physiologique des végétaux et des animaux*, vol. ii, pp. 73 *seq.* Paris.

[20] Sachs, J. (1875). *Text-Book of Botany*, p. 753. Translated (from the 3rd and 4th eds.) by A. W. Bennett and W. T. Thiselton Dyer. Oxford.

[21] Vries, H. de (1879). Über die inneren Vorgänge bei den Wachsthumskrümmungen mehrzelliger Organe. *Bot. Ztg*, 37, 830–8.

[22] Darwin, F. (1880). Über das Wachstum negativ heliotropischer Wurzeln im Licht und im Finstern. *Arb. Bot. Inst. Würzburg*, **2**, 521–8.

[23] Blaauw, A. H. (1919). Licht und Wachstum. III. (Die Erklärung des Phototropismus.) *Meded. LandbHoogesch., Wageningen*, **15**, 164–7; fig. 13.

[24] Nuernbergk, E. L. (1927). Untersuchungen über die Lichtverteilung in *Avena*-Koleoptilen und anderen phototropisch reizbaren Pflanzenorganen bei einseitigen Beleuchtung. *Bot. Abh.* H. 12, p. 130.

[25] Frank, A. B. (1870). *Die natürliche wagerechte Richtung von Pflanzenteilen.* Leipzig. (This work discussed the part played by geotropism and phototropism in maintaining the horizontal position of certain stems and other parts of plants. Since both gravity and light were found to contribute to the maintenance of horizontal growth, it became easy to imagine that their modes of action were in some way similar.)

[26] Sachs, J. (1887). *Lectures on the Physiology of Plants*, p. 692. Translated by H. Marshall Ward. Oxford.

[27] Wiesner, J. (1879). Die heliotropischen Erscheinungen im Pflanzenreiche. I. Theil. *Denkschr. Akad. Wiss. Wien*, **39**, 173–84.

—— (1882). Die heliotropischen Erscheinungen. II. Theil. *Denkschr. Akad. Wiss. Wien*, **43**, 13–16.

[28] —— (1879). Die heliotropischen Erscheinungen. I. Theil. *Denkschr. Akad. Wiss. Wien*, **39**, 198–201.

[29] Müller, N. J. C. (1872). Untersuchungen über die Krümmungen der Pflanzen gegen das Sonnenlicht. *Bot. Untersuchungen*, **3**, 57–82. Heidelberg.

[30] See, for example:
Spencer, H. (1862). *First Principles*, pp. 456 *seq.* London.

[31] Pfeffer, W. (1897–1904). *Pflanzenphysiologie*, vol. I, pp. 9–20. Leipzig.

[32] Dutrochet, H. J. (1824). *Recherches anatomiques sur la structure intime des animaux et des végétaux*, pp. 107, 117, 130. Paris.

[33] Darwin, F. (1891). On growth-curvatures in plants. *Rep. Brit. Ass.* p. 665.

[34] —— (ed.) (1887). *The Life and Letters of Charles Darwin*, vol. I, p. 98. London.

[35] B[lackman], F. F. (1932). Francis Darwin. *Proc. Roy. Soc.* B, **110**, Obituary Notices, p. viii.

[36] Ciesielski, T. (1872). Untersuchungen über die Abwärtskrümmung der Wurzel. *Beitr. Biol. Pfl.* **1**, 1–30.

[37] Avery, Jr., G. S. and Burkholder, P. R. (1936). Polarised growth and cell studies on the *Avena* coleoptile, phytohormone test object. *Bull. Torrey Bot. Cl.* **63**, 1–15.

[38] Darwin, C. assisted by Darwin, F. (1880). *The Power of Movement in Plants*, p. 468.

[39] —— (1880). *Ibid.* p. 486.

[40] —— (1880). *Ibid.* p. 487.

[41] Sachs, J. (1887). *Lectures on the Physiology of Plants*, p. 689.

[42] SACHS, J. (1887). *Ibid.* p. 677, note.

[43] GOEBEL, K. (1897). Julius Sachs. *Flora*, **84** (Ergänzungsband), 121.

[44] WIESNER, J. (1881). *Das Bewegungsvermögen der Pflanzen.* Vienna.

[45] DARWIN, F. and SEWARD, A. C. (eds.) (1903). *More Letters of Charles Darwin*, vol. II, p. 434. London.

[46] VÖCHTING, H. (1888). Über die Lichtstellung der Laubblätter. *Bot. Ztg*, **46**, 524–5.

[47] ROTHERT, W. (1894–96). Ueber Heliotropismus. *Beitr. Biol. Pfl.* **7**, 1–212.

[48] —— (1894–96). *Ibid.* 200–2.

[49] SIERP, H. and SEYBOLD, A. (1926). Untersuchungen über die Lichtempfindlichkeit der Spitze und des Stumpfes in der Koleoptile von *Avena sativa. Jb. wiss. Bot.* **65**, 592–610.

[50] LANGE, S. (1927). Die Verteilung der Lichtempfindlichkeit in der Spitze der Haferkoleoptile. *Jb. wiss. Bot.* **67**, 1–51.

[51] BLAAUW, A. H. (1908). Die Perzeption des Lichtes. *Rec. Trav. bot. néerl.* **5**, 257–77.

[52] PAYER, J. (1842). Mémoire sur la tendance des tiges vers la lumière. *C.R. Acad. Sci., Paris*, **15**, 1194–6.

[53] BACHMANN, FR. and BERGANN, FR. (1930). Über die Wertigkeit von Strahlen verschiedener Wellenlänge für die phototropische Reizung von *Avena sativa. Planta*, **10**, 744–55.

GALSTON, A. W. and BAKER, R. S. (1949). Studies on the physiology of light action. II. The photodynamic action of riboflavin. *Amer. J. Bot.* **36**, 778.

[54] KOHLBECKER, R. (1957). Die Abhängigkeit des Längenwachstums und der phototropischen Krümmungen von der Lichtqualität bei Keimwurzeln von *Sinapis alba. Z. Bot.* **45**, 507–24.

[55] ATKINS, G. A. (1936). The effect of pigment on phototropic response: a comparative study of reactions to monochromatic light. *Ann. Bot.* **50**, 197–218.

[56] PFEFFER, W. (1897–1904). *Pflanzenphysiologie*, vol. I, p. 12.

[57] BLAAUW, A. H. (1908). Die Perzeption des Lichtes. *Rec. Trav. bot. néerl.* **5**, 257–77.

[58] FRÖSCHEL, P. (1908). Untersuchungen über die heliotropische Präsentationszeit. I. *S.B. Akad. Wiss. Wien* (Abt. 1, *Math. Naturw. Kl.*), **117**, 235–56.

—— (1909). Untersuchungen über die heliotropische Präsentationszeit. II. *Ibid.* **118**, 1247–94.

[59] ARISZ, W. H. (1911). On the connection between stimulus and effect in phototropic curvatures of seedlings of *Avena sativa. Proc. Acad. Sci. Amst.* **13**, 1022–31.

Note the widely different estimates of the threshold obtained by:

NOACK, K. (1914). Die Bedeutung der schiefen Lichtrichtung für die Helioperzeption parallelotropen Organe. *Z. Bot.* **6**, 44.

GUTTENBERG, H. VON (1923). Studien über den Phototropismus der Pflanzen. *Beitr. allg. Bot.* **2**, 238 *seq.*

[60] WIESNER, J. (1882). Die heliotropischen Erscheinungen in Pflanzenreiche. II. Theil. *Denkschr. Akad. Wiss. Wien*, **43**, 13–16.

[61] Vogt, E. (1915). Über den Einfluss des Lichts auf das Wachstum der Koleoptile von *Avena sativa*. *Z. Bot.* **7**, 193–270.

[62] Blaauw, A. H. (1915). Licht und Wachstum. II. *Z. Bot.* **7**, 465–532.

[63] For example:
Noack, K. (1914). Die Bedeutung der schiefen Lichtrichtung. *Z. Bot.* **6**, 44.

[64] Sierp, H. (1921). Untersuchungen über die durch Licht und Dunkelheit hervorgerufenen Wachstumsreaktionen bei der Koleoptile von *Avena sativa* und ihr Zusammenhang mit den phototropischen Krümmungen. *Z. Bot.* **13**, 113–72.

[65] Dillewijn, C. van (1927). Die Lichtwachstumsreaktionen von *Avena*. *Rec. Trav. bot. néerl.* **24**, 400 *seq.*

[66] Arisz, W. H. (1915). Untersuchungen über den Phototropismus. *Rec. Trav. bot. néerl.* **12**, 44–216.

[67] Buy, H. G. du and Nuernbergk, E. (1934). Phototropismus und Wachstum der Pflanzen. II. *Ergebn. Biol.* **10**, 256–8.

[68] Arisz, W. H. (1915). Untersuchungen über den Phototropismus. *Rec. Trav. bot. néerl.* **12**, 143–5.

[69] Dillewijn, C. van (1927). Die Lichtwachstumsreaktionen von *Avena*. *Rec. Trav. bot. néerl.* **24**, 323–99.

[70] Sierp, H. (1921). *Z. Bot.* **13**, 155–70.

[71] Rothert, W. (1894–96). Ueber Heliotropismus. *Beitr. Biol. Pflanzen*, **7**, 634.

[72] Fitting, H. (1907). Die Leitung tropistischer Reize in parallelotropen Pflanzenteilen. *Jb. wiss. Bot.* **44**, 177–253.

[73] Boysen Jensen, P. (1911). La transmission de l'initiation phototropique dans l'*Avena*. *Overs. danske Vidensk. Selsk. Forh.* pp. 3–24.

[74] Wolk, P. C. van der (1911). Investigation of the transmission of light stimuli in the seedlings of *Avena*. *Proc. Acad. Sci. Amst.* **14**, 327–42.

[75] Purdy, Helen A. Studies on the path of transmission of phototropic and geotropic stimuli in the coleoptile of *Avena*. *K. danske vidensk. Selsk.* (*Biol. Medd.*), **3**, no. 8, 1–29.

[76] See, for example:
Wolk, P. C. van der (1911). *Proc. Acad. Sci. Amst.* **13**, 336–9.
Guttenberg, H. R. von (1913). Über akropetale heliotropische Reizleitung. *Jb. wiss. Bot.* **52**, 333–50.
Arisz, W. H. (1915). *Rec. Trav. bot. néerl.* **12**, 102–7.

[77] Rothert, W. (1894–96). *Beitr. Biol. Pflanzen*, **7**, 184–5.

[78] Boysen Jensen, P. (1936). *Growth Hormones in Plants*, p. 14. New York and London.

[79] —— (1911). *Overs. danske Vidensk. Selsk. Forh.* pp. 22–4.

[80] Fitting, H. (1909). Die Beeinflussung der Orchideenblüten durch die Bestäubung und durch andere Umstände. *Z. Bot.* **1**, 1–86.
—— (1910). Weitere entwicklungsphysiologische Untersuchungen an Orchideenbluten. *Z. Bot.* **2**, 225–67.

[81] Starling, E. H. (1905). The chemical correlation of the functions of the body. *Lancet*, **2**, 339–41, 423–5, 501–3, 579–83.

[82] Sachs, J. (1878–82). Stoff und Form der Pflanzenorgane. *Arb. Bot. Inst. Würzburg*, **2**, 452–88, 689–718.

[83] Paál, A. (1917–19). Über phototropische Reizleitung. *Jb. wiss. Bot.* **58**, 406–58.

[84] Compare, for example, the results reported in the following:
Stark, P. (1921). Studien über traumatotrope und haptotrope Reizleitungs vorgänge mit besonderer Berücksichtigung der Reizübertragung auf fremde Arten und Gattungen. *Jb. wiss. Bot.* **60**, 67–134.
Seubert, Elisabeth (1925). Über Wachstumsregulatoren in der Koleoptile von *Avena*. *Z. Bot.* **17**, 49–88.

[85] Brauner, L. (1922). Lichtkrümmung und Lichtwachstumsreaktion. *Z. Bot.* **14**, 497–547.

[86] Söding, H. (1925). Zur Kenntnis der Wuchshormone in der Haferkoleoptile. *Jb. wiss. Bot.* **64**, 587–603.

[87] Nielsen, N. (1924). Studies on the transmission of stimuli in the coleoptile of *Avena*. *Dansk bot. Ark.* **4**, nr. 8, 1–45.

[88] Priestley, J. H. (1926). Light and growth. IV. An examination of the phototropic mechanism concerned in the curvature of coleoptiles of the Gramineae. *New Phytol.* **25**, 227–47.

[89] See, for example:
Stark, P. and Dreschel, O. (1922). Phototropische Reizleitungsvorgänge bei Unterbrechung des organischen Zusammenhangs. *Jb. wiss. Bot.* **61**, 339–71.

[90] Lundegårdh, H. (1922–24). Ein Beitrag zur quantitativen Analyse des Phototropismus. *Ark. Bot.* **18**, no. 3, 32–43.
Pisek, A. (1926). Untersuchungen über den Autotropismus der Haferkoleoptile bei Lichtkrümmung, über Reizleitung und den Zusammenhang von Lichtwachstumsreaktion und Phototropismus. *Jb. wiss. Bot.* **65**, 460–501.
Beyer, A. (1927). Experimentelle Studien zur Blaauwschen Theorie. I. Die Wachstumverhältnisse bei der phototropischen Krümmung vorbelichteter *Avena*-Keimlinge. *Planta*, **4**, 411–36.

[91] Went, F. F. A. C. (1924). Communication on Miss A. Bakker's 'Investigations regarding the existence of separate zones of perception and reaction in the seedlings of the Paniceae'. *Proc. Acad. Sci. Amst.* **27**, 503–4.

[92] Went, F. W. (1927). On growth-accelerating substances in the coleoptile of *Avena sativa*. *Proc. Acad. Sci. Amst.* **30**, 10–19.

[93] —— (1927). *Ibid.* 11 *seq.*

[94] —— (1927). *Ibid.* 15.

[95] —— (1928). Wuchsstoff und Wachstum. *Rec. Trav. bot. néerl.* **25**, 1–116.

[96] Blackman, F. F. (1905). Optima and limiting factors. *Ann. Bot.* **19**, 281–95.

[97] Cholodny, N. (1927). Wuchshormone und Tropismen bei den Pflanzen. *Biol. Zbl.* **47**, 604–26.

[98] Audus, L. J. and Brownbridge, M. E. (1957). Studies on the geotropism of roots. I. Growth-rate distribution during response and the effects of applied auxins. *J. Exp. Bot.* **8**, 105–24.

[99] Seubert, Elisabeth (1925). *Z. Bot.* **17**, 49–88.

[100] Kögl, F. and Haagen Smit, A. J. (1931). Über die Chemie des Wuchsstoffs. *Proc. Acad. Sci. Amst.* **34**, 1411–16.

[101] KONIGSBERGER, V. J. and VERKAAIK, B. (1938). On phototropic curvatures in *Avena*, caused by photo-inactivation of auxin *a* via its lactone. *Rec. Trav. bot. néerl.* **35**, 1–13.

[102] KÖGL, F. and SCHURINGA, G. J. (1944). Über die Inaktivierung von Auxin-*a*-Lacton bei verschiedenen Wellenlängen und den Einfluss von Carotinoiden auf die Lichtreaktion. *Hoppe-Seyl. Z.* **280**, 148–61.

[103] WALD, G. and DU BUY, H. G. (1936). Pigments of the oat coleoptile. *Science*, **84**, 247.

[104] BÜNNING, E. (1937). Phototropismus und Carotinoide. II. Das Carotin der Reizaufnahmezonen von *Pilobolus*, *Phycomyces* und *Avena*. *Planta*, **27**, 148–58.

[105] REINERT, J. (1953). Über die Wirkung von Riboflavin und Carotin beim Phototropismus von *Avena*-Koleoptilen und bei anderen pflanzlichen Lichtreizreaktionen. *Z. Bot.* **41**, 103–22.

[106] HAAGEN-SMIT, A. J., DANDLIKER, W. B., WITTWER, S. H. and MURNEEK, A. E. (1946). Isolation of 3-indole-acetic acid from immature corn kernels. *Amer. J. Bot.* **33**, 118–20.

[107] WILDMAN, S. G. and BONNER, J. (1948). Observations on the chemical nature and formation of auxin in the *Avena* coleoptile. *Amer. J. Bot.* **35**, 740–46.

[108] GUTTENBERG, H. VON (1942). Über die Bildung und Aktivierung des Wuchsstoffes in den höheren Pflanzen. *Naturwissenschaften*, **30**, 109–12.

[109] DETTWEILER, C. (1942). Über den Einfluss des Heteroauxins auf die Wuchsstoffbildung in höheren Pflanzen. *Planta*, **33**, 258–77.

[110] GUTTENBERG, H. VON and BÜSCHEL, R. (1944). Über die Wirkung des Heteroauxins auf isolierte Wuchsstoffrei Pflanzenteile. *Planta*, **34**, 49–66.

[111] REINERT, J. (1950). Über den Wuchstoffgehalt der *Avena*-Koleoptilspitze und die chemische Natur des extrahierbaren Auxins. *Z. Naturforsch.* **5B**, 374–80.

[112] WIEDOW-PATZOLD, H.-L. and VON GUTTENBERG, H. (1957). Weitere Untersuchungen über den nativen Wuchsstoff. *Planta*, **49**, 588–97.
See also:
RAADTS, EDITH and SÖDING, H. (1957). Chromatographische Untersuchungen über die Wuchsstoffe der Haferkoleoptile. *Planta*, **49**, 47–60.

[113] STARK, P. and DRESCHEL, O. (1922). *Jb. wiss. Bot.* **61**, 354.

[114] SÖDING, H. (1935–36). Wirkt der Wuchsstoff artspezifisch? *Jb. wiss. Bot.* **82**, 534–54.

[115] LARSEN, P. (1951). Formation, occurrence, and inactivation of growth substances. *Annu. Rev. Pl. Physiol.* **2**, 175–9.
See also:
GORDON, S. A. (1954). Occurrence, formation, and inactivation of auxins. *Annu. Rev. Pl. Physiol.* **5**, 343–50.

[116] BONNER, J. and BANDURSKI, R. S. (1952). Studies of the physiology, pharmacology and biochemistry of the auxins. *Annu. Rev. Pl. Physiol.* **3**, 59.

[117] KEFFORD, N. P. (1955). The growth substances separated from plant extracts by chromatography. I. *J. Exp. Bot.* **6**, 129–51; II. *Ibid.* 245–55.

WAIN, R. L. and WIGHTMAN, F. (eds.) (1956). *The Chemistry and Mode of Action of Plant Growth Substances*. London.

[118] AUDUS, L. J. (1959). *Plant Growth Substances*, 2nd ed. London.

[119] WEIJ, H. G. VAN DER (1932). Der Mechanismus des Wuchsstofftransportes. *Rec. Trav. bot. néerl.* **29**, 464–78.

[120] BOTTELIER, H. P. (1934). Über den Einfluss äusserer Faktoren auf die Protoplasmaströmung in der *Avena*-Koleoptile. *Rec. Trav. bot. néerl.* **31**, 517–60.

[121] CLARK, W. G. (1938). Electrical polarity and auxin transport. *Pl. Physiol.* **13**, 529–52.

[122] BEYER, A. (1927–28). Beiträge zum Problem der Reizleitung. *Z. Bot.* **20**, 321–43.

[123] See, for example:

GUTTENBERG, H. VON (1950). Über das Vorkommen und die primäre Wirkung von Auxin und Heteroauxin. *Naturwissenschaften*, **37**, 65.

JACOBS, W. P. (1956). Internal factors controlling cell differentiation in the flowering plants. *Amer. Nat.* **90**, 163–9.

[124] STARK, P. and DRESCHEL, O. (1922). *Jahrb. wiss. Bot.* **61**, 346.

[125] OVERBEEK, J. VAN (1932). An analysis of phototropism in dicotyledons. *Proc. Acad. Sci. Amst.* **35**, 1325–35.

[126] BURKHOLDER, P. R. and JOHNSTON, E. S. (1937). Inactivation of plant growth substance by light. *Smithson. Misc. Coll.* **95**, no. 20, 1–14.

[127] OVERBEEK, J. VAN (1935). Wuchsstoff, Lichtwachsstumsreaktion und Phototropismus bei *Raphanus*. *Rec. Trav. bot. néerl.* **30**, 537–626.

[128] KONINGSBERGER, V. J. and VERKAAIK, B. (1938). *Rec. Trav. bot. néerl.* **35**, 1–13.

[129] STEWART, W. S. and WENT, F. W. (1940). Light stability of auxin in *Avena* coleoptiles. *Bot. Gaz.* **101**, 706–14.

[130] WILDMAN, S. G. and BONNER, J. (1948). *Amer. J. Bot.* **33**, 740–1.

[131] OPPENOORTH, JR., W. F. F. (1941). On the role of auxin in phototropism and light growth reactions of *Avena* coleoptiles. *Rec. Trav. bot. néerl.* **38**, 335–43.

[132] GALSTON, A. W. (1949). Riboflavin-sensitized photo-oxidation of indoleacetic acid and related compounds. *Proc. nat. Acad. Sci. Wash.* **35**, 10–17.

[133] GALSTON, A. W. and BAKER, R. S. (1949). Studies on the physiology of light action. II. The photo-dynamic action of riboflavin. *Amer. J. Bot.* **36**, 773–80.

[134] REINERT, J. (1953). *Z. Bot.* **41**, 110 *seq.*

See also:

SHROPSHIRE, JR., W. and WITHROW, R. B. (1958). Action spectrum of phototropic tip-curvature of *Avena*. *Pl. Physiol.* **33**, 360–5.

[135] See, for example:

NAVEZ, A. E. (1933). Growth-promoting substance and illumination. *Proc. nat. Acad. Sci. Wash.* **19**, 636–8.

[136] GUTTENBERG, H. VON and ZETSCHE, K. (1956). Der Einfluss des Lichtes auf die Auxinbildung und den Auxintransport. *Planta*, **48**, 99–134.

[137] OPPENOORTH, JR., W. F. F. (1941). *Rec. Trav. bot. néerl.* **38**, 335–43.

[138] Overbeek, J. van (1932). *Proc. Acad. Sci. Amst.* **35**, 1330 *seq.*

[139] Wilden, Marga (1939–40). Zur Analyse der positiven und negativen phototropischen Krümmungen. *Planta*, **30**, 286–8.

[140] Asana, R. D. (1938). On the relation between the distribution of auxin in the tip of the *Avena* coleoptile and the first negative phototropic curvature. *Ann. Bot., Lond.*, N.S., **2**, 955–7.

[141] Boysen Jensen, P. (1933). Über die durch einseitige Lichtwirkung hervorgerufene transversale Leitung des Wuchsstoffes in der *Avena*-Koleoptile. *Planta*, **19**, 335–44.

[142] —— (1928). Die phototropische Induktion in der Spitze der *Avena*-Koleoptile. *Ibid.* **5**, 464–77.

[143] Went, F. W. (1928). Die Erklärung des phototropischen Krümmungsverlaufs. *Rec. Trav. bot. néerl.* **25** A, 483–9.

[144] For example:
Galston, A. W. (1950). Phototropism. II. *Bot. Rev.* **16**, 361–78.

[145] Briggs, W. R., Tocher, R. D. and Wilson, J. F. (1957). Phototropic auxin redistribution in corn coleoptiles. *Science*, **126**, 210–12.

[146] Bunning, E., Reisener, H.-J., Weygand, F., Simon, H. and Klebe, J. F. (1956). Versuche mit radioaktiver Indolylessigsäure zur Prüfung der sogenannten Ablenkung des Wuchshormonstromes durch Licht. *Z. Naturforsch.* **11** B, 363–4.

[147] Went, F. W. (1927). *Proc. Acad. Sci. Amst.* **30**, 16–17.

[148] Buy, H. G. du (1933). Über Wachstum und Phototropismus von *Avena sativa*. *Rec. Trav. bot. néerl.* **30**, 870–7.

[149] Overbeek, J. van (1933). *Rec. Trav. bot. néerl.* **30**, 582–4.

[150] Meyer, J. and Pohl, R. (1956). Neue Erkenntnisse zum Problem der phototropen Krümmung bei höheren Pflanzen. *Naturwissenschaften*, **43**, 5.
Guttenberg, H. von and Zetsche, K. (1956). *Planta*, **48**, 99–134.
Niedergang-Kamien, E. and Leopold, A. C. (1957). Inhibitors of polar auxin transport. *Physiol. Plant.* **10**, 29–38.

[151] Huber, H. (1951). Über den Einfluss der Belichtung auf die Wuchsstoffempfindlichkeit der Keimstengel von *Cucumis sativus* L. *Ber. schweiz. bot. Ges.* **61**, 499–538.

[152] Laibach, F. (1938–39). Zur Frage der Inaktivierung des Wuchsstoffes durch Licht. *Ber. dtsch. bot. Ges.* **56**, 298–306.

[153] Cholodny, N. (1930). Mikropotometrische Untersuchungen über das Wachstum und die Tropismen der Koleoptile von *Avena sativa*. *Jb. wiss. Bot.* **73**, 720–58.

[154] Brauner, L. (1952). Induktion phototropischer Reaktionen durch ein künstliches Perzeptionsorgan. *Experientia*, **8**, 102–3.

[155] Buy, H. G. du and Nuernbergk, E. (1934). *Ergebn. Biol.* **10**, 256–8.

[156] Gessner, F. (1935–36). Phototropismus und Wanddehnbarkeit. *Jb. wiss. Bot.* **82**, 796–802.

[157] Overbeek, J. van (1939). Phototropism. *Bot. Rev.* **5**, 659.
See also:
Tagawa, T. and Bonner, J. (1957). Mechanical properties of the *Avena* coleoptile as related to auxin and to ionic interactions. *Plant. Physiol.* **32**, 207–12.

[158] Nuernbergk, E. L. (1927). *Bot. Abh.* H. 12, p. 130.

[159] Buy, H. G. du (1934). Der Phototropismus der *Avena*-Koleoptile und die Lichtabfallstheorie. *Ber. dtsch. bot. Ges.* **52**, 540 *seq.*

Brauner, L. (1955). Über die Funktion der Spitzenzone beim Phototropismus der *Avena*-Koleoptile. *Z. Bot.* **43**, 467–98.

Bünning, E., Dorn, J., Schneiderhöhn, G. and Thorning, I. (1953). Zur Funktion von Lactoflavin und Carotin beim Phototropismus und bei lichtbedingten Wachstumsbeeinflussungen. *Ber. dtsch. bot. Ges.* **66**, 333–40.

[160] Bünning, E. *et al.* (1953). *Ber. dtsch. bot. Ges.* **66**, 333–40.

Brauner, L. (1957). The perception of the phototropic stimulus in the oat coleoptile. *Symp. Soc. Exp. Biol.* **11**, 86–94.

[161] Schrank, A. R. (1946). Note on the effect of unilateral illumination on the transverse electrical polarity in the *Avena* coleoptile. *Plant Physiol.* **21**, 362–5.

[162] Clark, W. G. (1938). *Plant Physiol.* **13**, 529–52.

[163] Haig, C. (1935). The phototropic responses of *Avena* in relation to intensity and wave-length. *Biol. Bull., Woods Hole,* **69**, 305–24.

[164] See, for example:

Yamane, G. (1940). Über den positiven und negativen Phototropismus von Laubblättern der *Fatsia japonica* in Zusammenhang mit der Wuchsstoffwirkung. *Bot. Mag.* (*Tokyo*), **54**, 117–29.

The movements of the leaves of the 'compass plants' probably result from a combination of stimuli. See:

Dolk, H. E. (1931). The movement of leaves of the compass-plant *Lactuca scariola*. *Amer. J. Bot.* **18**, 195–204.

[165] Langham, D. G. (1941). The effect of light on growth habit of plants. *Amer. J. Bot.* **28**, 951–6.

2

PALAEONTOLOGY AND EVOLUTION

[1] Darwin, C. (1910). *The Origin of Species*, p. 248. Popular imp, 6th ed. London.

[2] In the British literature we have among others the following:

George, T. N. (1951). *Evolution in Outline*. London. (Short, but particularly useful.)

—— (1948). Evolution in fossil communities. *Proc. R. Phil. Soc. Glasg.* **73**, 23–42.

—— (1953–54). Fossils and the evolutionary process. *Advanc. Sci.* N.S., **10**, 132–44.

Swinnerton, H. H. (1939). Palaeontology and the mechanics of evolution. *Quart. J. Geol. Soc. Lond.* **95**, xxxiii–lxx.

—— (1940). The study of variation in fossils. *Ibid.* **96**, lxxvii–cxviii.

Notes and References: Chapter 2

DAVIES, A. M. (1937). *Evolution and its Modern Critics.* London.

BATHER, F. A. (1920). Fossils and life. *Rep. Brit. Ass.* pp. 61–86.

—— (1927). Biological classification. *Quart. J. Geol. Soc. Lond.* 83, lxii–civ.

WATTS, W. W. (1925). *Evolution in the Light of Modern Knowledge.* London.

In America, general works on evolution pay more attention to palaeontology than they do in Britain, the two following being specially important:

SIMPSON, G. G. (1953). *The Major Features of Evolution.* New York.

JEPSEN, G. L. and MAYR, E. (eds.) (1949). *Genetics, Paleontology and Evolution.* Princeton.

[3] GEIKIE, SIR ARCHIBALD (1909). *Charles Darwin as Geologist.* (Rede Lecture.) Cambridge.

[4] HUXLEY, T. H. (1893). *Darwiniana*, p. 225. London.

[5] LYELL, C. (1833). *Principles of Geology*, vol. III, pp. 1, 2. London.

[6] THOMAS, G. (1950). Processes of fossilization. *New Biol.* no. 8, 75–97. (One of the best discussions of fossilization.)

HAWKINS, H. L. (1920). *Invertebrate Palaeontology.* London.

WALTON, J. (1940). *Fossil Plants.* London.

[7] Well exemplified by graptolites. See:

ELLES, GERTRUDE L. (1944). The identification of graptolites. *Geol. Mag.* 81, 145–58.

[8] For a particularly clear and searching examination of the principles of stratigraphical geology, see

KRUMBEIN, W. C. and SLOSS, L. L. (1951). *Stratigraphy and Sedimentation.* San Francisco.

[9] There is no thoroughly satisfactory general treatise on the geology of Britain. The following should be consulted:

WOODWARD, H. B. (1887). *Geology of England and Wales*, 2nd ed. London. (Still, perhaps, the most attractive for southern Britain.)

JUKES-BROWNE, A. J. (1912). *The Student's Handbook of Stratigraphical Geology*, 2nd ed. London.

EVANS, J. W. and STUBBLEFIELD, C. J. (1929). *Handbook of the Geology of Great Britain.* London.

Excellent text-book expositions are to be found in:

STAMP, L. D. (1934). *An Introduction to Stratigraphy*, 2nd ed. London.

WELLS, A. K. (1959). *Outline of Historical Geology*, 4th ed. London.

A picture of the changing scene whereon the history of life in the British area has unfolded itself will be found in:

WILLS, L. J. (1951). *Palaeogeographical Atlas of the British Isles.* London.

For an authoritative and up-to-date account of the geology of Britain, reference must be made to the indispensable GEOLOGICAL SURVEY (1935→). *British Regional Geology.* 18 vols. London.

[10] For an example of a discussion of this problem, with special reference to the Lias, see:

TRUEMAN, A. E. (1923). Some theoretical aspects of correlation. *Proc. Geol. Ass., Lond.*, 34, 193–206.

STAMP, L. D. (1925). Some practical aspects of correlation: a criticism. *Proc. Geol. Ass., Lond.*, 36, 11–27.

[11] Very effective use of this graphical device is made by:

SIMPSON, G. G. (1950). *The Meaning of Evolution.* New Haven and Oxford.

—— (1953). *Life of the Past.* New Haven and London.

ROMER, A. S. (1945). *Vertebrate Paleontology*, 2nd ed. Chicago.

[12] Summaries for the groups, in turn, are given in the general systematic works, such as:

ZITTEL, K. A. (1913–32). *Textbook of Palaeontology*, 2nd ed. London.

WOODS, H. (1946). *Palaeontology: Invertebrate*, 8th ed. Cambridge.

MOORE, R. C. (ed.) (1953–). *Treatise on Invertebrate Paleontology.* New York.

For accounts of the whole history of life, see:

NICHOLSON, H. A. (1877). *Ancient Life History of the Earth.* London.

OAKLEY, K. P. and MUIR-WOOD, HELEN (1949). *Succession of Life through Geological Time.* London (British Museum).

[13] ARKELL, W. J. (1935–48). Ammonites of the English Corallian Beds. *Palaeontogr. Soc.* (*Monogr.*), pp. 380 *seq.*

[14] See particularly:

TRUEMAN, A. E. (1924). The species-concept in palaeontology. *Geol. Mag.* 61, 355–60.

—— (1940). The meaning of orthogenesis. *Trans. Geol. Soc. Glasg.* 20, 77–95.

[15] An authority who has always paid much attention to separate structural features is Swinnerton; see, for instance:

SWINNERTON, H. H. (1932). Unit characters. *Biol. Rev.* 7, 321–35.

[16] For some discussion of this matter, see:

GEORGE, T. N. (1933). Palingenesis and palaeontology. *Biol. Rev.* 8, 107–35.

SWINNERTON, H. H. (1938). Development and evolution. *Rep. Brit. Ass.* pp. 57–84.

See also the works of L. F. Spath referred to hereafter.

[17] NEAVERSON, E. (1955). *Stratigraphical Palaeontology*, 2nd ed. London.

[18] LANG, W. H. and COOKSON, ISABEL C. (1935). On a flora, including vascular land plants, associated with *Monograptus*, in rocks of Silurian age, from Victoria, Australia. *Phil. Trans.* B, 224, 421–49.

[19] HOWELL, B. F. and STUBBLEFIELD, C. J. (1950). A revision of the fauna of the North Welsh *Conocoryphe viola* beds implying a Lower Cambrian age. *Geol. Mag.* 87, 1–16.

[20] HARKNESS, R. and HICKS, H. (1871). On the ancient rocks of the St David's Promontory, South Wales, and their fossil contents. *Quart. J. Geol. Soc. Lond.* 27, 389.

JONES, O. T. (1930). The Cambrian rocks. *Proc. Geol. Ass., Lond.*, 41, 250–6.

[21] CALLAWAY, C. (1878). On some small Trilobites from the Cambrian rocks of Comley (Shropshire). *Quart. J. Geol. Soc. Lond.* 34, 759.

COBBOLD, E. S. (1910). On the Quartzites of Shropshire. *Quart. J. Geol. Soc. Lond.* 66, 19–51.

Cobbold, E. S. (1920). The Cambrian horizons of Comley (Shropshire) and their Brachiopoda, Pteropoda, Gasteropoda, etc. *Quart. J. Geol. Soc. Lond.* **76**, 325–86.

—— (1925). Notes on the Cambrian area of Comley. *Proc. Geol. Ass., Lond.*, **36**, 367–74.

—— (1927). The structure of the Cambrian area of Comley. *Quart. J. Geol. Soc. Lond.* **83**, 551–73.

—— (1931). Additional fossils from the Cambrian rocks of Comley, Shropshire. *Ibid.* **87**, 459–512.

—— (1936). The Conchostraca of the Cambrian area of Comley, Shropshire, with a note on a new variety of *Atops reticulatus* (Walcott). *Ibid.* **92**, 221–35.

Cobbold, E. S. and Pocock, R. W. (1934). The Cambrian area of Rushton (Shropshire). *Phil. Trans.* B. **223**, 305–409.

Lapworth, C. (1888). Discovery of the *Olenellus* fauna in Britain. *Geol. Mag.* **25**, 484–7.

—— (1891). On *Olenellus callavei* and its geological relationships. *Ibid.* **28**, 529–36.

Raw, F. (1931). Notes on some sections of Lower Cambrian rocks from Comley, Shropshire. *Quart. J. Geol. Soc. Lond.* **87**, 503- 12.

—— (1936). Mesonacidae of Comley in Shropshire, with a discussion of classification within the family. *Ibid.* **92**, 236–93.

[22] Groom, T. (1899). The geological structure of the Southern Malverns, and of the adjacent district to the west. *Quart. J. Geol. Soc. Lond.* **55**, 129–69.

—— (1902). The sequence of the Cambrian and associated beds of the Malvern Hills. *Ibid.* **58**, 90.

—— in Monckton, H. W. and Herries, R. S. (eds.) (1910). *Geology in the Field*, pp. 701–3. London.

Matley, C. A. (1902). On the Cambrian Brachiopoda of the Malvern Hills. *Quart. J. Geol. Soc. Lond.* **58**, 135–49.

[23] Cobbold, E. S. (1919). Cambrian Hyolithidae, etc., from Hartshill in the Nuneaton district, Warwickshire. *Geol. Mag.* **56**, 149–58.

[24] Geikie, A. (1891). Discovery of the *Olenellus*-zone in the Northwest Highlands. *Geol. Mag.* **28**, 498–9.

Peach, B. N. (1912). The relation between the Cambrian faunas of Scotland and North America. *Rep. Brit. Ass.* pp. 448–59.

Peach, B. N., Horne, J. *et al.* (1907). The Geological structure of the Northwest Highlands of Scotland. *Mem. Geol. Surv. Gt. Brit.* chaps. 23 *et seq.* Glasgow.

[25] Cox, A. H. and Wells, A. K. (1927). The geology of the Dolgelly district Merionethshire. *Proc. Geol. Ass., Lond.*, **38**, 273.

Matley, C. A. and Wilson, T. S. (1946). The Harlech Dome, north of the Barmouth estuary. *Quart. J. Geol. Soc. Lond.* **102**, 13.

Nicholas, T. C. (1915). St Tudwal's peninsula. *Quart. J. Geol. Soc. Lond.* **71**, 83–143, 451–72.

Salter, J. W. and Etheridge, R. (1880). The fossils of N. Wales. *Mem. Geol. Surv. Gt. Brit.* p. 337. London.

[26] Harkness, R. and Hicks, H. (1871). *Quart. J. Geol. Soc. Lond.* **27**, 384–404.

JONES, O. T. (1930). *Proc. Geol. Ass., Lond.*, **41**, 253.
—— (1930). Caerbwdy to Porth-y-Rhaw. *Ibid.* **41**, 416–18.
SALTER, J. W. (1863). On the discovery of *Paradoxides* in Britain. *Quart. J. Geol. Soc. Lond.* **19**, 274–7.
—— (1864). On some new fossils from the Lingula-flags of Wales. *Ibid.* **20**, 233–41.

[27] COBBOLD, E. S. (1911). Trilobites from the *Paradoxides* Beds of Comley (Shropshire). *Quart. J. Geol. Soc. Lond.* **67**, 282–300.
—— (1920). *Ibid.* **76**, 325–86.
—— (1925). *Proc. Geol. Ass., Lond.*, **36**, 367–74.
—— (1927). *Quart. J. Geol. Soc. Lond.* **83**, 551–73.
COBBOLD, E. S. and POCOCK, R. W. (1934). *Phil. Trans.* B, **223**, 305.

[28] ILLING, V. C. (1915). The Paradoxidian fauna of a part of the Stockingford Shales. *Quart. J. Geol. Soc. Lond.* **71**, 386–450.

[29] BELT, T. (1867). On the Lingula-flags or Ffestiniog Group of the Dolgelly district. *Geol. Mag.* **4**, 538–43.
COX, A. H. and WELLS, A. K. (1927). *Proc. Geol. Ass., Lond.*, **38**, 275.
FEARNSIDES, W. G. (1905). On the geology of Arenig Fawr and Moel Llyfnant. *Quart. J. Geol. Soc. Lond.* **61**, 612.
—— (1910). The Tremadoc slates and associated rocks of south-east Caernarvonshire. *Ibid.* **66**, 142–88.
FEARNSIDES, W. G. and DAVIES, W. (1943). The Geology of Dendraeth. The country between Troeth Mawr and Troeth Bach, Merioneth. *Quart. J. Geol. Soc. Lond.* **99**, 249.
LAKE, P. and REYNOLDS, S. H. (1896). The Lingula-flags and igneous rocks of the neighbourhood of Dolgelly. *Quart. J. Geol. Soc. Lond.* **52**, 511–22.
MATLEY, C. A. and WILSON, T. S. (1946). The Harlech Dome. *Quart. J. Geol. Soc. Lond.* **102**, 14–40.
SALTER, J. W. and ETHERIDGE, R. (1880). *Mem. Geol. Surv. Gt. Brit.* p. 342. London.
WELLS, A. K. (1925). The geology of the Rhobell Fawr district (Merioneth). *Quart. J. Geol. Soc. Lond.* **81**, 465.

[30] THOMAS, H. H. (1914). Haverfordwest. *Mem. Geol. Surv. Gr. Brit.* p. 12. London.

[31] COBBOLD, E. S. and POCOCK, R. W. (1934). *Phil. Trans.* B, **223**, 317, 391.
STUBBLEFIELD, C. J. (1929). A new Upper Cambrian section in S. Shropshire. *Summ. Prog. Geol. Surv.* pp. 55–62. London.

[32] GROOM, T. T. (1902). The sequence of the Cambrian and associated Beds of the Malvern Hills. *Quart. J. Geol. Soc. Lond.* **58**, 98.
—— in MONCKTON, H. W. and HERRIES, R. S. (eds.) (1910). *Geology in the Field*, p. 703.

[33] ILLING, V. C. (1913). (1) Notes on certain Trilobites found in the Stockingford Shales. (2) Recent discoveries in the Stockingford Shales near Nuneaton. *Geol. Mag.* **50**, 452.
LAPWORTH, C. (1898). Geology of the Birmingham district. *Proc. Geol. Ass., Lond.*, **15**, 346.

[34] FEARNSIDES, W. G. (1905). On the geology of Arenig Fawr and Moel Llyfnant. *Quart. J. Geol. Soc. Lond.* **61**, 614.

FEARNSIDES, W. G. (1910). The Tremadoc slates and associated rocks of southeast Caernarvonshire. *Quart. J. Geol. Soc. Lond.* **66**, 155.

FEARNSIDES, W. G. and DAVIES, W. (1943) *Quart. J. Geol. Soc. Lond.* **99**, 249.

SALTER, J. W. and ETHERIDGE, R. (1880). *Mem. Geol. Surv. Gt. Brit.* p. 350. London.

WELLS, A. K. (1925). *Quart. J. Geol. Soc. Lond.* **81**, 465.

COX, A. H. and WELLS, A. K. (1920). The Lower Palaeozoic rocks of the Arthog-Dolgelly district (Merionethshire). *Quart. J. Geol. Soc. Lond.* **76**, 261.

—— (1927). *Proc. Geol. Ass., Lond.*, **38**, 275, 320.

[35] CALLAWAY, C. (1877). On a new area of Upper Cambrian rocks in south Shropshire, with a description of a new fauna. *Quart. J. Geol. Soc. Lond.* **33**, 652–72.

STUBBLEFIELD, C. J. (1929). *Summ. Prog. geol. Surv., Lond.*, p. 58.

STUBBLEFIELD, C. J. and BULMAN, O. M. B. (1927). The Shineton Shales of the Wrekin district. *Quart. J. Geol. Soc. Lond.* **83**, 96–146.

[36] WHITTARD, W. F. (1931). The Shelve country, Shropshire. *Proc. Geol. Ass., Lond.*, **42**, 324.

[37] GROOM, T. T. (1902). *Quart. J. Geol. Soc. Lond.* **58**, 105.

—— in MONCKTON, H. W. and HERRIES, R. S. (eds.) (1910). *Geology in the Field*, p. 703.

[38] COX, A. H. (1912). On an inlier of Longmyndian and Cambrian rocks at Pedwardine (Herefordshire). *Quart. J. Geol. Soc. Lond.* **68**, 369.

[39] SMITH, S. and STUBBLEFIELD, C. J. (1933). On the occurrence of Tremadoc shales in the Tortweth inlier (Gloucestershire). *Quart. J. Geol. Soc. Lond.* **89**, 357–78.

[40] LAPWORTH, C. (1898). *Proc. Geol. Ass., Lond.*, **15**, 347.

[41] NEAVERSON, E. (1955). *Stratigraphical Palaeontology*, p. 503.

[42] This and the following quotations are from:

WHITTARD, W. F. (1953). The enigma of the earliest fossils. *Proc. Bristol Nat. Soc.* **28**, 289–304.

[43] CUSHMAN, J. A. (1950). *Foraminifera.* Cambridge.

[44] LANG, W. D. (1923). Trends in British Carboniferous corals. *Proc. Geol. Ass., Lond.*, **34**, 120–36.

—— (1938). Some further considerations on trends in corals. *Ibid.* **49**, 148–59.

[45] SWINNERTON, H. H. (1923). *Outlines of Palaeontology*, 1st ed. p. 58. London.

[46] HAWKINS, H. L. (1943). Evolution and habit among the Echinoidea. *Quart. J. Geol. Soc. Lond.* **99**, lii–lxxv.

The same author has discussed possible evolutionary trends in the apical disc in:

—— (1912). On the evolution of the apical system in the Holectypoida. *Geol. Mag.* **49**, 8–16.

—— (1917). Morphological studies in the Echinoidea Holectypoida and their allies. *Ibid.* **54**, 249–56.

—— (1920). *Invertebrate Palaeontology*, p. 94. London.

Possible evolutionary trends in the ambulacrum are discussed in:
—— (1920). The ambulacrum of the Echinoidea Holectypoida. *Phil. Trans.* B, **209**, 377–480.

[47] Cooper, G. A. and Williams, A. (1952). Significance of the stratigraphic distribution of brachiopods. *J. Paleont.* **26**, 326–37.

[48] Zittel, K. A. (1900). *Textbook of Palaeontology*, vol. I, p. 533. London. The sentence was probably written by Hyatt who was largely responsible for the section on the Nautiloidea in this work.

[49] Spath, L. F. (1933). Evolution of the Cephalopoda. *Biol. Rev.* **8**, 418–62.

[50] Spath, L. F. (1923–43). Ammonoidea of the Gault. *Palaeontogr. Soc.* (*Monogr.*), 1–787.
—— (1938). *The Ammonites of the Liassic Family Liparoceratidae.* London (British Museum).

[51] Swinnerton, H. H. (1947). *Outlines of Palaeontology*, 3rd ed. p. 227.

[52] Elles, Gertrude L. (1922). The Graptolite faunas of the British Isles. *Proc. Geol. Ass., Lond.*, **33**, 168–200.

[53] This work is brought together in:
Moore, R. C. (ed.) (1955). *Treatise on Invertebrate Paleontology: Graptolithina.* New York.
Bulman, O. M. B. (1958). The sequence of Graptolite faunas. *Palaeontology*, **1**, 159–173.

[54] Consult, for example:
Swinton, W. E. (1948). *The Corridor of Life.* London.
Simpson, G. G. (1950). *The Meaning of Evolution*, chaps. 4–7. New Haven and Oxford.
Colbert, E. H. (1955). *Evolution of the Vertebrates.* New York and London.

[55] Le Gros Clark, Sir Wilfred (1956). Review of *Evolution of the Vertebrates*, by E. H. Colbert. *Sci. Progr. Twent. Cent.* **44**, 357.

[56] de Beer, Sir Gavin (1954). *Archaeopteryx* and Evolution. *Advance. Sci., Lond.*, N.S., **11**, 160–70. The quotation given is slightly abridged from the original presidential address (Zoology section).

[57] Edwards, W. N. (1935). *Guide to the Fossil Plants.* London (British Museum).

[58] Seward, A. C. (1931). *Plant Life through the Ages.* Cambridge.

[59] Scott, D. H. (1924). *Extinct Plants and Problems of Evolution.* London.

[60] Florin, R. (1951). Evolution in Cordaites and Conifers. *Acta Horti Berg.* **15**, 285–388.

[61] Carruthers, R. G. (1910). The evolution of *Zaphrentis delanouei*. *Quart. J. Geol. Soc. Lond.* **66**, 523–38. This paper contains the original work and, although itself excellently short and to the point, has been much summarized and analysed. Some additional diagnoses and names are given in:
Hudson, R. G. S. (1940). *Zaphrentis carruthersi* etc. *Proc. Yorks. Geol. Soc.* **24**, 290–311.
Hill, Dorothy (1940). Carboniferous rugose corals of Scotland. *Palaeontogr. Soc.* p. 144.
A useful diagram will be found in:
Swinnerton, H. H. (1947). *Outlines of Palaeontology*, 3rd ed. p. 53.

Notes and References: Chapter 2

An analysis of the facts from the point of view of principles of nomenclature will be found in:

SYLVESTER-BRADLEY, P. C. (1951). Subspecies in palaeontology. *Geol. Mag.* **88**, 88–102.

[62] HILL, DOROTHY (1940). *Palaeontogr. Soc.* p. 145.

[63] SWINNERTON, H. H. (1938). *Rep. Brit. Ass.* p. 62.

[64] VAUGHAN, A. (1915). Correlation of Dinantian and Avonian. *Quart. J. Geol. Soc. Lond.* **71**, 36.

[65] HILL, DOROTHY (1936). British Silurian rugose corals with acanthine septa. *Phil. Trans.* B, **226**, 189–217.

[66] HILL, DOROTHY and BUTLER, A. J. (1935). *Cymatelasma*, a new genus of Silurian rugose corals. *Geol. Mag.* **73**, 516–27.

[67] LANG, W. D. (1938). *Proc. Geol. Ass., Lond.*, **49**, 148–59.

[68] ROWE, A. W. (1899). Analysis of the genus *Micraster*. *Quart. J. Geol. Soc. Lond.* **55**, 494–547.

CHATWIN, C. P. (1924). Dartford. *Mem. Geol. Surv. Gt. Brit.* p. 18. London.

KERMACK, K. A. (1954). Biometrical study of *Micraster coranguinum* etc. *Phil. Trans.* B, **237**, 375–428.

[69] SWINNERTON, H. H. (1947). *Outlines of Palaeontology*, p. 61.

[70] JONES, O. T. (1928). *Plectambonites* and some allied genera. *Mem. Geol. Surv. Gt. Brit. Palaeontology*, **1**, 367.

[71] ALEXANDER, FRANCES E. S. (1948). *Anomia reticularis* Linnaeus. *Quart. J. Geol. Soc. Lond.* **104**, 207–20.

[72] PRENTICE, J. E. (1949). The *hemisphaericus*-like gigantellids of the southern Pennines. *Proc. Yorks. Geol. Soc.* **27**, 247–69.

[73] LANG, W. D. (1923). Evolution: a resultant. *Proc. Geol. Ass., Lond.*, **34**, 7–20.

ARKELL, W. J. (1933). Oysters of the Fuller's Earth. *Proc. Cotteswold Nat. Fld. Cl.* **25**, 21–68.

[74] TRUEMAN, A. E. (1922). The use of *Gryphaea* in the correlation of the Lower Lias. *Geol. Mag.* **59**, 256–68.

Further material and discussion is given in the following:

TRUEMAN, A. E. (1940). *Trans. Geol. Soc. Glasg.* **20**, 81.

MACLENNAN, R. M. and TRUEMAN, A. E. (1942). *Gryphaea incurva* from the Lower Lias of Loch Aline. *Proc. Roy. Soc. Edinb.* B, **61**, 211–32.

SWINNERTON, H. H. (1939). *Quart. J. Geol. Soc. Lond.* **95**, xxxiii–lxx.

—— (1940). *Ibid.* **96**, lxxvii–cxviii.

[75] GEORGE, T. N. (1951). *Evolution in Outline*, p. 74.

[76] WOODS, H. (1912). Evolution of *Inoceramus* in the Cretaceous period. *Quart. J. Geol. Soc. Lond.* **68**, 1–20.

[77] TRUEMAN, A. E. and WEIR, J. (1946). British Carboniferous non-marine Lamellibranchia. *Palaeontogr. Soc.* xxix.

[78] GEORGE, T. N. (1948). *Proc. Roy. Phil. Soc. Glasg.* **73**, 22–42.

[79] EAGAR, R. M. C. (1947). Non-marine lamellibranch succession etc. *Phil. Trans.* B, **233**, 48.

[80] Well shown in the figure in:

WRAY, D. A. (1947). *British Regional Geology. The Pennines*, p. 41. London.

[81] A[rkell], W. J. (1931). Review of *Statistisch-Biostratigraphische Untersuchungen an Mitteljurassischen Ammoniten etc.*, by Roland Brinkmann. *Geol. Mag.* **68**, 373–6.
[82] Swinnerton, H. H. (1947). *Outlines of Palaeontology*, pp. 245–7.
[83] Challinor, J. (1945). A graptolite lineage from North Cardiganshire. *Geol. Mag.* **82**, 97–106.
[84] Simpson, G. G. (1953). *Life of the Past*, p. 263.
[85] Holmes, A. (1947). The construction of a geological time-scale. *Trans. Geol. Soc. Glasg.* **21**, 145.

3

NATURAL SELECTION

[1] Lush, J. L. (1954). Rates of genetic changes in populations of farm animals. *Caryologia*, **6** (Suppl.), 589–99.
[2] Wright, S. (1949). In Jepsen, G. L. and Mayr, E. (eds.), *Genetics, Paleontology and Evolution*. Princeton. Pp. 364–89.
[3] Haldane, J. B. S. (1924). A mathematical theory of natural and artificial selection. *Trans. Camb. Phil. Soc.* **23**, 19–41.
[4] Weldon, W. F. R. (1901). A first study of natural selection in *Clausilia laminata* (Montagu). *Biometrika*, **1**, 109–24.
[5] Mørch, E. T. (1941). *Chondrodystrophic Dwarfs in Denmark*. Copenhagen.
[6] Gunther, M. and Penrose, L. S. (1935). The genetics of epiloia. *J. Genet.* **31**, 413–30.
[7] Haldane, J. B. S. (1946). The mutation rate of the gene for haemophilia, and its segregation ratios in males and females. *Ann. Eugen., Lond.*, **13**, 262–71.
[8] Allison, A. C. (1954). Protection afforded by sickle-cell trait against subtertian malarial infection. *Brit. Med. J.* **1**, 290–4.
[9] Dobzhansky, Th. (1951). *Genetics and the Origin of Species*. New York.
[10] Karn, M. N. and Penrose, L. S. (1951). Birth weight and gestation time in relation to maternal age, parity and infant survival. *Ann. Eugen., Lond.*, **16**, 147–64.
[11] Simpson, G. G. (1953). *The Major Features of Evolution*. New York.
[12] Waddington, C. H. (1953). Genetic assimilation of an acquired character. *Evolution*, **7**, 118–26.
[13] Robson, E. B. (1955). Birth weight in cousins. *Ann. Hum. Genet., Lond.*, **19**, 262–8.
[14] Rendel, J. M. (1943). Variations in the weights of hatched and unhatched ducks' eggs. *Biometrika*, **33**, 48–58.
[15] Falconer, D. S. (1954). Asymmetrical responses in selection experiments. In Buzzati-Traverso, A. A. (ed.), *Symposium on Genetics of Population Structure*. Publications Internat. Union Biol. Sciences, Series B, no. 15, pp. 16–41. Naples.

Notes and References: Chapter 3

[16] KETTLEWELL, H. B. D. (1956). Further selection experiments on industrial melanism in the *Lepidoptera*. *Heredity*, **10**, 287–301.

[17] ——(1958). A survey of the frequencies of *Biston bitularia* L. (Lep.) and its melanic forms in Great Britain. *Ibid.* **12**, 51–72.

[18] CAIN, A. J. and SHEPPARD, P. N. (1952). The effects of natural selection on body colour in the land snail *Cepaea nemoralis*. *Heredity*, **6**, 217–31.

[19] SHEPPARD, P. M. (1951). Fluctuations in the selective value of certain phenotypes in the polymorphic land snail *Cepaea nemoralis* (L.). *Heredity*, **5**, 125–34.

[20] RENDEL, J. M. (1945). Genetics and cytology of *Drosophila subobscura*. II. Normal and selective matings in *Drosophila subobscura*. *J. Genet.* **46**, 287–302.

[21] MAYNARD SMITH, J. (1956). Fertility, mating behaviour and sexual selection in *Drosophila subobscura*. *J. Genet.* **54**, 261–74.

[22] MÜNTZING, A. (1932). Cytogenetic investigations on synthetic *Galeopsis tetrahit*. *Hereditas*, **16**, 105–54.

[23] STEBBINS, G. L. (1950). *Variation and Evolution in Plants*. London.

[24] ZEUNER, F. E. (1945). *The Pleistocene Period. Its Climate, Chronology and Faunal Successions*. London.

[25] FISHER, R. A. (1930). *The Genetical Theory of Natural Selection*. Oxford.

[26] MAYR, E. (1942). *Systematics and the Origin of Species*. New York.

[27] HARLAND, S. C. (1936). The genetical conception of the species. *Biol. Rev.* **11**, 83–112.

[28] THORPE, W. H. (1938). Further experiments in olfactory conditioning in a parasitic insect. The nature of the conditioning process. *Proc. Roy. Soc.* B, **126**, 370–97.

[29] KOOPMAN, K. F. (1950). Natural selection for reproductive isolation between *Drosophila pseudoobscura* and *Drosphila persimilis*. *Evolution*, **4**, 135–48.

[30] TRYON, R. C. (1931). Studies in individual differences in maze ability. II. The determination of individual differences by age, weight, sex and pigmentation. *J. Comp. Psychol.* **12**, 1–22. III. The community of function between two maze abilities (95–116). IV. The constancy of individual differences: correlation between learning and relearning (303–46). V. Luminosity and visual acuity as systematic causes of individual differences and an hypothesis of maze ability (401–20).

[31] KRECHEVSKY, I. (1933). The hereditary nature of hypotheses. *J. Comp. Psychol.* **16**, 99–116.

[32] SPURWAY, H. (1949). Remarks on Vavilov's law of homologous variation. *Ric. sci.* **19** (Suppl.), 3–9.

[33] Many of the topics of this chapter are dealt with more fully in: MAYNARD SMITH, J. (1958). *The Theory of Evolution*. London.

4

DEVELOPMENTS IN THE STUDY OF ANIMAL COMMUNICATION

[1] FRISCH, K. VON (1950). *Bees, their Vision, Chemical Senses and Language.* New York.
—— (1954). *The Dancing Bees.* London.
[2] NEUHAUS, W. (1956). Die Unterscheidungsfähigkeit des Hundes für Duftgemische. *Z. vergl. Physiol.* 39, 25–48.
[3] SCHMID, B. (1935). Über die Ermittelung des menschlichen und tierischen Individualgeruches durch den Hund. *Z. vergl. Physiol.* 22, 524–38.
[4] GOZ, H. (1954). Über den Art- und Individualgeruch bei Fischen. *Z. vergl. Physiol.* 29, 1–45.
[5] NIXON, H. L. and RIBBANDS, C. R. (1952). Food transmission in the honeybee community. *Proc. Roy. Soc.* B, 140, 43–50.
[6] KALMUS, H. and RIBBANDS, C. R. (1952). The origin of the odours by which honey bees distinguish their companions. *Proc. Roy. Soc.* B, 140, 50–9.
[7] RENNER, M. (1955). Neue Untersuchungen über die physiologische Wirkung des Duftorganes der Honigbiene. *Naturwissenschaften*, 21, 589.
[8] FRISCH, K. VON (1941). Über einen Schreckstoff der Fischhaut und seine biologische Bedeutung. *Z. verg. Physiol.* 29, 46–145.
[9] SCHUTZ, F. (1956). Vergleichende Untersuchungen über die Schreckreaktion bei Fischen und deren Verbreitung. *Z. vergl. Physiol.* 38, 84–135.
[10] EIBL-EIBESFELDT, I. (1949). Über das Vorkommen von Schreckstoffen bei Erdkrötenquappen. *Experientia*, 5, 236.
[11] HEINTZ, E. (1954). Actions répulsives exercées sur divers animaux par des substances contenues dans le peau ou le corps d'animaux de même espèce. *C.R. Soc. Biol., Paris*, 148, 585, 717.
[12] FRISCH, K. VON (1923). Über die 'Sprache' der Bienen. *Zool. Jb.* (Abt. 3), 40, 1–186.
[13] HAAS, A. (1949). Arttypische Flugbahnen von Hummelmänchen. *Z. vergl. Physiol.* 31, 281–307.
[14] SHAFFER, B. M. (1957). Aspects of cellular aggregation in cellular slime moulds. I. Orientation and chemotaxis. *Amer. Nat.* 91, 19–35.
[15] TINBERGEN, N. and TER PELKWIJK, J. J. (1938). De kleine Watersalamander. *Lev. Nat.* 43, 232–7.
[16] SCHWINCK, I. (1954). Experimentelle Untersuchungen über Geruchssinn und Strömungswahrnehmung in der Orientierung bei Nachtschmetterlingen. *Z. vergl. Physiol.* 37, 19–56.
[17] KETTLEWELL, B. D. H. (1946). Female assembling scents. *Entomologist*, 79, 8–14.
[18] SCHAFFER, J. (1940). *Die Hautdrüsenorgane der Säugetiere.* Berlin and Vienna.

[19] CRANE, J. (1955). Imaginal behaviour of a Trinidad butterfly, *Heliconius erato hydrara* Hewston, with special reference to the social use of colour. *Zoologica*, **40**, 167–95.

[20] HALDANE, J. B. S. (1955). Animal communication and the origin of human language. *Sci. Progr. Twent. Cent.* **43**, 385–401.

[21] McDERMOTT, F. A. (1917). Observations on the light emission of American Lampyridae. *Canad. Ent.* **49**, 53–61.

[22] BUCK, J. B. (1937). Studies on the firefly. II. The signal system and colour vision in *Photinus pyralis*. *Physiol. Zoöl.* **10**, 412–19.

[23] LIESENFELD, F. J. (1956). Untersuchungen am Netz und über der Erschütterungssin von *Zygiella x-notata* (Cl.) (Araneidae). *Z. vergl. Physiol.* **38**, 563–92.

[24] BRISTOWE, W. S. (1941). *The Comity of Spiders*. II. London.

CRANE, J. (1949). Comparative biology of Salticid spiders at Rancho Grande. IV. An analysis of display. *Zoologica*, **34**, 159–214.

DREES, O. (1952). Untersuchungen über die angeborenen Verhaltenweisen bei Springspinnen (Salticidae). *Z. Tierpsychol.* **9**, 169–207.

[25] ILSE, D. (1941). The colour vision of insects. *Proc. Roy. Phil. Soc. Glasg.* **65**, 68–82.

[26] HELLWIG, H. and LUDWIG, W. (1951). Versuche zum Frage der Arterkennung bei Insekten. *Z. Tierpsychol.* **9**, 456–62.

[27] ALTEVOGT, R. (1957). Untersuchungen zur Biologie, Ökologie und Physiologie indischer Winkerkrabben. *Z. morph. Ökol. Tiere*, **46**, 1–110.

CRANE, J. (1943). Display, breeding and relationships of Fiddler Crabs (Brachyura, genus *Uca*) in the north eastern United States. *Zoologica*, **28**, 217–23.

[28] KITZLER, G. (1941). Die Paarungsbiologie einiger Eidechsen. *Z. Tierpsychol.* **4**, 353–402.

[29] TINBERGEN, N. (1953). *Social Behaviour in Animals*. London.

[30] NOBLE, G. K. and CURTIS, B. (1939). The social behaviour of the jewel fish, *Hemichromis bimaculatus* Gill. *Bull. Amer. Mus. Nat. Hist.* **76**, 1–46.

[31] BAERENDS, G. P. and BAERENDS-VAN ROON, J. M. (1950). An introduction to the study of the ethology of Cichlid fishes. *Behaviour* (Suppl.), **1**.

[32] CINAT TOMSON, G. H. (1923). Die geschlechtliche Zuchtwahl beim Wellensittich (*Melopsittacus undulatus* Shaw). *Biol. Zbl.* **46**, 543–52.

[33] MARLER, P. (1955). Studies of fighting in chaffinches. (2). The effect on dominance relations of disguising females as males. *Brit. J. Anim. Behav.* **3**, 137–46.

[34] MORRIS, D. (1957). Typical intensity and its relation to the problem of ritualisation. *Behaviour*, **11**, 1–12.

[35] MOYNIHAN, M. (1955). Some aspects of reproductive behaviour in the Black-headed Gull (*Larus r. ridibundus*) and related species. *Behaviour* (Suppl.), **4**.

[36] LORENZ, K. (1941). Vergleichende Bewegungsstudien an Anatinen. *J. Ornithol.* **89**, 194–294.

WEIDMANN, U. (1956). Verhaltsstudien an der Stockente (*Anas platyrhynchos* L.). I. Das Aktionssystem. *Z. Tierpsychol.* **13**, 208–71.

[37] GUHL, A. M. and ORTMAN, M. M. (1953). Visual patterns in the recognition of individuals among chickens. *Condor*, **55**, 287–98.

NOBLE, G. K. and CURTIS, B. (1939). *Bull. Amer. Mus. Nat. Hist.* **76**, 1–46.

[38] NICE, M. M. (1943). Studies in the life history of the Song Sparrow. II. Behaviour of the Song Sparrow and other Passerines. *Trans. Linn. Soc. N.Y.* **4**, 1–328.

[39] SCHLOETH, R. (1956). Zur Psychologie der Begegnung zwischen Tiere. *Behaviour*, **10**, 1–79.

[40] EIBL-EIBESFELDT, I. (1953). Zur Ethologie des Hamsters (*Cricetus cricetus* L.). *Z. Tierpsychol.* **10**, 204–54.

[41] SCHLOETH, R. (1956). *Behaviour*, **10**, 1–79.

[42] EIBL-EIBESFELDT, I. (1956). Einige Bemerkungen über den Ursprung von Ausdrucksbewegungen bei Säugetieren. *Z. Säugetierkunde*, **21**, 29–43.

[43] LEYHAUSEN, P. (1956). Das Verhalten der Katzen (Felidae). *Handbuch Zool.* Bd. VIII, 10 (21), pp. 1–34. Berlin.

[44] SCHENKEL, R. (1947). Ausdruckstudien an Wölfen. *Behaviour*, **1**, 81–129.

[45] KOHTS, N. (1935). Infant ape and human child (instincts, emotions, plays and habits) (Russian with English summary). *Sci. Mem. Mus. Darwin., Moscow*, **3**, 1–596.

[46] FROIS-WITTMANN, J. (1930). The judgement of facial expression. *J. Exp. Psychol.* **13**, 113–51.

[47] ROTH, L. M. (1948). A study of mosquito behaviour. An experimental study of the sexual behaviour of *Aedes aegypti* (L). *Am. Midl. Nat.* **40**, 265–352.

[48] WOODS, E. F. (1956). Queen piping. *Bee World*, **37**, 185–95, 216–19.

[49] FABER, A. (1953). *Laut- und Gebärdensprache bei Insekten.* Stuttgart.

JACOBS, W. (1953). Verhaltensbiologische Studien an Feldheuschrecken. *Tierpsychol.* (Suppl.), 1.

PIERCE, G. (1949). *The Songs of Insects.* Cambridge, Mass.

[50] PRINGLE, J. W. S. (1954). A physiological analysis of cicada song. *J. Exp. Biol.* **31**, 525–60.

[51] FISH, M. P. (1956). Animal sounds in the sea. *Sci. Amer.* **194**, 93–102.

[52] BARBER, S. B. and MOWRAY, W. H. (1956). Mechanism of sound production in the sculpin. *Science*, **124**, 219–20.

[53] FISH, M. P., KELSEY, A. S. and MOWRAY, W. H. (1952). Studies on the production of underwater sound by North Atlantic coastal fishes. *J. Mar. Res.* **11**, 180–93.

[54] PUMPHREY, R. J. (1940). Hearing in insects. *Biol. Rev.* **15**, 107–32.

[55] MARLER, P. (1955). Characteristics of some animal calls. *Nature, Lond.*, **176**, 6–8.

[56] HOWARD, H. E. (1920). *Territory in Bird Life.* London.

[57] STEVENS, S. S. and DAVIS, H. (1938). *Hearing.* New York.

[58] SCHWARTZKOPFF, J. (1955). On the hearing of birds. *Auk*, **72**, 340–47.

[59] Marler, P. (1956). Behaviour of the chaffinch, *Fringilla coelebs. Behaviour* (Suppl.), **5**.

[60] Busnel, M. C. (1953). Contribution à l'étude des émissions acoustiques des Orthoptères. *Ann. l'I.N.R.A.* **3**, 333–421.

[61] Pumphrey, R. J. (1940). *Biol. Rev.* **15**, 107–32.

[62] Haskell, P. T. (1957). Sound in the insect world. *New Biol.* no. 23, pp. 29–47.

[63] Huber, F. (1956). Heuschrecken- und Grillenlaute und ihre Bedeutung. *Naturwissenschaften*, **14**, 317–21.

[64] Pringle, J. W. S. (1954). *J. Exp. Biol.* **31**, 525–60.

[65] Huber, F. (1956). *Naturwissenschaften*, **14**, 317–21.

[66] White, G. (1789). *The Natural History and Antiquities of Selborne.* London.

[67] Hartshorne, C. (1956). The monotony threshold in bird song. *Auk*, **73**, 176–92.

Saunders, A. A. (1951). *A Guide to Bird Songs.* New York.

[68] Nice, M. M. (1943). *Trans. Linn. Soc. N.Y.* **4**, 1–328.

[69] Borror, D. J. (1956). Variation in Carolina wren songs. *Auk*, **73**, 211–29.

[70] Lack, D. (1954). *The Natural Regulation of Animal Numbers.* Oxford.

[71] Baerends, G. P. (1950). Specialisations in organs and movements with a releasing function. *Symp. Soc. Exp. Biol.* **4**, 337–60.

[72] Sauer, F. (1955). Über Variationen der Artgesänge bei Grasmücken. *J. Ornithol.* **96**, 129–46.

[73] Messmer, E. (1956). Die Entwicklung der Lautäusserungen und einiger Verhaltensweisen der Amsel (*Turdus merula merula* L.) unter natürlichen Bedingungen und nach Einzelaufsucht in schalldichten Räumen. *Z. Tierpsychol.* **13**, 341–441.

[74] Poulsen, H. (1951). Inheritance and learning in the song of the chaffinch (*Fringilla coelebs* L.). *Behaviour*, **3**, 216–28.

[75] Thorpe, W. H. (1956). *Learning and Instinct in Animals.* London.

[76] Scott, W. E. D. (1901). Data on song in birds: observations on the song of Baltimore orioles in captivity. *Science*, **14**, 522–6.

[77] Lanyon, W. E. (1957). The comparative biology of the Meadowlarks (*Sturnella*) in Wisconsin. *Publ. Nuttall Ornithol. Cl.* **1**, 1–67.

[78] Poulsen, H. (1951). *Behaviour*, **3**, 216–28.

—— (1954). On the song of the Linnet (*Carduelis cannabina* (L.)). *Dansk. orn. Foren. Tidsskr.* **48**, 32–7.

[79] Nice, M. M. (1943). *Trans. Linn. Soc. N.Y.* **4**, 1–328.

[80] Thorpe, W. H. (1954). The process of song learning in the chaffinch, as studied by means of the sound spectrograph. *Nature, Lond.*, **173**, 465.

[81] Marler, P. (1956). *Behaviour* (Suppl.), **5**.

[82] Messmer, E. (1956). *Z. Tierpsychol.* **13**, 341–441.

[83] Stadler, H. (1930). Vogeldialekt. *Alauda*, **2** (Suppl.), 1–66.

[84] Dilger, W. C. (1956). Hostile behaviour and reproductive isolating mechanisms in the avian genera *Catharus* and *Hyocichla*. *Auk*, **73**, 313–53.

[85] Stein, R. C. (1956). A comparative study of advertising song in the *Hyocichla* thrushes. *Auk*, **73**, 508–12.

Notes and References: Chapter 4

[86] MARLER, P. (1952). Variation in the song of the chaffinch, *Fringilla coelebs*. *Ibis*, **94**, 458–72.

[87] BENSON, C. W. (1948). Geographical voice variation in African birds. *Ibis*, 90, 48–71.

[88] LACK, D. and SOUTHERN, H. N. (1949). Birds on Tenerife. *Ibis*, **91**, 607–26.

[89] MARLER, P. and BOATMAN, D. J. (1951). Observations on the birds of Pico, Azores. *Ibis*, **93**, 90–9.

[90] POULSEN, H. (1951). *Behaviour*, **3**, 216–28.

THORPE, W. H. (1954). The process of song learning in the chaffinch, as studied by means of the sound spectrograph. *Nature, Lond.*, **173**, 465.

[91] MARLER, P. and BOATMAN, D. J. (1951). *Ibis*. **93**, 90–9.

[92] LYNES, H. (1914). Remarks on the geographical distribution of the chiffchaff and willow warbler. *Ibis*, **2**, 304–14.

[93] PROMPTOV, A. N. and LUKINA, E. V. (1945). Conditioned reflectory differentiation in Passeres and its biological value. *C.R.* (*Dokl.*) *Acad. Sci. U.R.S.S.* **46**, 382–4.

[94] MARLER, P. (1957). Specific distinctiveness in the communication signals of birds. *Behaviour*, **11**, 13–39.

[95] MAYR, E., ANDREW, R. J. and HINDE, R. A. (1956). Die systematische Stellung der Gattung *Fringilla*. *J. Ornithol.* **97**, 258–73.

[96] MARLER, P. (1956). The voice of the chaffinch and its function as a language. *Ibis*, **98**, 231–61.

[97] CULLEN, E. (1957). Adaptations in the Kittiwake to cliff nesting. *Ibis*, **99**, 275–302.

[98] GIBB, J. (1954). Feeding ecology of tits, with notes on treecreeper and goldcrest. *Ibis*, **96**, 513–43.

[99] LORENZ, K. (1935). Der Kumpan in der Umwelt des Vogels. *J. Ornithol.* **79**, 67–127.

—— (1950). The comparative method in studying innate behaviour patterns. *Symp. Soc. Exp. Biol.* **4**, 221–68.

[100] TINBERGEN, N. (1952). Derived activities, their causation, biological significance, origin and emancipation during evolution. *Quart. Rev. Biol.* **27**, 1–32.

[101] CRAIG, W. (1921–22). A note on Darwin's work on the expression of the emotions in man and the animals. *J. Abnorm.* (*Soc.*) *Psychol.* **16**, 356–66.

[102] KRAMER, S. (1957). Personal communication on insect courtship.

[103] HALDANE, J. B. S. (1954). La signalisation animale. *Ann. Biol.* **30**, 89–98.

—— (1956). Les aspects physico-chimiques des instincts. In *L'instinct dans le comportement des animaux et l'homme*. Paris.

[104] EIBL-EIBESFELDT, I. (1956). *Z. Säugetierk.* **21**, 29–43.

[105] SCHAFFER, J. (1940). *Die Hautdrüsenorgane der Säugetiere.*

[106] DAANJE, A. (1950). On the locomotory movements in birds and the intention movements derived from them. *Behaviour*, **3**, 48–98.

EIBL-EIBESFELDT, I. (1956). *Z. Säugetierk.* **21**, 29–43.

MORRIS, D. (1956). The feather postures of birds and the problem of the origin of social signals. *Behaviour*, **9**, 75–113.

TINBERGEN, N. (1952). *Quart. Rev. Biol.* **27**, 1–32.
[107] YOUNG, P. T. (1943). *Emotion in Man and Animal*. New York.
[108] MORRIS, D. (1956). *Behaviour*, **9**, 75–113.
[109] HINGSTON, R. W. G. (1933). *Animal Colour and Adornment*. London.
MORRIS, D. (1956). *Behaviour*, **9**, 75–113.
[110] HUXLEY, J. S. (1923). Courtship activities in the red-throated diver (*Colymbus stellatus* Pontopp.); together with a discussion of the evolution of courtship in birds. *J. Linn. Soc.* (*Zool.*), **35**, 253–92.
[111] LACK, D. (1941). Some aspects of instinctive behaviour and display in birds. *Ibis*, **5**, 407–41.
[112] EIBL-EIBESFELDT, I. (1956). *Z. Säugetierk.* **21**, 29–43.
[113] HUXLEY, J. S. (1923). *J. Linn. Soc.* (*Zool.*), **35**, 253–92.
[114] TINBERGEN, N. (1940). Die Übersprungbewegungen. *Z. Tierpsychol.* **4**, 1–40.
[115] LORENZ, K. (1941). *J. Ornithol.* **89**, 194–294.
[116] TINBERGEN, N. (1940). *Z. Tierpsychol.* **4**, 1–40.
—— (1952). *Quart. Rev. Biol.* **27**, 1–32.
[117] TINBERGEN, N. and MOYNIHAN, M. (1952). Head flagging in the black-headed gull. *Brit. Birds*, **45**, 19–22.
[118] MOYNIHAN, M. (1955*b*). Remarks on the original sources of displays. *Auk*, **72**, 240–6.
[119] LORENZ, K. (1952). *King Solomon's Ring*. London.
[120] HINDE, R. A. (1955–56). A comparative study of the courtship of certain finches (Fringillidae). *Ibis*, **97**, 706–45; *ibid.* **98**, 1–23.
[121] MORRIS, D. (1956). *Behaviour*, **9**, 75–113.
[122] HINDE, R. A. (1955–56). *Ibis*, **97**, 706–45; *ibid.* **98**, 1–23.
[123] ANDREW, R. J. (1956). Some remarks on behaviour in conflict situations with special reference to *Emberiza* spp. *Brit. J. Anim. Behav.* **4**, 41–4.
[124] FRISCH, K. VON (1954). *The Dancing Bees*. London.
[125] BOUTAN, L. (1913). Le Pseudo-langage. Observations effectuées sur un anthropoide: le Gibbon (*Hylobates leucogenys*-Ogilby). *Act. Soc. Linn. Bordeaux*, **47**, 5–81.
[126] CARPENTER, C. R. (1934). A field study of the behaviour and social relations of howler monkeys. *Comp. Psychol. Monogr.* **10**, 1–168.
[127] KING, J. A. (1955). Social behaviour, social organisation and population dynamics in a black-tailed prariedog town in the Black Hills of South Dakota. *Contr. Lab. Vert. Zool.* **67**. Ann Arbor.
[128] SAUER, F. (1954). Die Entwicklung der Lautäusserungen vom Ei ab Schalldicht gehaltener Dorngrasmücken (*Sylvia c. communis* Latham) im Vergleich mit später isolierten und mit wildlebenden Artgenossen. *Z. Tierpsychol.* **11**, 10–93.
[129] NICE, M. M. (1943). *Trans. Linn Soc. N.Y.* **4**, 1–328.
[130] ARMSTRONG, E. A. (1955). *The Wren*. London.
[131] MARLER, P. (1956). *Ibis*, **98**, 231–61.
[132] BARRINGTON, D. (1773). Experiments and observations on the singing of birds. *Phil. Trans.* **63**, 249–91.
[133] YERKES, R. M. and YERKES, A. W. (1929). *The Great Apes*. Yale.
[134] HALDANE, J. B. S. (1952). *Rationalist Ann.* pp. 37–45.

[135] PUMPHREY, R. J. (1951). The Origin of Language. An inaugural lecture. University of Liverpool.

[136] FRISCH, K. VON (1950). *Bees*. New York.
—— (1954). *The Dancing Bees*. London.

[137] KOEHLER, O. (1956). Thinking without words. *Proc. 14th Int. Congr. Zool.* p. 75. (Copenhagen, 1953.)

5

CROSS- AND SELF-FERTILIZATION IN PLANTS

[1] SPRENGEL, C. K. (1793). *Das entdeckte Geheimniss der Natur im Bau und in der Befruchtung der Blumen*. Berlin.

[2] DARWIN, C. (1868). *The Variation of Animals and Plants under Domestication*. London.

[3] WEISMANN, A. (1883). *Über die Vererbung*. Jena. (Translation by A. E. Shipley (1889). Oxford.)

[4] —— (1885). *Die Kontinuität des Keimplasmas als Grundlage einer Theorie der Vererbung*. Jena. (Translation by S. Schönland (1889). Oxford.)

[5] HERTWIG, O. (1876). *Beiträge zur Kenntniss der Bildung, Befruchtung und Theilung des thierischen Eies*. Leipzig.

[6] BENEDEN, E. VAN (1883). Recherches sur la maturation de l'œuf, la fécondation et la division cellulaire. *Arch. Biol.* **4**.

[7] STRASBURGER, E. (1884). *Neue Untersuchungen über den Befruchtungsvorgang bei den Phanerogamen als Grundlage für eine Theorie der Zeugung*. Jena.

[8] WALDEYER, W. (1888). Über Karyokinese und ihre Beziehung zu den Befruchtungsorganen. *Arch. mikr. Anat.* **32**, 1–122.

[9] HERTWIG, O. (1885). *Das Problem der Befruchtung und der Isotropie des Eies*. Jena.

[10] BOVERI, T. (1889). Ein geschlechtlich erzeugter Organismus ohne mütterlich Eigenschaften. *S.B. Ges. Morph. Physiol. Münch.* **5**. (Translation by T. H. Morgan (1893). *Amer. Nat.* **27**, 222–32.)

[11] WEISMANN, A. (1890). Bemerkungen zu einigen Tages-Problemen. *Biol. Zbl.* **10**, 1–12, 33–44. (Translation by A. E. Shipley in Poulton, E. B. and Shipley, A. E. (eds.), *Essays upon Heredity and Kindred Biological Problems by A. Weismann*, vol. II, pp. 71–97. Oxford, 1892.)

[12] NÄGELI, K. (1884). *Mechanisch-physiologische Theorie der Abstammungslehre*. München und Leipzig.

[13] MENDEL, G. (1866). Versuche über Pflanzenhybriden. *Verh. naturf. Ver. Brünn*, **4**, 3–44. (Reprinted in *Flora, Jena*, **89**, 364–403 (1901). Translation by the Royal Horticultural Society in Bateson, 1909 (Ref. 16).)

[14] MENDEL, G. (1869). Über einige aus künstlicher Befruchtung gewonnene *Hieracium*-Bastarde. *Verh. naturf. Ver. Brünn*, **8**, 26–31. (Translation by the Royal Horticultural Society in Bateson, 1909 (Ref. 16).)

[15] DARWIN, C. (1859). ***On the Origin of Species by Means of Natural Selection, or the Preservation of Favoured Races in the Struggle for Life.*** London.

[16] BATESON, W. (1909). ***Mendel's Principles of Heredity.*** Cambridge.

[17] VRIES, H. DE (1889). ***Intracellulare Pangenesis.*** Jena. (Translation by C. S. Gager (1910). Chicago.)

[18] BATESON, W. (1894). ***Materials for the Study of Variation treated with Especial Regard to Discontinuity in the Origin of Species.*** London.

[19] VRIES, H. DE (1900). Das Spaltungsgesetz der Bastarde. *Ber. dtsch. bot. Ges.* **18**, 83–90.

[20] CORRENS, C. (1900). G. Mendel's Regel über das Verhalten der Nachkommenschaft der Rassenbastarde. *Ber. dtsch. bot. Ges.* **18**, 158–68.

[21] TSCHERMAK, E. VON (1900). Über künstliche Kreuzung bei *Pisum sativum. Ber. dtsch. bot. Ges.* **18**, 232–9.

[22] JOHANNSEN, W. (1909). ***Elemente der exakten Erblichkeitslehre mit Grundzügen der biologischen Variationsstatistik.*** Jena.

[23] MORGAN, T. H., STURTEVANT, A. H., MULLER, H. J. and BRIDGES, C. B. (1915). ***The Mechanism of Mendelian Heredity.*** New York.

[24] FISHER, R. A. (1930). ***The Genetical Theory of Natural Selection.*** Oxford.

[25] HARDY, G. H. (1908). Mendelian proportions in a mixed population. *Science*, **28**, 49–50.

[26] WEINBERG, W. (1908). Über den Nachweis der Vererbung beim Menschen. *Jh. Ver. vaterl. Naturk. Württemb.* **64**, 368–82.

[27] VRIES, H. DE (1901–3). ***Die Mutationstheorie.*** Leipzig. (Translation by J. B. Farmer and A. D. Darbishire (1910–11). London.)

[28] MULLER, H. J. (1927). Artificial transmutation of the gene. *Science*, **66**, 84–7.

[29] AUERBACH, C. and ROBSON, J. M. (1947). The production of mutations by chemical substances. *Proc. Roy. Soc. Edinb.* B, **62**, 271–83.

[30] LOCK, R. H. (1906). ***Recent Progress in the Study of Variation, Heredity and Evolution.*** London.

See, for example:

[31] KÖLREUTER, J. G. (1766). ***Dritte Fortsetzung der vorläufigen Nachricht von einigen das Geschlecht der Pflanzen betreffenden Versuchen und Beobachtungen.*** Leipzig.

[32] EAST, E. M. (1909). The distinction between development and heredity in inbreeding. *Amer. Nat.* **43**, 173–81.

[33] SHULL, G. H. (1911). The genotypes of maize. *Amer. Nat.* **45**, 234–52.

[34] —— (1914). Duplicate genes for capsule form in *Bursa bursa-pastoris. Z. indukt. Abstamm.- u. VererbLehre*, **12**, 97–149.

[35] EAST, E. M. and JONES, D. F. (1919). ***Inbreeding and Outbreeding: their Genetic and Sociological Significance.*** Philadelphia.

[36] BATESON, W., SAUNDERS, E. R. and PUNNETT, R. C. (1906). Experimental studies in the physiology of heredity. *Rep. Evol. Comm. Roy. Soc.* **3**, 8–11, 31–7.

[37] BEADLE, G. W. and COONRADT, V. L. (1944). Heterocaryosis in *Neurospora crassa. Genetics*, **29**, 291–308.

[38] ZALOKAR, M. (1948). The *p*-aminobenzoic acid requirement of the 'sulfonamide-requiring' mutant strain of *Neurospora. Proc. Nat. Acad. Sci., Wash.*, **34**, 32–6.

[39] Emerson, S. (1948). A physiological basis for some suppressor mutations and possibly for one gene heterosis. *Proc. Nat. Acad. Sci., Wash.*, **34**, 72–4.

[40] —— in J. W. Gowan (ed.) (1952). *Heterosis.* Iowa.

[41] Mather, K. (1955). The genetical basis of heterosis. *Proc. Roy. Soc.* B, **144**, 143–50.

[42] Jinks, J. L. (1955). A survey of the genetical basis of heterosis in a variety of diallel crosses. *Heredity*, **9**, 223–38.

[43] Lerner, I. M. (1954). *Genetic Homeostasis.* Edinburgh.

[44] Haldane, J. B. S. (1955). On the biochemistry of heterosis, and the stabilization of polymorphism. *Proc. Roy. Soc.* B, **144**, 217–20.

[45] Lewis, D. (1955). Gene interaction, environment and hybrid vigour. *Proc. Roy. Soc.* B, **144**, 178–85.

[46] Jinks, J. L. and Mather, K. (1955). Stability in development of heterozygotes and homozygotes. *Proc. Roy. Soc.* B, **143**, 561–78.

[47] Müller, F. (1868). Notizen über die Geschlechtsverhältnisse brasilianischer Pflanzen. *Bot. Ztg.* **26**, 113–16.

[48] Correns, C. (1913). Selbsterilität und Individualstoffe. *Biol. Zbl.* **33**, 389–423.

[49] East, E. M. and Mangelsdorf, A. J. (1925). A new interpretation of the hereditary behaviour of self-sterile plants. *Proc. Nat. Acad. Sci., Wash.*, **11**, 166–83.

[50] Filzer, P. (1926). Die Selbststerilität von *Veronica syriaca. Z. indukt. Abstamm.- u. VererbLehre*, **41**, 137–97.

[51] Lewis, D. (1954). Comparative incompatibility in Angiosperms and Fungi. *Advanc. Genet.* **6**, 235–85.

[52] Lehmann, E. (1944). *Veronica filiformis* Sm., eine selbsterile Pflanze. *Jb. wiss. Bot.* **91**, 395–403.

[53] Gerstel, D. U. (1950). Self-incompatibility studies in Guayule. II. Inheritance. *Genetics*, **35**, 482–506.

[54] Hughes, M. B. and Babcock, E. B. (1950). Self-incompatibility in *Crepis foetida* L. subsp. *rhoeadifolia. Genetics*, **35**, 570–88.

[55] Crowe, L. K. (1954). Incompatibility in *Cosmos bipinnatus. Heredity*, **8**, 1–11.

[56] Bateman, A. J. (1954). Self-incompatibility systems in Angiosperms. II. *Iberis amara. Heredity*, **8**, 305–32.

[57] —— (1955). Self-incompatibility systems in Angiosperms. III. Cruciferae. *Ibid.* **9**, 53–68.

[58] Darwin, C. (1876). *The Effects of Cross- and Self-Fertilization in the Vegetable Kingdom.* London.

[59] Bateman, A. J. (1956). Cryptic self-incompatibility in the Wallflower: *Cheiranthus cheiri* L. *Heredity*, **10**, 257–61.

[60] East, E. M. (1940). The distribution of self-sterility in the flowering plants. *Proc. Amer. Phil. Soc.* **82**, 449–518.

[61] Fryxell, P. A. (1957). Mode of reproduction of higher plants. *Bot. Rev.* **23**, 135–233.

[62] East, E. M. (1929). Self-sterility. *Bibliogr. genet.* **5**, 331–68.

[63] Lewis, D. (1952). Serological reactions of pollen incompatibility substances. *Proc. Roy. Soc.* B, **140**, 127–35.

[64] Lewis, D. (1943). The physiology of incompatibility in plants. I. The effect of temperature. *Proc. Roy. Soc.* B, **131**, 13–26.

[65] Linskens, H. F. (1955). Physiologische Untersuchungen der Pollenschlauch-Hemmung selbststeriler Petunien. *Z. Bot.* **43**, 1–44.

[66] Darwin, C. (1877). *The Different Forms of Flowers on Plants of the Same Species*. London.

[67] See, for example:
Lewis, D. (1949). Incompatibility in flowering plants. *Biol. Rev.* **24**, 472–96.

[68] Crosby, J. L. (1949). Selection of an unfavourable gene-complex. *Evolution*, **3**, 212–30.

[69] Barlow, N. Unpublished. See [72].

[70] Ubisch, G. von (1921). Zur Genetik der trimorphen Heterostylie sowie einige Bemerkungen zur dimorphen Heterostylie. *Biol. Zbl.* **41**, 88–96.

[71] Barlow, N. (1923). Inheritance of the three forms in trimorphic species. *J. Genet.* **13**, 133–46.

[72] East, E. M. (1927). The inheritance of heterostyly in *Lythrum salicaria*. *Genetics*, **12**, 393–414.

[73] Fisher, R. A. and Mather, K. (1943). The inheritance of style length in *Lythrum salicaria*. *Ann. Eugen., Lond.*, **12**, 1–23.

[74] Fyfe, V. C. (1950). The genetics of tristyly in *Oxalis valdiviensis*. *Heredity*, **4**, 365–71.

[75] Ubisch, G. von (1926). Koppelung von Farbe und Heterostylie bei *Oxalis rosea*. *Biol. Zbl.* **46**, 633–45.

[76] Fyfe, V. C. (1956). Two modes of inheritance of the short-styled form in the 'genus' *Oxalis*. *Nature, Lond.*, **177**, 942–3.

[77] Lewis, D. (1943). The physiology of incompatibility in plants. II. *Linum grandiflorum*. *Ann. Bot., Lond.*, N.S., **7**, 115–22.

[78] —— (1956). Incompatibility and plant breeding. *Brookhaven Symp. Biol.* **9**, 89–100.

[79] Valentine, D. H. (1954). Seed-incompatibility. *VIIIe Congr. int. Bot., Paris*, **10**, 170–1.

[80] Baur, E. (1912). Ein Fall von geschlechtsbegrenzter Vererbung bei *Melandrium album*. *Z. indukt. Abstamm.- u. VererbLehre*, **8**, 335–6.

[81] Shull, G. H. (1914). Sex-limited inheritance in *Lychnis dioica* L. *Z. indukt. Abstamm.- u. VererbLehre*, **12**, 265–302.

[82] Warmke, H. E. (1946). Sex determination and sex balance in *Melandrium*. *Amer. J. Bot.* **33**, 648–60.

[83] Lewis, D. (1942). The evolution of sex in flowering plants. *Biol. Rev.* **17**, 46–67.

[84] Janick, J. and Stephenson, E. C. (1955). Genetics of the monoecious character in spinach. *Genetics*, **40**, 429–37.

[85] Correns, C. (1928). Bestimmung, Vererbung und Verteilung des Geschlechtes bei den höheren Pflanzen. *Handb. Vererbungsw.* **2**.

[86] East, E. M. (1934). The nucleus-plasma problem. *Amer. Nat.* **68**, 289–303, 402–39.

[87] Lewis, D. and Crowe, L. K. (1956). The genetics and evolution of gynodioecy. *Evolution*, **10**, 115–25.

[88] Wilkie, D. (1956). Incompatibility in bracken. *Heredity*, **10**, 247–56.

Notes and References: Chapter 5

[89] ALLEN, C. E. (1917). A chromosome difference correlated with sex differences in *Sphaerocarpos*. *Science*, **46**, 466–7.

[90] —— (1919). The basis of sex inheritance in *Sphaerocarpos*. *Proc. Amer. Phil. Soc.* **58**, 289–316.

[91] WHITEHOUSE, H. L. K. (1954). Incompatibility in Fungi. *VIIIe Congr. int. Bot., Paris*, **10**, 152–60.

[92] —— (1949). Multiple-allelomorph heterothallism in the Fungi. *New Phytol.* **48**, 212–44.

[93] MATHER, K. (1942). Heterothally as an outbreeding mechanism in Fungi. *Nature, Lond.*, **149**, 54–6.

[94] —— (1944). Genetical control of incompatibility in Angiosperms and Fungi. *Ibid.* **153**, 392–4.

[95] CROWE, L. K. (1955). The evolution of incompatibility in species of *Oenothera*. *Heredity*, **9**, 293–322.

[96] DARLINGTON, C. D. (1939). *The Evolution of Genetic Systems*. Cambridge.

[97] WILLIAMS, W. (1951). Genetics of incompatibility in Alsike Clover, *Trifolium hybridum*. *Heredity*, **5**, 51–73.

[98] KIRCHNER, O. (1905). Über die Wirkung der Selbstbestäubung bei den Papilionaceen. *Naturw. Z. Land- u. Forstw.* **3**, 97.

[99] MUNTZING, A. (1936). The evolutionary significance of autopolyploidy. *Hereditas*, **21**, 263–378.

[100] STEBBINS, G. L. (1950). *Variation and Evolution in Plants*. London.

[101] BAKER, H. G. (1953). Race formation and reproductive method in flowering plants. *Symp. Soc. Exp. Biol.* **7**, 114–45.

[102] —— (1953). Dimorphism and monomorphism in the Plumbaginaceae. II. Pollen and stigmata in the genus *Limonium*. *Ann. Bot., Lond.*, N.S., **17**, 433–45.

[103] —— (1954). Dimorphism and incompatibility in the Plumbaginaceae. *VIIIe Congr. int. Bot., Paris*, **10**, 133–4.

[104] CROSBY, J. L. (1954). Populations and evolution: the significance of incompatibility. *VIIIe Congr. int. Bot., Paris*, **10**, 163–9.

[105] MATHER, K. and DE WINTON, D. (1941). Adaptation and counter-adaptation of the breeding system in *Primula*. *Ann. Bot., Lond.*, N.S., **5**, 297–311.

[106] MATHER, K. (1943). Specific differences in *Petunia*. I. Incompatibility. *J. Genet.* **45**, 215–35.

[107] —— (1943). Polygenic inheritance and natural selection. *Biol. Rev.* **18**, 32–64.

[108] —— (1940). Outbreeding and separation of the sexes. *Nature, Lond.*, **145**, 484–6.

[109] WHITEHOUSE, H. L. K. (1949). Heterothallism and sex in the Fungi. *Biol. Rev.* **24**, 411–47.

[110] BURNETT, J. H. (1956). The mating systems of Fungi. I. *New Phytol.* **55**, 50–90.

[111] DODGE, B. O. (1927). Nuclear phenomena associated with heterothallism and homothallism in the ascomycete *Neurospora*. *J. Agric. Res.* **35**, 289–305.

[112] FISCHER, G. W. (1953). Evidence of complex physiological heterothallism in *Ustilago striiformis hordei*. *Res. Stud. St. Coll. Wash.* **21**, 57–72.

[113] Bauch, R. (1932). Die Sexualität von *Ustilago scorzonerae* und *U. zeae. Phytopath. Z.* **5**, 315–21.

[114] Rowell, J. B. (1955). Functional role of compatibility factors and an *in vitro* test for sexual compatibility with haploid lines of *Ustilago zeae. Phytopathology*, **45**, 370–4.

[115] Ephrussi, B. (1953). *Nucleo-Cytoplasmic Relations in Micro-Organisms.* Oxford.

[116] Sonneborn, T. M. (1947). Recent advances in the genetics of *Paramecium* and *Euplotes. Advanc. Genet.* **1**, 263–358.

[117] Beale, G. H. (1954). *The Genetics of* Paramecium aurelia. Cambridge.

[118] Mather, K. (1949). *Biometrical Genetics. The Study of Continuous Variation.* London.

[119] Lundqvist, A. (1954). Studies on self-sterility in rye, *Secale cereale* L. *Hereditas*, **40**, 278–94.

[120] Knight, R. and Rogers, H. H. (1955). Incompatibility in *Theobroma cacao. Heredity*, **9**, 69–77.

[121] Whitehouse, H. L. K. (1950). Multiple-allelomorph incompatibility of pollen and style in the evolution of the Angiosperms. *Ann. Bot., Lond.*, N.S., **14**, 199–216.

6

BUFFON, LAMARCK AND DARWIN

[1] Willis, R. (1847). *The Works of William Harvey, M.D.*, p. 41. London.

[2] —— (1847). *Ibid.* pp. 42–5.

[3] Galen, C. *De Usu partium*, lib. vi, cap. 10.

[4] Kühn, C. G. (1822). *Medicorum Graecorum Opera.* Lipsiae. *Claudii Galeni Opera omnia*, III, pp. 455–7.

[5] Cesalpino, A. (1593). *Andreae Caesalpini Aretini Quaestionum Peripateticarum Lib. V, Q. Medicarum Lib. II, De Medicament. facultatibus Lib. II.* Venetiis. *Q. Med. Liber Second.*, Q. 17, pp. 234–5.
—— (1593). *Q. Peripat. Liber Quint.*, Q. 3, pp. 115 *et seq.*

[6] Woodger, J. H. (1952). *Biology and Language.* Cambridge. Appendix A, pp. 75–92. An analysis of Harvey's *De Motu cordis et sanguinis.*

[7] Sinnott, E. W., Dunn, L. C. and Dobzhansky, Th. (1952). *Principles of Genetics.* New York. Appendix, pp. 463–93. *Experiments in Plant-Hybridization*, by Gregor Mendel.

[8] Bayon, H. P. (1938, 1939). William Harvey, physician and biologist: his precursors, opponents and successors. *Ann. Sci.* **3**, 59–118, 435–56; *ibid.* **4**, 65–106, 329–89.

[9] Lyell, Sir Charles (1851). Anniversary address of the President. *Quart. J. Geol. Soc. Lond.* **7**, xxv-lxxvi.

[10] Lamarck, J. B. P. A. (1809). *Philosophie Zoologique, ou Exposition des Considérations relatives à l'histoire naturelle des Animaux, etc., etc.* Paris. (This work is subsequently referred to as *Philosophie Zoologique.*)

[11] Comte, A. I. (1838). *Cours de philosophie positive*, vol. iii, Leçon 42, pp. 554–5. Paris.

[12] Buffon, G. L. L. Comte de (1753). *Histoire naturelle, générale et particulière, avec la description du Cabinet du Roi*, v, pp. xii–xv. Paris. (This author is subsequently referred to as 'Buffon' and this work as *Hist. nat.*).

[13] Buffon, H. N. de (1860). *Correspondance inédite de Buffon*, vol. ii, p. 68. Paris.

[14] Russell, Bertrand (1951). *A Critical Exposition of the Philosophy of Leibniz*, 4th impr., 2nd ed., pp. 20 (note), 66, 67, 223. London.

[15] Latta, R. (1898). *Leibniz. The Monadology and other Philosophical Writings*, p. 64. Oxford.

[16] Buffon (1749). *Hist. nat.* vol. i, p. 11.

[17] Linnaeus, C. (1751). *Philosophia Botanica*, Sectio 77. Stockholm.

[18] Hermann, J. (1783). *Tabula affinitatum animalium, etc.* Argentorati.

[19] Lovejoy, A. O. (1936). *The Great Chain of Being, A Study of the History of an Idea.* Cambridge, Mass.

[20] Daudin, H. (1926). *Les Méthodes de classification et l'idée de série en botanique et en zoologie de Linné à Lamarck, 1740–1790.* Paris.

[21] Comte, A. I. (1838). *Cours de philosophie positive*, Leçons 40–2.

[22] Aristotle, *De Anima*, Book iii, 9; 432 b 20.

[23] Barber, W. H. (1955). *Leibniz in France from Arnauld to Voltaire.* Oxford.

[24] Assezat, J. (ed.) (1875). *Œuvres complètes de Diderot*, vol. i, pp. 275–342. Paris.

[25] Ravier, E. (1937). *Bibliographie des Œuvres de Leibniz*, p. 264. Paris.

[26] Leibniz, G. W. (1749). *Summi polyhistoris Godefridi Guilielmi Leibnitii Protogaea etc.* p. 41. Goettingae.

[27] Buffon (1749). *Hist. nat.* i, ii and iii.

[28] Guyenot, E. (1941). *Les Sciences de la Vie aux XVIIe et XVIIIe Siècles, l'Idée d'Evolution*, p. 394. Paris.

[29] Rostand, J. (1932). *L'Evolution des Espèces*, pp. 57–8. Paris.

[30] Zirkle, C. (1941). Natural Selection before the 'Origin of Species'. *Proc. Amer. Phil. Soc.* **84**, 71–123.

[31] Buffon (1749). *Hist. nat.* vol. ii ,pp. 40–1.

[32] —— (1749). *Ibid.* p. 38.

[33] —— (1749). *Ibid.* p. 39.

[34] —— (1749). *Ibid.* p. 23.

[35] —— (1776). *Hist. nat.* (Suppl. iii), p. 32.

[36] Wilkie, J. S. (1956). The idea of evolution in the writings of Buffon. I. *Ann. Sci.* **12**, 48–62. II. *Ibid.* **12**, 212–27. III. *Ibid.* **12**, 256–66.

[37] Lanessan, J. L. de (1889). Buffon et Darwin. *Rev. Scient.* **1**, 385–91; **2**, 425–32.

[38] Perrier, E. (1886). *La Philosophie Zoologique avant Darwin*, 2nd ed. Paris.

[39] Rostand, J. (1931). *L'Etat présent du Transformisme.* Paris.
—— (1932). *L'Evolution des Espèces.* Paris.

[40] Butler, Samuel (1911). *Evolution, Old and New*, 3rd ed. London.

[41] Ostoya, P. (1951). *Les Théories de l'Evolution*, p. 47. Paris. (Lists other French authors sharing this opinion.)

[42] Rostand, J. (1932). *L'Evolution des Espèces*, p. 54; but see also p. 60, footnote.
[43] Flourens, P. (1860). *Des Manuscrits de Buffon avec des fac-similé de Buffon et de ses collaborateurs*. Paris.
[44] Buffon (1749). *Hist. nat.* vol. I, p. 12.
[45] Locke, John. *An Essay concerning Human Understanding*, book III, ch. 6, section 12.
[46] Buffon, H. N. de (1860). *Correspondance inédit de Buffon*, vol. I, p. 46.
[47] Buffon, G. L. L. (1753). *Hist. nat.* vol. IV, p. 387.
[48] Maupertuis, P. L. (1953). *Les Œuvres de Mr de Maupertuis*, vol. II, pp. 378–92. Berlin.
[49] —— *Ibid.* p. 386.
[50] —— *Ibid.* p. 388.
[51] Buffon (1753). *Hist. nat.* vol. IV, pp. 389–90.
[52] Baumann (1751). *Dissertatio inauguralis, de universali naturae systemate, pro gradu doctoris habita.* Erlangen.
[53] Maupertuis, P. L. (1756). *Œuvres de Mr de Maupertuis, à Lyons*, vol. II, pp. 137–68, Système de la Nature.
[54] —— (1756). *Ibid.* pp. 150–1.
[55] Darwin, Charles (1880). *The Origin of Species*, 6th ed. p. xv. London.
[56] Wells, W. C. (1818). *Two Essays: one upon Single Vision with two eyes; the other on Dew; etc., etc.*, pp. 435–6. London.
[57] Butler, Samuel (1911). *Evolution, Old and New*, 3rd ed. London.
[58] Buffon (1753). *Hist. nat.* vol. IV, p. 377.
[59] Ostoya, P. (1951). *Les Théories de l'Evolution.* Paris, p. 52.
[60] Buffon (1778). *Hist. nat.* (Suppl. V), pp. 1–254, Epoques de la Nature.
[61] —— (1761). *Ibid.* vol. IX, pp. 106–9.
[62] —— (1776). *Ibid.* (Suppl. III), pp. 3, 5.
[63] —— (1766). *Ibid.* vol. XIV, pp. 342–3.
[64] —— (1749). *Ibid.* vol. IV, p. 382.
[65] Wilkie, J. S. (1956). *Ann. Sci.* 12, 48–62.
[66] Buffon (1749). *Hist. nat.* vol. IV, p. 386.
[67] —— (1776). *Ibid.* (Suppl. III), p. 3.
[68] —— (1766). *Ibid.* vol. XIV, pp. 372–3.
[69] —— (1766). *Ibid.* pp. 371–2.
[70] —— (1766). *Ibid.* p. 325.
[71] —— (1755). *Ibid.* vol. V, p. 217.
[72] —— (1755). *Ibid.* pp. 226–7.
[73] —— (1766). *Ibid.* vol. XIV, p. 317.
[74] —— (1766). *Ibid.* p. 334.
[75] —— (1749). *Ibid.* vol. I, p. 43.
[76] —— (1764). *Ibid.* vol. XII, pp. iii–xvi.
[77] —— (1765). *Ibid.* vol. XIII, pp. i–xx.
[78] Lamarck, J. B. P. A. (1815). *Histoire Naturelle des Animaux sans Vertèbres.* vol. I, p. 296. Paris. (The title page bears 'Par M. le Chevalier de Lamarck'.) (This author is subsequently referred to as 'Lamarck' and this work as *Animaux sans Vertèbres.*)

[79] —— (1809). *Philosophie Zoologique.* Paris.
[80] —— (1815). *Animaux sans Vertèbres.* Paris.
[81] —— (1835). *Histoire Naturelle des Animaux sans Vertèbres.* Paris. (The title-page bears 'Par J. B. P. A. de Lamarck'. There are notes by G. P. Deshayes and H. Milne Edwards.)
[82] —— (1815). *Animaux sans Vertèbres,* vol. I, pp. 8, 80.
[83] —— (1815). *Ibid.* p. 53.
[84] —— (1815). *Ibid.* p. 52.
[85] —— (1815). *Ibid.* p. 131, for example.
[86] —— (1815). *Ibid.* p. 153, for example.
[87] —— (1809). *Philosophie Zoologique,* vol. II, pp. 462–3.
[88] —— (1815). *Animaux sans Vertèbres,* vol. I, p. 461; and, in the edition of 1835, vol. I, p. 323.
[89] —— (1815). *Ibid.* p. 201.
[90] —— (1815). *Ibid.* pp. 181–2.
[91] —— (1815). *Ibid.* p. 185.
[92] —— (1809). *Philosophie Zoologique,* pp. 130–4.
[93] —— (1809). *Ibid.* p. 159.
[94] —— (1809). *Ibid.* p. 169.
[95] —— (1809). *Ibid.* pp. 185, 212, 217.
[96] —— (1809). *Ibid.* pp. 220–1.
[97] CUVIER, G. (1800). *Leçons d'Anatomie Comparée, de G. Cuvier,* An VIII, i, pp. 59–60. Paris.
[98] LAMARCK (1815). *Animaux sans Vertèbres,* vol. I, pp. 132–3.
[99] —— (1815). *Ibid.* pp. 136–7.
[100] —— (1815). *Ibid.* p. 147.
[101] —— (1809). *Philosophie Zoologique,* p. 172.
[102] —— (1815). *Animaux sans Vertèbres,* vol. I, pp. 311, 331.
[103] —— (1815). *Ibid.* p. 322.
[104] —— (1815). *Ibid.* p. 194.
[105] —— (1809). *Philosophie Zoologique,* pp. 75–81.
[106] —— (1809). *Ibid.* pp. 100–1.
[107] —— (1815). *Animaux sans Vertèbres,* vol. I, p. 198.
[108] —— (1815). *Ibid.* p. 200.

For EU product safety concerns, contact us at Calle de José Abascal, 56–1°,
28003 Madrid, Spain or eugpsr@cambridge.org.

www.ingramcontent.com/pod-product-compliance
Lightning Source LLC
LaVergne TN
LVHW090305210926
848480LV00001B/140